Memory in a Social Context

Takashi Tsukiura • Satoshi Umeda
Editors

Memory in a Social Context

Brain, Mind, and Society

Springer

Editors
Takashi Tsukiura
Department of Cognitive and Behavioral Sciences, Graduate School of Human and Environmental Studies
Kyoto University
Kyoto, Japan

Satoshi Umeda
Department of Psychology
Keio University
Tokyo, Japan

ISBN 978-4-431-56589-5 ISBN 978-4-431-56591-8 (eBook)
https://doi.org/10.1007/978-4-431-56591-8

Library of Congress Control Number: 2017960167

Printed on acid-free paper

This Springer imprint is published by Springer Nature
The registered company is Springer Japan KK
The registered company address is: Chiyoda First Bldg. East, 8-1 Nishi-Kanda Chiyoda-ku, 101-0065 Tokyo, Japan

Preface

Memory is one of the important cognitive functions in human beings, and research of human memory traditionally has been investigated in the field of experimental psychology. A number of psychological researchers have made important findings about human memory and have provided critical theories to explain human memory processes, but this approach is hitting a brick wall. For example, in an experimental psychological approach, human memory functions are examined in a very controlled environment or in a laboratory-based approach, but the evidence obtained from that approach may not necessarily reflect real-life events in our mind. In addition, findings from experimental psychology have often ignored the link with biological structures, or the brain. In these situations, we have to think about efficient solutions for steps to progress in human memory studies. One solution is a cognitive neuroscience approach, in which functional neuroimaging techniques have enabled us to view how memory processes are represented in our brain. In addition, as another solution, we should extend the traditional concept of human memory into a wider framework by reconsidering memory functions in a social context. These advanced approaches may help us to understand how "social memory" is represented in the human brain and is processed in real-life situations. The purpose of this book is to provide new points of view of human memory in the link among mind, brain, and society.

Part I is "Basic Approach for Human Memory". In this part, two chapters summarize basic psychological and neuroscientific mechanisms of human memory in traditional approaches. In addition, they introduce future directions and advanced techniques to understand human memory. For the comprehension of psychological and brain-based mechanisms of complex social memories, these chapters create an important milestone for all readers.

Part II is "Interacting Mechanisms of Human Memory with Social Cognition". Human memory is modulated by interacting with many socio-emotional factors such as simple emotion, reward, or the self-referential process. In this part, eight interesting chapters discuss cognitive and social neuroscience findings that investigate memory in human beings in real-life societies of the past, present, and future. The framework of interacting mechanisms between human memory and socio-emotional processes will help readers to design newly developed experiments to examine how general cognitive functions as well as memory are affected by socio-emotional factors in a social context.

Part III is "Lifelong Effects on Memory and Social Cognition". Our brain and mind are changed by healthy aging or development from birth to death. In addition, emotional events experienced in childhood have large and successive impacts on human memory. In this part, four chapters report important reviews of cognitive neuroscience findings that investigate how the age-dependent changes of the brain and mind modulate memory processes in a social context and how the traumatic events in our lives affect memory and memory-related brain mechanisms.

Part IV is "Toward the Application of Memory Research into Our Society". Human cognitive functions including memory are affected by social interactions in our society, which is constructed of assemblies of cognitive functions in human beings as social entities. Thus, interactions between basic memory research and real-life memory activity is critical to understanding the real roles of human memory systems in our daily lives and to applying scientific findings of basic memory research to improving our lifestyles. In this part of the book, four researchers introduce examples of the application of basic memory research in our society based on the challenges faced in their research.

We hope that with this book all readers will understand the current status of social memory research in the multiple approaches of human memory experiments, and will be able to construct a new framework to find future directions for social memory research. In addition, we believe that research findings in social memory will be helpful in making people happier in their daily lives.

Finally, we would like to express thanks to all contributors to this book. In addition, we would like to acknowledge Dr. Yasutaka Okazaki and Dr. Yuko Matsumoto at Springer Japan and Ms. Ramabrabha Selvaraj at SPi Global.

Kyoto, Japan Takashi Tsukiura
Tokyo, Japan Satoshi Umeda

Contents

Part I
Basic Approach for Human Memory

Chapter 1
Experimental Psychology Approaches to Human Memory

Hiroyuki Shimizu

Abstract This chapter describes the experimental approaches to human memory in cognitive psychology. To date, various research methods have been developed and refined to investigate issues of human memory. These efforts have included various aspects of memory processes, differential aspects of memory of diverse participants, and investigations of memory disorder in different developmental or aging stages with typical or atypical ability. This chapter comprises three sections. The first section provides a basic outline of memory experiments for individuals, proposed as a tetrahedral model by Jenkins (Four points to remember: a tetrahedral model of memory experiments. In: Cermak LS, Craik FIM (eds) Levels of processing in human memory. Erlbaum, Hillsdale, pp 429–446, 1979). The second section examines the four main sets of variables (participants, materials, orienting tasks, and criterial tasks) in memory experiments that most influential studies have reported. The third section discusses other sets of variables (events in the retention interval, inside and outside laboratory, and metacognitive judgments) from recent studies that require consideration. Finally, the chapter provides a concluding comment on future directions with regard to developing and refining memory experiments.

Keywords Human memory • Memory experiment • Methodology • Jenkin's tetrahedral model • Experimental variable • Laboratory • Everyday memory • Metacognitive judgment

1.1 Introduction

Memory is considered to be a higher-ordered cognitive function that human beings have acquired through a process of long-standing evolution. Although not only human beings but also other species have memory function, human memory is extraordinarily complicated and mutually related to the other cognitive functions

H. Shimizu (✉)
Kobe Gakuin University, Kobe, Japan
e-mail: shimizu@human.kobegakuin.ac.jp

T. Tsukiura, S. Umeda (eds.), *Memory in a Social Context*,
https://doi.org/10.1007/978-4-431-56591-8_1

such as perception, thinking, inference, decision-making, and language. Therefore, in the field of psychology of human memory, numerous and diverse research methods have been developed and refined. These efforts have included various aspects of memory processes, differential aspects of memory of diverse participants, and investigations of memory disorder in different developmental or aging stages with typical or atypical ability. In particular, since the late 1970s, much attention has been paid to behavior and phenomenon of memory in everyday situations and the studies on so-called everyday memory have been developed using a variety of research methods. Many studies employing methods other than experiments, such as psychological questionnaires, testing, interviews, and observations, have been reported in psychological journals on memory. As some kinds of memory behavior and phenomena associated with everyday memory are less likely to occur in laboratory settings, the limitations and constraints of laboratory research have been acknowledged and the significance of personal contexts and variations in the sociocultural background of participants have been emphasized (e.g., Neisser 1982). Nevertheless, there are still many interesting things to be revealed using experiments in laboratory settings with causally analytical approaches to human memory rather than observational or correlational approaches, such as interviews and questionnaires in natural settings. Furthermore, many memory researchers have shown an interest in and attempted to introduce advanced technology (e.g., neuroimaging) to investigate mechanisms or phenomena of memory, or collaborated with researchers beyond the boundaries of their fields (e.g., Nyberg and Cabeza 2000).

Generally speaking, human memory is a mental ability that allows people to hold the things that they have experienced in the past and subsequently reproduce them in some way. In cognitive psychology, especially human information-processing perspectives, memory is usually explained as a process in which information from the outside world is encoded, stored, and retrieved. These three phases are thought to be a basic process of human memory, but modern cognitive psychology pursues the more complex and detailed mechanisms or phenomena of memory. It is necessary to use special research paradigms to examine specific memory mechanisms. In summary, there is a wide range of research methods available for investigating memory according to research questions that guide particular studies. The empirical scientific research methods on memory primarily include experiments, questionnaires, interviews, assessments, observations, etc. In accordance with the nature of research questions and hypotheses, best methods, or sometimes the best combination of two or more methods, should be chosen. This chapter describes the experimental psychology approaches to human memory in three sections. The first section provides a basic outline of memory experiments for the individuals, proposed as the tetrahedral model by Jenkins (1979). The second section examines the four main sets of variables (participants, materials, orienting tasks, and criterial tasks) in memory experiments that most influential studies have reported. The third section discusses other sets of variables (events in retention interval, inside and outside laboratory, and metacognitive judgments) from recent studies that require consideration. Finally, the chapter provides a concluding comment on future directions with regard to developing and refining memory experiments.

1.2 Basic Theoretical Framework of Memory Experiments

Most memory experiments, and the other psychological experiments, follow a basic scheme or procedural pattern. This includes formulating hypothesis, selecting participants, assigning participants to groups, and determining experimental variables. Following these basic steps, the experimenters manipulate stimulus variables and/or organismic variables as independent variables in relation to the research questions and hypotheses, and measure the specific responses and behavior considered as a dependent variable. In addition, factors other than independent variables that potentially affect the dependent variable are called extraneous variables, which should be appropriately controlled for to avoid the impact of the extraneous variables on the measured performance. These are sometimes referred to as "threats to validity." There are five main techniques researchers use to control extraneous variables: (a) elimination, (b) constancy of conditions, (c) balancing, (d) counterbalancing, and (e) randomization. In short, at first, researchers try to remove the extraneous variable itself, and if this proves to be difficult to do, they need to keep the influence of the extraneous variables constant for all conditions. Moreover, when researchers cannot keep this constant, they attempt to balance or counterbalance it. When researchers cannot control the extraneous variables sufficiently using the four techniques identified, they must randomize the likelihood of inevitable occurrence of the influence of extraneous variables (see McGuigan 1996).

What and how researchers determine and treat as independent, dependent, or extraneous variables depends on the research questions and hypotheses to be tested. As mentioned above, a useful basic framework for comprehensively understanding experimental variables in memory research is the well-known tetrahedral model of memory experiments proposed by Jenkins (1979), a modified version of which appears in Fig. 1.1. In this model, all variables relating to memory experiments are divided into four sets of variables, corresponding to four vertices constituting a three-dimensional structure (i.e., a tetrahedron): (a) types of participants being tested, (b) types of materials being studied and tested, (c) orienting tasks instructed to the participants in an encoding phase, and (d) criterial tasks (tests) given to the participants in a retrieval phase. That is, in this model, "the memory phenomena that we see depend on what kinds of subjects we study, what kinds of acquisition conditions we provide, what kinds of materials we choose to work with, and what kinds of criterial measures we obtain. Furthermore, the dependencies themselves are complex; the variables interact vigorously with one another" (Jenkins 1979, p. 431). Each vertex represents a set of variables in which memory researchers may be interested, each edge and each plane represent two-way and three-way interactions respectively, and the whole tetrahedron represents the four-way interaction of all the variables. If researchers are interested in the relations between the two sets of variables (or relations among the three sets of variables), they focus on the edges (or plains) of this tetrahedral model.

Although Roediger (2008) subsequently argued that there are in fact more variables that researchers need to consider, this original model provides a comprehensive understanding of the structure in memory experiments. For example, as introduced in Kantowitz et al. (2008), in the levels-of-processing experiment (Craik and Tulving

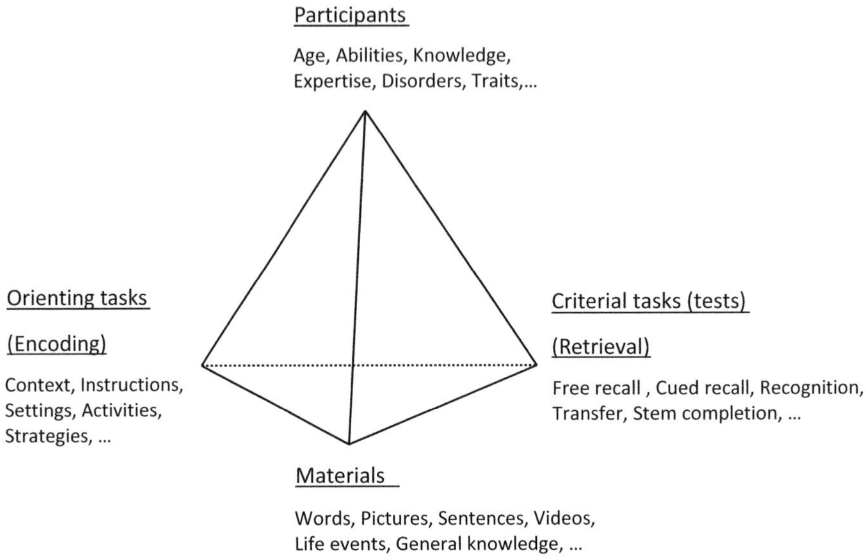

Fig. 1.1 A modified version of Jenkins' tetrahedral model of memory experiments. *Note*. From Jenkins et al., and Roediger 2008. Copyright © 2008 by Annual Reviews. Adapted with permission

1975), the researchers focused on how the orienting task (the questions orienting the participants to process the words in different ways in relation to levels of processing) influenced the performance in unexpected tests of recall. The other two sets of variables, not being of interest, were not manipulated as independent or dependent variables, but controlled as extraneous variables. Jenkins (1979) exemplified a well-known experiment by Flavell et al. (1970) as focusing on the edge of interaction between participants and orienting tasks. Flavell et al. (1970) showed that young children did not exhibit any different behaviors when they were given instructions to "learn" compared with when they were given instructions to "look at." Furthermore, even when they were given instructions to "rehearse" and then they repeated the words over and over, such instructions did not lead them to review the preceding words they might still remember. On the other hand, older children were shown to follow the more effective rehearsal strategies. Therefore, even if "the same" orienting tasks are assigned to participants with different ages, the participants have different orienting activities and memory performances.

1.3 Four Principal Sets of Variables in Memory Experiments

Jenkins' tetrahedral model is effective for covering a wide range of structure in memory experiments, but several supplements and additions have evolved since this model was published nearly four decades ago. These have included, (a) participants, (b) materials, (c) orienting tasks, and (d) criteria tasks.

1.3.1 Participants

An advantageous point of studying memory in humans is that the participants usually have the ability to understand verbal instructions from experimenters, hold their own experiences, and are able to orally and literally describe them. However, when researchers are interested in memory in infants and young children, and in people with physical and intellectual difficulties, the memory experiments need to be modified to adapt to the participants' abilities. Obviously, the critical personal variables are participants' age, expertise, and brain damage. In particular, when we focus on a single case with specific expertise or brain damage, special techniques help researchers to capture the personal traits, even if there is no difficulty with verbal comprehension and speech ability. Takahashi et al. (2006), for example, focused on a Japanese memorist who set a world record for reciting the first 40,000 digits of pi. The experimenters carried out a series of memory tests on him and compared his performance with a control group consisting of 18 Japanese male participants matched for age and educational level. Resulting data showed that he performed better on tasks with digits, but not better for studying word lists and story recall, indicating that his overall performance did not suggest that he had superior cognitive abilities. Thus, researchers need to compare the performance in various tasks between a critical case and a control group under the same conditions, to characterize the individual's specific memory ability.

1.3.2 Materials

It is widely recognized that Hermann Ebbinghaus first introduced experimental methods in the study of memory, and developed nonsense syllables (meaningless combinations of letters) as materials of homogeneous difficulty for serial learning to control the variability of familiar words and eliminate the influence of past experiences and preexisting knowledge of the participants before the memory experiments (Ebbinghaus 1885/1964). The norms of association or meaningfulness values of nonsense syllables have been studied since an earlier work in the 1920s (e.g., Glaze 1928; Noble 1952; Jenkins 1985), and English and other language versions were published (e.g., Umemoto 1969). Also, the recent tendency in the utilization of ecologically valid materials such as faces and scenes is currently obvious, because the concerns in memory research have been expanded from laboratory settings to everyday settings. Today, there are various archives and databases including many sets of syllables, words, sentences, texts, pictures, videos, etc., rated and standardized in diverse fashions, that can be immensely useful as stimuli or materials in memory research.

Another extension of the variability in materials can be seen in the studies on false memory using the Deese/Roediger–McDermott (DRM) paradigm (Deese 1959; Roediger and McDermott 1995), in which participants are presented with lists of associatively related words, all of which are related to a general topic or theme that is represented by a word that did not appear in the original list. For example, if the list contained only the words *nap, pillow, dream, bed*, and *snore*, participants correctly remember not only some of those words, but also falsely remember the theme, or critical lure, *sleep*. The likelihood of recalling the critical lures that were not presented in the original list was as high as that of recalling items that were actually presented in the middle positions of the list (Roediger and McDermott 1995). Therefore, researchers have recently been interested in not only memory for the materials that were presented, but also memory of information that is associatively related to what was presented but did not occur; namely, false memories.

1.3.3 Orienting Tasks

As mentioned above, orienting tasks are related to a manipulation of the processing that occurs in the encoding phase of a given material. Lockhart (2000) argued that encoding is jointly determined by three broad factors: induced cognitive state, task demands, and the nature of the stimulus material. In particular, we note the development of memory research on the significance of manipulation to induce specific cognitive states, and the importance of task demands, representatively shown in the levels-of-processing effects from the study by Craik and Tulving (1975) previously discussed.

Additionally, the influences of events occurring immediately *before* and *after* the encoding of the material to be remembered have been repeatedly reported in many studies. As for the influences of events occurring immediately before encoding, the findings of the classic study on the phenomenon known as "release from proactive interference (PI release)" are significant (Wickens 1972). In a typical study of PI release, participants studied words that all come from the same category (e.g., *food-stuffs*) in the first three trials. In the fourth trial, the category was changed and the words refer to another category (e.g., *parts of the body*). The participants usually exhibit a decline in recall performance across the three trials as a result of semantic interference from previous trials (i.e., proactive interference). This probably means that, after several trials, the participants have some difficulty discriminating between words presented in the current trial and words presented in previous trials. However, when words from a different category were presented for study, the level of recall performance was recovered to the same level as in previous trials. These findings can be exemplified as an effect of events occurring just before the encoding of items by inducing participants to a cognitive state.

With regard to the influences of events occurring immediately after encoding, the recent studies on retrieval-induced forgetting indicate the importance of experiences immediately after the encoding of items (Anderson et al. 1994). Retrieval-induced forgetting is usually studied with a retrieval practice paradigm, in which participants first study simple verbal categories (e.g., *fruits, drinks*) for a later memory test, and then they are asked to repeatedly recall some of the examples (e.g., *orange*) by category-stem cueing (e.g., *fruits-or__*) that they just studied, from some of the categories. After this "retrieval practice," a final test is given in which they are asked to recall all examples from every category. The results showed that the performance of cued recall of the remaining unpracticed examples in the practiced category (e.g., *fruits-banana*), compared with unpracticed items from baseline categories that were also studied, but none of whose examples receive retrieval practice (e.g., *drinks-wine*), was lower, suggesting that the practicing of retrieval might cause forgetting. Therefore, retrieval-induced forgetting suggests that experiences of practicing or reviewing the parts of materials, immediately after initial encoding of the materials, would impair recall of the parts that were also studied but not practiced or reviewed.

1.3.4 Criterial Tasks

Roediger (2008) argued that, since Jenkins' model was proposed, a number of criteria tasks have been employed to measure retention, from classic tasks such as free recall, serial recall, paired associate learning, and various recognition procedures, to newer tasks such as primed completion of word fragments or answering general knowledge questions. Similar to encoding, retrieval can be measured intentionally (when participants are asked to remember events) or incidentally (when the impact of previous experience is assessed by reflecting through transfer or priming). The distinction between intentional and incidental tasks corresponds to that between explicit and implicit tests of memory, which are also sometimes called direct and indirect memory tests (Jacoby 1984; Schacter 1987). For example, we can use a method of testing of recall or recognition with verbal instructions asking participants to study and remember the material in the context of research on explicit memory in which the conscious and intentional recollection of previous experiences and information was done. On the other hand, to examine the participants' implicit memory implying a type of memory without conscious awareness of previous experiences, performance on a task requesting the participants to guess, judge or generate something for the presented items between the conditions of participants having and not having previous specific experiences should be compared.

Furthermore, Lockhart (2000) added prospective remembering to the criterial tasks in relation to a retrieval phase of memory. In prospective memory tasks, participants are asked to recall and perform actions at a designated future time (Einstein and McDaniel 1996). The prospective memory tasks are roughly divided into time-based and event-based tasks. In time-based prospective remembering, participants are asked to perform an action at a given time in the future (e.g., checking the chicken in the pot in 10 min), whereas in event-based prospective remembering, an action is asked to be performed contingent on the occurrence of the other event (e.g., remembering to post the letter when the participants first encounter a post box, a post office, or postman). That is, the criterial tasks are not always related to remembering the past experiences at the time the experimenter instructed the participants to remember, but sometimes focus on when and how spontaneously and successfully the participants generate actions at the future time of retrieval.

1.4 Other Sets of Variables to Be Considered

Although all memory experiments can be characterized in an extended version of Jenkins' (1979) tetrahedral model, the subsequent studies in the decades following the development of this model showed other sets of variables that should be taken into account. Roediger (2008) pointed out that another factor affecting many phenomena of memory is whether a manipulation of conditions occurs in randomized, within-subjects designs rather than between-subjects (or within-subject, blocked) designs. Accordingly, other sets of variables should also be considered.

1.4.1 Events in Retention Interval

First, the issues of variations in conditions during the retention interval should be addressed in relation to the intermediate phase of memory processes between encoding and retrieval (i.e., the storage phase). Lockhart (2000) pointed out that two major conditions of interest have been the duration of the retention interval and the nature of events that occur during the retrieval interval. However, the latter has recently attracted more attention than the former, because the duration can be manipulated or controlled more easily than the events during the interval. As far as the influences of events that occur during the retention intervals are concerned, one of the classic examples of methods is rehearsal prevention, stemming from a so-called Brown–Peterson paradigm (Brown 1958; Peterson and Peterson 1959) for examining the limits of short-term memory. In this procedure, participants are presented with a sequence of items (e.g., three words or a consonant trigram), followed by a distractor task-filling retention interval of varying duration (3–18 s), such as counting backward in 3s from 999 until recall of the study items in the correct serial order is instructed. When the distractor task sufficiently occupies processing capacity during the retention interval so that any rehearsal of the items to be remembered is impossible, recall performance of the items was impaired. That is, the longer the participants performed backward counting, the less likely they were to recall the items. These findings indicate that preventing the usage of additional cognitive resources during the retention interval has an effect on recall performance in the criterial tasks.

Another example of memory research emphasizing the importance of experiences in a storage phase of memory processes comes from the study on misinformation effects, the finding that exposure to misleading post-event information can lead eyewitnesses to report items and events that they never actually saw. In the experimental paradigm developed by Loftus and colleagues (e.g., Loftus and Palmer 1974), participants are presented with a slide sequence depicting a complex and forensically relevant event, such as a traffic accident or theft. Immediately, participants are questioned about the event they witnessed. The critical manipulation is that the questioning includes leading or misleading information. Finally, participants are tested on their memory for the witnessed event. The dependent variable of interest is the extent to which misled participants incorporate the misleading suggestions into their eyewitness reports, compared with a control group of participants who were not misled. A series of experiments revealed that misleading post-event questions caused a variety of distortions in eyewitness reports (Loftus 1979). The misinformation effect can be considered an example of retroactive interference in traditional studies, indicating that information presented later interferes with the retention of previously encoded information (Barnes and Underwood 1959). The new information that people receive works retroactively to distort memory of the original event. Thus, the studies on the misinformation effects indicate that human memory is influenced by the events occurring during retention intervals in the storage phase between encoding and retrieval of the materials.

Moreover, Lockhart (2000) argued that another retention interval condition is derived from the directed forgetting paradigm, in which, for example, participants are presented with a list of words to remember, but in advance, some of the presented items were designated as items to be remembered and others as items to be forgotten (Bjork 1970). Because the designation unexpectedly and surprisingly follows the item presentation with delay, at the time of study, participants have no information about the items to be remembered or to be forgotten. Any differences in subsequent remembering cannot be attributed to differences in encoding, but only to events during the retention interval. The results showed that recall of the items designated to be forgotten was poorer than recall of the items designated to be remembered. The directed forgetting paradigm has been used to study inhibitory processes in memory (Bjork 1989).

1.4.2 Inside and Outside the Laboratory

There has been a debate about the advantages and disadvantages of laboratory approaches and everyday memory approaches, and which is more appropriate for memory research in terms of generalizability and ecological validity (e.g., Bruce 1985; Banaji and Crowder 1989). After all, whether memory researchers adopt laboratory settings or natural settings depends on the purposes of research, study research questions, and the constraints of participants' traits. As an example, some memory researchers studying eyewitness testimony and prospective memory in everyday situations often prefer field experiments in natural settings, and report significant findings that may be unattainable by conventional experiments in laboratory settings (e.g., Brigham et al. 1982; Kim and Mayhorn 2008). In this sense, a setting or situation in which experiments are administered is considered a crucial factor in memory experiments.

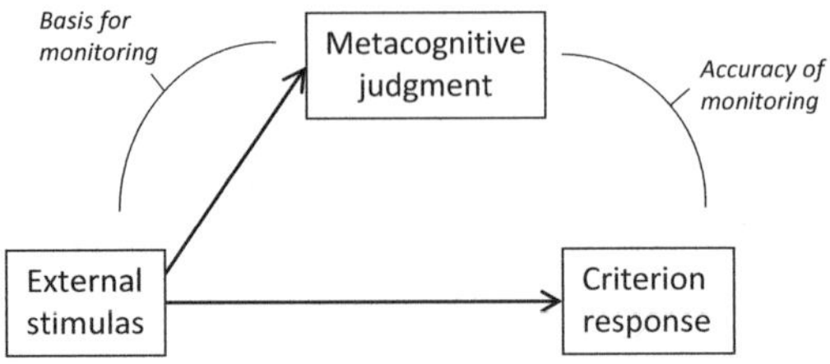

Fig. 1.2 Schematic representation of the distinction between an independent variable's effect on the magnitude of a metacognitive monitoring judgment (i.e., a basis of the judgment) vs the degree of relationship between the judgment and criterion performance (i.e., accuracy of the judgment). *Note*. From Nelson 1996. Copyright ©1996 by the American Psychological Association. Adapted with permission

1.4.3 Metacognitive Judgments

Since Nelson and Narens (1990) proposed a theoretical framework on metamemory (a part of metacognition), referring to people's knowledge about memory and their monitoring and control of memory processes, metamemory research has become increasingly popular (e.g., Dunlosky and Metcalfe 2009). When researchers focus on metacognitive monitoring, including ease-of-learning judgments, judgments of learning, feeling-of-knowing judgments, etc., it is difficult to place such metacognitive judgments within the sets of variables discussed so far in this chapter. This is because metacognitive judgments stem from external stimuli, and do not necessarily correspond to the criterion responses. Nelson (1996) emphasized that it is worthwhile distinguishing between two kinds of effects that the manipulation of the external stimuli can have (Fig. 1.2). One effect is on the magnitude of the metacognitive judgment, whereas the other effect is on the degree of the relationship between the metacognitive judgment and the criterion responses (i.e., an effect on the degree of metacognitive accuracy). For example, an intentional/incidental learning instruction does not affect later recall, but does enhance individuals' accuracy at judging what they have learned (Mazzoni and Nelson 1995).

1.5 Concluding Comments

This chapter discussed the issues of variables to be considered in memory experiments, based on the basic conceptual framework that Jenkins (1979) proposed. Memory research has steadily developed, and in the process of development, many new ideas and techniques have been offered in relation to the methods employed in memory studies. Broadly speaking, one of the major trends in modern memory research is focusing on the unobservable and conscious/unconscious elements of memory processes in various naturalistic contexts and situations. However, to establish generalizable truths of human memory, experimental methods are essential for clarifying the causal relationships of variables in memory phenomena. At the same time, the recent development of methods in memory questionnaires and interview techniques, conventionally aimed at confirming correlational relationships of multiple factors with multivariate analysis methods (e.g., factor analyses), has attempted to reveal causal relationships of factors (e.g., a structural equation modeling). The empirical findings based on the data reflecting participants' introspection or consciousness through questionnaires and interviews are so interesting and stimulating that the values of these findings determined through experimental methods have been shared among memory researchers. Also, neuropsychology and neuroimaging technology currently play an increasingly significant role in memory research, and these trends require further development and refinement of experimental methodology on human memory by widely covering not only behavioral data, but also metacognitive, introspective, and/or conscious/unconscious data in laboratory settings.

Acknowledgments This research was partially supported by JSPS KAKENHI Grant Number 17 K04510.

References

Anderson MC, Bjork RA, Bjork EL (1994) Remembering can cause forgetting: retrieval dynamics in long-term memory. J Exp Psychol Learn Mem Cogn 20(5):1063–1087
Banaji MR, Crowder RG (1989) The bankruptcy of everyday memory. Am Psychol 44(9):1185–1193
Barnes JM, Underwood BJ (1959) Fate of first-list association in transfer theory. J Exp Psychol 58(2):97–105
Bjork RA (1970) Positive forgetting: the noninterference of items intentionally forgotten. J Verbal Learn Verbal Behav 9(3):255–268
Bjork RA (1989) Retrieval inhibition as an adaptive mechanism in human memory. In: Roediger HL III, Craik FIM (eds) Varieties of memory and consciousness: essays in honour of Endel Tulving. Erlbaum, Hillsdale, pp 309–330
Brigham JC, Maass A, Snyder LD, Spaulding K (1982) Accuracy of eyewitness identification in a field setting. J Pers Soc Psychol 42(4):673–681
Brown J (1958) Some tests of the decay theory of immediate memory. Q J Exp Psychol 10(1):12–21
Bruce D (1985) The how and why of ecological memory. J Exp Psychol Gen 114(1):78–90
Craik FIM, Tulving E (1975) Depth of processing and the retention of words in episodic memory. J Exp Psychol Gen 104(3):268–294
Deese J (1959) On the prediction of occurrence of particular verbal intrusions in immediate recall. J Exp Psychol 58(1):17–22
Dunlosky J, Metcalfe J (2009) Metacognition. Sage, Thousand Oaks
Ebbinghaus H (1885/1964) Memory: a contribution to experimental psychology. Dover, New York (Originally published 1885)
Einstein GO, McDaniel MA (1996) Retrieval processes in prospective memory: theoretical approaches and some new empirical findings. In: Brandimonte M, Einstein G, McDaniel M (eds) Prospective memory: theory and applications. Lawrence Erlbaum Associates, Hillsdale, pp 115–142
Flavell JH, Friedrichs AG, Hoyt JD (1970) Developmental changes in memorization processes. Cogn Psychol 1(4):324–340
Glaze JA (1928) The association value of non-sense syllables. Pedagog Semin J Gen Psychol 35:255–269

Jacoby LL (1984) Incidental versus intentional retrieval: remembering and awareness as separate issues. In: Squires LR, Butters N (eds) Neuropsychology of memory. Guilford, New York, pp 145–156

Jenkins JJ (1979) Four points to remember: a tetrahedral model of memory experiments. In: Cermak LS, Craik FIM (eds) Levels of processing in human memory. Erlbaum, Hillsdale, pp 429–446

Jenkins JJ (1985) Nonsense syllables: comprehending the "almost incomprehensible variation". J Exp Psychol Learn Mem Cogn 11(3):455–460

Kantowitz BH, Roediger HL III, Elmes DG (2008) Experimental psychology, 9th edn. Wadsworth, Belmont

Kim P, Mayhorn CB (2008) Exploring students' prospective memory inside and outside the lab. Am J Psychol 121(2):241–254

Lockhart RS (2000) Methods of memory research. In: Tulving E, Craik FIM (eds) The Oxford handbook of memory. Oxford University Press, New York, pp 45–57

Loftus EF (1979) Eyewitness testimony. Harvard University Press, Cambridge, MA

Loftus EF, Palmer JE (1974) Reconstruction of automobile destruction: an example of the interaction between language and memory. J Verbal Learn Verbal Behav 13(5):585–589

Mazzoni G, Nelson TO (1995) Judgments of learning are affected by the kind of encoding in ways that cannot be attributed to the level of recall. J Exp Psychol Learn Mem Cogn 21(5):1263–1274

McGuigan FJ (1996) Experimental psychology: methods of research, 7th edn. Prentice Hall, Englewood Cliffs

Neisser U (1982) Memory observed. W. H. Freeman, San Francisco

Nelson (1996) Consciousness and metacognition. Am Psychol 51(2):102–116

Nelson TO, Narens L (1990) Metamemory: a theoretical framework and new findings. In: Bower GH (ed) The psychology of learning and motivation, vol 26. Academic, New York, pp 125–173

Noble CE (1952) An analysis of meaning. Psychol Rev 59(6):421–430

Nyberg L, Cabeza R (2000) Brain imaging of memory. In: Tulving E, Craik FIM (eds) The Oxford handbook of memory. Oxford University Press, New York, pp 501–519

Peterson LR, Peterson MJ (1959) Short-term retention of individual verbal items. J Exp Psychol 58(3):193–198

Roediger HL (2008) Relativity of remembering: why the laws of memory vanished. Annu Rev Psychol 59:225–254

Roediger HL III, McDermott KB (1995) Creating false memories: remembering words not presented on lists. J Exp Psychol Learn Mem Cogn 21(4):803–814

Schacter DL (1987) Implicit memory: history and current status. J Exp Psychol Learn Mem Cogn 13(3):501–518

Takahashi M, Shimizu H, Saito S, Tomoyori H (2006) One percent ability and ninety-nine percent perspiration: a study of a Japanese memorist. J Exp Psychol Learn Mem Cogn 32(5):1195–1200

Umemoto T (1969) The norms of association values in Japan. University of Tokyo Press, Tokyo. (梅本堯夫 (1969). 連想基準表 東京大学出版会)

Wickens DD (1972) Characteristics of word encoding. In: Melton AW, Martin E (eds) Coding processes in human memory. Winston & Sons, Washington, DC, pp 191–215

Chapter 2
Functional Neuroimaging Approaches to Human Memory

Junichi Chikazoe and Seiki Konishi

Abstract Historically, studies of brain-damaged humans and experimentally lesioned animals have provided abundant evidence regarding neural underpinning for episodic memory. These studies have revealed that the medial temporal lobe, including the hippocampus, plays a critical role in different memory stages including encoding, consolidation, and retrieval. Furthermore, these studies demonstrated differences in the level of impairment caused by a lesion across memory stages, suggesting that each stage might recruit different brain regions, although some of them might overlap. The advancement of neuroimaging methods such as functional magnetic resonance imaging (fMRI) enabled investigation of the thorough coverage of brain regions without invasion in healthy human subjects. In this chapter, we offer comprehensive and concise commentaries on functional neuroimaging approaches to reveal mechanisms underlying memory encoding, retrieval and consolidation. We further describe novel approaches such as multi-voxel pattern analysis used for decoding of memory representations and real-time fMRI that could show causality beyond correlation.

Keywords Memory encoding • Memory retrieval • Memory consolidation • Multi-voxel pattern analysis • Real-time fMRI • Memory representation

2.1 Introduction

In our daily life, we experience various events: seeing a cherry blossom, feeling a dull pain in a knee, listening to birdsong in the forest, etc. Even during the plainest day, some subtle signature of such events may be memorable. What determines whether an experience is later remembered or forgotten? Abundant evidence from behavioral neuropsychological studies has demonstrated that each stage of the

J. Chikazoe (✉)
Section of Brain Function Information, National Institute for Physiological Sciences, Myodaiji, Okazaki 444-8585, Japan
e-mail: j.chikazoe@gmail.com

S. Konishi
Department of Neurophysiology, Juntendo University School of Medicine, Tokyo, Japan

T. Tsukiura, S. Umeda (eds.), *Memory in a Social Context*,
https://doi.org/10.1007/978-4-431-56591-8_2

memory process is supported by partially overlapped but considerably distinct brain regions (Squire 1982). For example, neuropsychological studies revealed that patients with medial temporal lobe (MTL) damage had severe difficulty in encoding memory for novel events, whereas they could recall events that occurred before the traumatic event/surgery to some extent (Scoville and Milner 1957). Furthermore, memory loss caused by hippocampal damage showed a characteristic feature: the patients had difficulty in recalling events that had occurred recently, but they could recall older events (Scoville and Milner 1957). This feature, which is called "temporal gradient," suggested that memories from the hippocampus might be moved to the neo-cortex in a more stabilized form of storage (memory consolidation) (Squire and Alvarez 1995). Although evidence from patients with brain damage is strong, they lack spatial resolution: in most cases, the damaged region is not localized, and it sometimes involves a broad range of brain areas that may have different functions. Moreover, the precise influence of a lesion on each memory stage (i.e., encoding, consolidation, and retrieval) is difficult to specify, which makes the role of a brain region for each memory stage ambiguous. In contrast, neuroimaging techniques such as functional magnetic resonance imaging (fMRI) has higher spatial resolution whereby functional organization on a subcentimeter scale can be detected (Hirose et al. 2009). Importantly, fMRI allows separation across memory stages. In this chapter, we describe a variety of task designs and analyses aimed at unveiling neural correlates of memory at each stage. We further refer to recent techniques such as controlling participants' memory performance by modifying the timing of stimulus presentation based on online analysis of neuroimaging data, which is called "real-time fMRI."

2.2 Encoding

Neural correlates underlying memory encoding have been intensively investigated for decades. An experimental paradigm, the so-called subsequent memory paradigm, enabled comparing brain activity for items that are remembered or forgotten in a memory test, thus establishing a direct link between brain activity and successful memory formation (Dolcos et al. 2012; Wagner et al. 1998). In this paradigm, brain activity during the encoding phase is recorded and then trials were sorted, based on whether to be remembered or forgotten in the later memory test. By calculating the difference in brain activity associated with the later remembered and forgotten items, one can detect brain regions that mediate processes required for successful memory encoding (Paller et al. 1987). The fMRI signal that is greater for the items later remembered than for those later forgotten indicates the contribution of those regions to successful encoding (a subsequent memory [SM] effect). With this paradigm, neuroimaging studies revealed the SM effects in multiple brain regions, including the prefrontal cortex and medial temporal lobe (MTL) (Fernandez and Tendolkar 2001; Simons and Spiers 2003; Wagner et al. 1998) along with the fusiform cortex (Dickerson et al. 2007; Garoff et al. 2005; Kim and Cabeza 2007) and posterior parietal cortex (Sommer et al. 2005; Uncapher and Rugg 2009).

Interestingly, several previous studies (Daselaar et al. 2004; Kim et al. 2010) reported that the subsequent memory performance was predicted by the deactivation of the default-mode network (a negative subsequent memory) (Duverne et al. 2009; Otten and Rugg 2001), which consists of the anterior and posterior midline cortex, the temporoparietal junction, and the superior frontal cortex (Raichle et al. 2001). A meta-analysis of 74 fMRI studies statistically confirmed those results and further revealed that the activation of the inferior frontal cortex was lateralized to the left hemisphere, whereas the other regions showed bilateral activation (Kim 2011). In this meta-analysis, the hemispheric laterality of the deactivation was found in the temporoparietal junction: temporoparietal deactivation was stronger in the right than in the left hemisphere. The meaning of the negative SM effects (i.e., forgotten > remembered) remains unclear. One possible explanation is that it may indicate greater activity interfering with successful encoding. Indeed, the default-mode network is known to be associated with mind-wandering during the task (Mason et al. 2007), which may interfere with the memory encoding of external events. Another possibility is that it may reflect a momentary lapse of attention. Previous attention studies demonstrated that less efficient stimulus processing during attentional lapses was characterized by less deactivation of the default-mode network (Weissman et al. 2006). From this perspective, greater deactivation predicting items remembered later may reflect the extent of attention allocated to the item.

2.3 Retrieval

Neural correlates related to memory retrieval can be investigated using two standard approaches. It depends on the processes of interest which approach is preferable. The first approach is aimed at revealing "retrieval success," which refers to processes that are selectively engaged when a retrieval attempt is successful (Rugg et al. 1996). The second approach is based on a "dual-process" model of memory (Yonelinas 2002). This model posits that a retrieval cue can elicit two qualitatively distinct kinds of mnemonic information, that is, recollection and familiarity (Rugg and Vilberg 2013). Recollection signal provides information about qualitative aspects of the past event, such as context or source of information whereas familiarity signal can support simple judgments of occurrence of the event. We demonstrate how to specify neural correlates of memory retrieval in addition to these two approaches.

2.3.1 Retrieval Success

Neuroimaging studies can specify brain regions associated with a certain cognitive component by contrasting at least two conditions. For example, to retrieve memory from a visual word cue, visual and language systems should be recruited besides memory systems. If we assume that these systems would not interact with each other, we can hypothetically obtain a brain activation map of memory retrieval by subtracting the control condition (visual + language) from the target condition (visual + language + memory retrieval). In the case of the memory test, there are four conditions: hit (remembered old item), miss (forgotten old item), false alarm (new item incorrectly labeled as old), and correct rejection (new time correctly labeled as new). Apparently the "hit" condition is associated with successful retrieval; however, which condition to use for the control condition may depend on the research question. For example, to directly compare brain responses for the same stimuli between encoding and retrieval, some studies used the "miss" condition for the control (Huijbers et al. 2013). However, most studies used the "correct rejection" condition for the control, as this condition is thought to share similar cognitive components with "hit" except retrieval success (Tulving 1999). Interestingly, the contrast "hit versus correct rejection" has consistently revealed the involvement of the parietal lobe, including the lateral posterior parietal cortex, precuneus, and the mid-cingulate cortex (Kahn et al. 2004; Konishi et al. 2000; Wheeler and Buckner 2004). This result was surprising as the parietal lobe was not thought to support declarative memory previously (Wagner et al. 2005); however, it has been consistently reported by fMRI studies of memory retrieval from various laboratories (Konishi et al. 2000; McDermott et al. 2000; Spaniol et al. 2009; Wheeler and Buckner 2004). Resting state functional connectivity studies have consistently reported that these three regions form a functional network. In resting state functional connectivity analysis, low-frequency (<0.1 Hz) correlations across brain regions were computed, which revealed that those three regions showed high correlations during rest; in other words, in the absence of explicit task conditions. The roles of the parietal cortex in memory have been discussed in several review papers (Cabeza et al. 2012; Wagner et al. 2005). However, its actual role is still controversial (Gilmore et al. 2015).

2.3.2 Dual Process

A retrieval cue can elicit two qualitatively distinct pieces of mnemonic information, that is, the process of recognizing the item based on retrieval of specific contextual details, such as when or where an item was presented (recollection) and the process of recognizing an item based on its perceived memory strength without retrieval of any specific details (familiarity) about the previously studied episode (Diana et al. 2007). This view was first introduced by Atkinson and Mandler respectively (Atkinson et al. 1974; Atkinson and Juola 1973; Mandler 1979; Mandler et al. 1969), and has been developed over the decades (Yonelinas 2002). Although there are several variations such as the "Atkinson model," the "Mandler model," and the "Yonelinas model," this view is generally called the "dual-process" theory (Yonelinas 2002). Neuroimaging studies have played a critical role for the development of "dual-process" theory by providing the evidence for functional–anatomical dissociation of these two processes. To identify neural correlates of recollection and familiarity, previous neuroimaging studies employed several methods (Diana et al. 2007). For example, the remember/know procedure requires participants to assess whether a given item elicits contextual recollection or whether the recognition of the item is based on familiarity in the absence of recollection (Henson et al. 1999). Another method is collecting rating scores for recognition confidence. Recollection is often accompanied by highly confident recognition owing to the retrieval of specific details of the studied event, whereas familiarity is expected to contribute across a wide range of confidence responses, reflecting the level of familiarity strength (Daselaar et al. 2006). The other methods are testing associative memory (e.g., a pair of two stimuli) (Kirwan and Stark 2004) or source memory (e.g. during which task was the word presented?) (Davachi et al. 2003; Kensinger and Schacter 2006). These studies revealed that the parahippocampal cortex and hippocampus are associated with recollection, whereas the perirhinal cortex supports familiarity.

2.3.3 Other Retrieval Studies

Most studies of memory retrieval focused on retrieval success or the dual process theory. However, there are several important studies aimed at revealing the memory processes of humans. Konishi and his colleagues devised a new memory paradigm in which participants performed a memory test for paired Fourier figures studied ~8 weeks before the test and those studied immediately before the test. They revealed a significant signal increase in the right hippocampus during retrieval of newly studied pairs relative to initially studied pairs. In contrast, a significant signal increase was observed in the anterior temporal cortex during retrieval of the initially studied pairs. These results provided evidence for the formation of temporal neocortical representation for stable long-term memory (Yamashita et al. 2009). Other studies employed multivoxel pattern analysis (MVPA) by which representational similarity across encoding and retrieval was examined (Polyn et al. 2005; Ritchey et al. 2013). For that purpose, neuroimaging data were collected for both the encoding and the retrieval of visual stimuli, and correlations of activation patterns were computed. It was revealed that memory success tracked fluctuations in encoding-retrieval similarity in frontal and posterior cortices (Ritchey et al. 2013). By examining category-specific activation patterns (i.e. faces, locations and objects) during a free recall task, Norman and his colleague revealed that the reappearance of a given category's activity pattern correlated with verbal recalls and preceded the recall event by several seconds. This result was consistent with the hypothesis that category-specific activity is cueing the memory system to retrieve studied items (Polyn et al. 2005).

2.4 Consolidation

Long-term storage of episodic memories is hypothesized to result from the information transfer from the hippocampus to the neocortex over time. This information restructuring should be accompanied by the change in the neural circuits and is thought to be mediated by reactivation of the same patterns of neural activity that are present during a previous experience (Sutherland and McNaughton 2000; Tambini et al. 2010). This process has been termed consolidation and numerous rodent studies have provided evidence that reactivation during awake–rest/sleep support memory (Buzsaki 1996; Girardeau et al. 2009; Jadhav et al. 2012; Wilson and McNaughton 1994). In humans, little has been known about the consolidation process until quite recently, although neuroimaging studies of memory have focused on the encoding and retrieval processes. This is at least partly due to the methodological difficulty: the traditional univariate analysis needs knowledge of the event timing to model neurovascular responses measured by fMRI. By investigating the change in resting state blood oxygen level-dependent (BOLD) correlation across brain regions before and after stimulus presentation, Davachi and colleagues demonstrated that memory performance could be predicted by the increase in BOLD correlation between the hippocampus and the lateral occipital complex (Tambini et al. 2010). In this study, resting states before and after stimulus presentation were scanned, from which BOLD correlations between the hippocampus and cortices were computed. After visual stimuli were presented in a 20-min session, resting states were scanned again. Participants performed a memory test about what they saw during the experiment outside of the scanner after completing the scanning session. It was revealed that high subsequent memory was predicted by the increase in BOLD correlation between the hippocampus and the lateral occipital complex during rest following presentation of visual stimuli. More recently, using MVPA, association between reactivation after stimulus presentation and subsequent memory performance was investigated on a trial-by-trial basis (Deuker et al. 2013; Staresina et al. 2013). The basic procedure of these studies is training the classifier based on the fMRI data during stimulus presentation, and then applying it to the fMRI data during rest or a low cognitive demand task (i.e., self-paced odd/even judgment) following the stimulus presentation session. These studies consistently reported that visual stimuli were spontaneously replayed without any sensory cues and, more importantly, such replay/reactivation of the occipital/temporal cortices predicted which visual stimuli were subsequently remembered.

2.5 Real-Time fMRI

One of the most important challenges in neuroscience is showing causality of neural activity in behavior. The efficiency of memory formation varies over time for any individual, which may be, at least partly, explained by fluctuation of the brain state.

The subsequent memory paradigm referred to above shows an association between brain activity and memory performance. However, in recent neuroimaging studies researchers have tried to provide evidence for the causality. In Yoo et al. (2012), activation of the parahippocampal cortex was monitored in real-time, and scene presentations were triggered when participants entered "good" or "bad" brain states for the learning of novel scenes (Yoo et al. 2012). They revealed that subsequent recognition memory was more accurate for scenes presented in "good" than in "bad" brain states, suggesting that using real-time neuroimaging, education and training can be accelerated. Real-time fMRI is used not only for triggering the stimulus presentation, but also for delivering the information about brain states to participants during the experiment, which is called "neurofeedback." Neurofeedback is a common approach in the field of electroencephalography (Hammond 2011), in which participants can learn how to control their brain states by receiving the real-time feedback about their brain states. The difference between real-time EEG and fMRI is spatial resolution. The real-time fMRI enabled region-specific feedback to be given. It has been revealed by various previous studies that real-time fMRI neurofeedback is effective in eliciting the self-control of localized BOLD signals (Sulzer et al. 2013). A recent study investigated the utility of real-time fMRI neurofeedback as a tool to enhance human cognition by training healthy individuals to consciously control activity in the left dorsolateral prefrontal cortex (DLPFC) (Sherwood et al. 2016). In this study, half of participants received real-time fMRI neurofeedback from the left DLPFC in five training sessions across 2 weeks, whereas the other half in the control group underwent similar training outside the MRI and without real-time fMRI neurofeedback. Working memory performance was evaluated on the first and last days, using computerized tests. The group with real-time fMRI neurofeedback showed a significant increase in the ability to self-regulate the BOLD signal in the left DLPFC. Furthermore, the group with real-time fMRI neurofeedback showed improved task performance relative to the control group. These studies suggested that, by combining real-time fMRI, the causality of activity of the specific brain region in the subsequent memory performance could be tested. Furthermore, we may potentially manipulate the ability to learn by neurofeedback training, although this field of studies currently lacks strong evidence. Thus, further examination, including reproducibility, is required.

2.6 Concluding Remarks

Neuroimaging studies have significantly contributed to the further understanding of the human memory system. A recent technical advancement in MVPA enabled decoding of brain states from neuroimaging data (Haynes and Rees 2006). MVPA improves signal sensitivity by taking advantage of multiple voxel information; however, it is not merely an issue of sensitivity, but rather, it also accompanies a type of paradigm shift: brain activation revealed by univariate analysis indicates involvement, whereas pattern information indicates representational content (Mur et al.

2009). In the field of memory studies, this means that neural representations could be decoded on a trial-by-trial basis. As described above, this allows the influence of spontaneous recall on subsequent memory performance to be examined (Deuker et al. 2013; Staresina et al. 2013). Combining real-time fMRI with MVPA may be promising (Stoeckel et al. 2014), although relatively low temporal resolution (~4-s delay to peak activity from the onset) should be solved by analysis advancement or avoided using the experimental design.

References

Atkinson RC, Juola JF (1973) Factors influencing speed and accuracy of word recognition. Atten Perform IV:583–612

Atkinson RC, Herrmann DJ, Wescourt KT (1974) Search processes in recognition memory. In: Solso R (ed) Theories in cognitive psychology: the Loyola symposium. Halsted Press, New York, pp 193–238

Buzsaki G (1996) The hippocampo-neocortical dialogue. Cereb Cortex 6(2):81–92

Cabeza R, Ciaramelli E, Moscovitch M (2012) Cognitive contributions of the ventral parietal cortex: an integrative theoretical account. Trends Cogn Sci 16(6):338–352

Daselaar SM, Prince SE, Cabeza R (2004) When less means more: deactivations during encoding that predict subsequent memory. NeuroImage 23(3):921–927

Daselaar SM, Fleck MS, Cabeza R (2006) Triple dissociation in the medial temporal lobes: recollection, familiarity, and novelty. J Neurophysiol 96(4):1902–1911

Davachi L, Mitchell JP, Wagner AD (2003) Multiple routes to memory: distinct medial temporal lobe processes build item and source memories. Proc Natl Acad Sci U S A 100(4):2157–2162

Deuker L, Olligs J, Fell J, Kranz TA, Mormann F, Montag C, Reuter M, Elger CE, Axmacher N (2013) Memory consolidation by replay of stimulus-specific neural activity. J Neurosci 33(49):19373–19383

Diana RA, Yonelinas AP, Ranganath C (2007) Imaging recollection and familiarity in the medial temporal lobe: a three-component model. Trends Cogn Sci 11(9):379–386

Dickerson BC, Miller SL, Greve DN, Dale AM, Albert MS, Schacter DL, Sperling RA (2007) Prefrontal-hippocampal-fusiform activity during encoding predicts intraindividual differences in free recall ability: an event-related functional-anatomic MRI study. Hippocampus 17(11):1060–1070

Dolcos F, Denkova E, Dolcos S (2012) Neural correlates of emotional memories: a review of evidence from brain imaging studies. Psychologia 55(2):80–111

Duverne S, Motamedinia S, Rugg MD (2009) The relationship between aging, performance, and the neural correlates of successful memory encoding. Cereb Cortex 19(3):733–744

Fernandez G, Tendolkar I (2001) Integrated brain activity in medial temporal and prefrontal areas predicts subsequent memory performance: human declarative memory formation at the system level. Brain Res Bull 55(1):1–9

Garoff RJ, Slotnick SD, Schacter DL (2005) The neural origins of specific and general memory: the role of the fusiform cortex. Neuropsychologia 43(6):847–859

Gilmore AW, Nelson SM, McDermott KB (2015) A parietal memory network revealed by multiple MRI methods. Trends Cogn Sci 19(9):534–543

Girardeau G, Benchenane K, Wiener SI, Buzsaki G, Zugaro MB (2009) Selective suppression of hippocampal ripples impairs spatial memory. Nat Neurosci 12(10):1222–1223

Hammond DC (2011) What is neurofeedback: an update. J Neurother 15(4):305–336

Haynes JD, Rees G (2006) Decoding mental states from brain activity in humans. Nat Rev Neurosci 7(7):523–534

Henson RN, Rugg MD, Shallice T, Josephs O, Dolan RJ (1999) Recollection and familiarity in recognition memory: an event-related functional magnetic resonance imaging study. J Neurosci 19(10):3962–3972
Hirose S, Chikazoe J, Jimura K, Yamashita K, Miyashita Y, Konishi S (2009) Sub-centimeter scale functional organization in human inferior frontal gyrus. NeuroImage 47(2):442–450
Huijbers W, Schultz AP, Vannini P, McLaren DG, Wigman SE, Ward AM, Hedden T, Sperling RA (2013) The encoding/retrieval flip: interactions between memory performance and memory stage and relationship to intrinsic cortical networks. J Cogn Neurosci 25(7):1163–1179
Jadhav SP, Kemere C, German PW, Frank LM (2012) Awake hippocampal sharp-wave ripples support spatial memory. Science 336(6087):1454–1458
Kahn I, Davachi L, Wagner AD (2004) Functional-neuroanatomic correlates of recollection: implications for models of recognition memory. J Neurosci 24(17):4172–4180
Kensinger EA, Schacter DL (2006) Amygdala activity is associated with the successful encoding of item, but not source, information for positive and negative stimuli. J Neurosci 26(9):2564–2570
Kim H (2011) Neural activity that predicts subsequent memory and forgetting: a meta-analysis of 74 fMRI studies. NeuroImage 54(3):2446–2461
Kim H, Cabeza R (2007) Trusting our memories: dissociating the neural correlates of confidence in veridical versus illusory memories. J Neurosci 27(45):12190–12197
Kim H, Daselaar SM, Cabeza R (2010) Overlapping brain activity between episodic memory encoding and retrieval: roles of the task-positive and task-negative networks. NeuroImage 49(1):1045–1054
Kirwan CB, Stark CE (2004) Medial temporal lobe activation during encoding and retrieval of novel face-name pairs. Hippocampus 14(7):919–930
Konishi S, Wheeler ME, Donaldson DI, Buckner RL (2000) Neural correlates of episodic retrieval success. NeuroImage 12(3):276–286
Mandler G (1979) Organization and repetition: organizational principles with special reference to rote learning. In: Nilsson L (ed) Perspectives on memory research. Wiley, New York, pp 293–327
Mandler G, Pearlstone Z, Koopmans HS (1969) Effects of organization and semantic similarity on recall and recognition. J Verbal Learn Verbal Behav 8(3):410–423
Mason MF, Norton MI, Van Horn JD, Wegner DM, Grafton ST, Macrae CN (2007) Wandering minds: the default network and stimulus-independent thought. Science 315(5810):393–395
McDermott KB, Jones TC, Petersen SE, Lageman SK, Roediger HL 3rd (2000) Retrieval success is accompanied by enhanced activation in anterior prefrontal cortex during recognition memory: an event-related fMRI study. J Cogn Neurosci 12(6):965–976
Mur M, Bandettini PA, Kriegeskorte N (2009) Revealing representational content with pattern-information fMRI – an introductory guide. Soc Cogn Affect Neurosci 4(1):101–109
Otten LJ, Rugg MD (2001) When more means less: neural activity related to unsuccessful memory encoding. Curr Biol 11(19):1528–1530
Paller KA, Kutas M, Mayes AR (1987) Neural correlates of encoding in an incidental learning paradigm. Electroencephalogr Clin Neurophysiol 67(4):360–371
Polyn SM, Natu VS, Cohen JD, Norman KA (2005) Category-specific cortical activity precedes retrieval during memory search. Science 310(5756):1963–1966
Raichle ME, MacLeod AM, Snyder AZ, Powers WJ, Gusnard DA, Shulman GL (2001) A default mode of brain function. Proc Nat Acad Sci U S A 98(2):676–682
Ritchey M, Wing EA, LaBar KS, Cabeza R (2013) Neural similarity between encoding and retrieval is related to memory via hippocampal interactions. Cereb Cortex 23(12):2818–2828
Rugg MD, Vilberg KL (2013) Brain networks underlying episodic memory retrieval. Curr Opin Neurobiol 23(2):255–260
Rugg MD, Fletcher PC, Frith CD, Frackowiak RS, Dolan RJ (1996) Differential activation of the prefrontal cortex in successful and unsuccessful memory retrieval. Brain 119(Pt 6):2073–2083
Scoville WB, Milner B (1957) Loss of recent memory after bilateral hippocampal lesions. J Neurol Neurosurg Psychiatry 20(1):11–21

Sherwood MS, Kane JH, Weisend MP, Parker JG (2016) Enhanced control of dorsolateral prefrontal cortex neurophysiology with real-time functional magnetic resonance imaging (rt-fMRI) neurofeedback training and working memory practice. NeuroImage 124(Pt A):214–223

Simons JS, Spiers HJ (2003) Prefrontal and medial temporal lobe interactions in long-term memory. Nat Rev Neurosci 4(8):637–648

Sommer T, Rose M, Weiller C, Buchel C (2005) Contributions of occipital, parietal and parahippocampal cortex to encoding of object-location associations. Neuropsychologia 43(5):732–743

Spaniol J, Davidson PS, Kim AS, Han H, Moscovitch M, Grady CL (2009) Event-related fMRI studies of episodic encoding and retrieval: meta-analyses using activation likelihood estimation. Neuropsychologia 47(8–9):1765–1779

Squire LR (1982) The neuropsychology of human memory. Annu Rev Neurosci 5:241–273

Squire LR, Alvarez P (1995) Retrograde amnesia and memory consolidation: a neurobiological perspective. Curr Opin Neurobiol 5(2):169–177

Staresina BP, Alink A, Kriegeskorte N, Henson RN (2013) Awake reactivation predicts memory in humans. Proc Natl Acad Sci U S A 110(52):21159–21164

Stoeckel LE, Garrison KA, Ghosh S, Wighton P, Hanlon CA, Gilman JM, Greer S, Turk-Browne NB, deBettencourt MT, Scheinost D, Craddock C, Thompson T, Calderon V, Bauer CC, George M, Breiter HC, Whitfield-Gabrieli S, Gabrieli JD, LaConte SM, Hirshberg L, Brewer JA, Hampson M, Van Der Kouwe A, Mackey S, Evins AE (2014) Optimizing real time fMRI neurofeedback for therapeutic discovery and development. Neuroimage: Clinical 5:245–255

Sulzer J, Haller S, Scharnowski F, Weiskopf N, Birbaumer N, Blefari ML, Bruehl AB, Cohen LG, DeCharms RC, Gassert R, Goebel R, Herwig U, LaConte S, Linden D, Luft A, Seifritz E, Sitaram R (2013) Real-time fMRI neurofeedback: progress and challenges. NeuroImage 76:386–399

Sutherland GR, McNaughton B (2000) Memory trace reactivation in hippocampal and neocortical neuronal ensembles. Curr Opin Neurobiol 10(2):180–186

Tambini A, Ketz N, Davachi L (2010) Enhanced brain correlations during rest are related to memory for recent experiences. Neuron 65(2):280–290

Tulving E (1999) Memory, consciousness and the brain: the Tallinn conference. Psychology Press, Philadelphia

Uncapher MR, Rugg MD (2009) Selecting for memory? The influence of selective attention on the mnemonic binding of contextual information. J Neurosci 29(25):8270–8279

Wagner AD, Schacter DL, Rotte M, Koutstaal W, Maril A, Dale AM, Rosen BR, Buckner RL (1998) Building memories: remembering and forgetting of verbal experiences as predicted by brain activity. Science 281(5380):1188–1191

Wagner AD, Shannon BJ, Kahn I, Buckner RL (2005) Parietal lobe contributions to episodic memory retrieval. Trends Cogn Sci 9(9):445–453

Weissman DH, Roberts KC, Visscher KM, Woldorff MG (2006) The neural bases of momentary lapses in attention. Nat Neurosci 9(7):971–978

Wheeler ME, Buckner RL (2004) Functional-anatomic correlates of remembering and knowing. NeuroImage 21(4):1337–1349

Wilson MA, McNaughton BL (1994) Reactivation of hippocampal ensemble memories during sleep. Science 265(5172):676–679

Yamashita K, Hirose S, Kunimatsu A, Aoki S, Chikazoe J, Jimura K, Masutani Y, Abe O, Ohtomo K, Miyashita Y, Konishi S (2009) Formation of long-term memory representation in human temporal cortex related to pictorial paired associates. J Neurosci 29(33):10335–10340

Yonelinas AP (2002) The nature of recollection and familiarity: a review of 30 years of research. J Mem Lang 46(3):441–517

Yoo JJ, Hinds O, Ofen N, Thompson TW, Whitfield-Gabrieli S, Triantafyllou C, Gabrieli JD (2012) When the brain is prepared to learn: enhancing human learning using real-time fMRI. NeuroImage 59(1):846–852

Part II
Interacting Mechanisms of Human Memory with Social Cognition

Chapter 3
Psychological and Neural Functions of Working Memory in Social Contexts

Yukio Maehara

Abstract In everyday life, we need to temporally maintain and manipulate various pieces of information in our mind to achieve social and interpersonal goals. Such a temporal memory system that supports goal-directed behavior is called working memory. This article reviews how working memory functions in social contexts from both a psychological and a neuroscience perspective. Many neuroimaging studies have revealed that the frontoparietal network (FPN) is activated when working memory keeps information available independent of the information modality and that the default mode network (DMN) is deactivated when working memory is employed. Thus, the FPN and DMN often show an inverse pattern of activation. Psychological studies investigating the contributions of working memory to mind reading or "theory of mind" (ToM) show that working memory plays a role in keeping available information that is necessary for reasoning about the mental states of others. Therefore, a psychological perspective assumes that FPN activation accompanied by DMN deactivation might support ToM. However, the DMN is also involved in representing self-generated thoughts that are not perceived by our senses, including the mental states of others. Indeed, the DMN includes a set of brain regions that are indispensable to social cognition. Therefore, a neuroscience perspective enables us to assume that balancing and switching between the FPN, which is associated with working memory, and the DMN, which is associated with representing social information, might optimize social cognition. Finally, anticipated avenues for future research on working memory in social cognition are suggested.

Keywords Working memory • Attention • Frontoparietal network • Mind wandering • Default mode network • Theory of mind • Social cognition • Salience network • Joint cognition • Executive function

---

Y. Maehara (✉)
Faculty of Education, Nagasaki University, 1-14 Bunkyo-machi, Nagasaki 852-8521, Japan
e-mail: ymaehara@nagasaki-u.ac.jp

T. Tsukiura, S. Umeda (eds.), *Memory in a Social Context*,
https://doi.org/10.1007/978-4-431-56591-8_3

3.1 Introduction

When constructing a bookshelf with sections, we maintain the features of the boards depicted in the instructions while searching for the left-side and bottom boards so that we do not assemble the bookshelf incorrectly. When preparing a family dinner, we recall the recipe for another dish and maintain what we should do next in memory while we are boiling water and cutting vegetables so that we can cook several dishes at the same time. When shopping for clothes, we retain the colors and designs of clothes we have liked and imagine possible combinations with other clothes in our wardrobe while moving from store to store to decide what we should buy. These examples show that many everyday activities require us to temporarily retain some pieces of information in parallel with other information processing to make later decisions or actions to achieve a goal.

Undoubtedly, temporal maintenance of information for use in subsequent goal-directed cognitive activities is an indispensable ability in our daily life. Such an active, temporal memory function is generally called working memory (Baddeley 2012; Baddeley and Hitch 1974). As long as we live in a human society, we need to regulate our behaviors in pursuit of desirable social and interpersonal goals. These goals may be to understand correctly and respond adequately to what others say, to treat everyone kindly without discrimination, and to develop and maintain personal relationships with those around us. For instance, during conversations we must find and maintain essential points when listening to what others are saying so that we give adequate responses and ask questions when it is our turn to talk. Working memory may be needed for most social activities. However, to date, psychological and neural studies of working memory in social situations are not vigorous or systematic relative to research examining working memory in more purely cognitive domains, such as language comprehension and logical reasoning. Determining the specific contributions of working memory in social and interpersonal contexts may provide valuable insights into educational practice and cognitive interventions to improve social skills in individuals with disorders of social cognition and working memory, including major depression (Baune et al. 2014; Mulders et al. 2015), anxiety disorder (Koban and Pourtois 2014; Ouimet et al. 2009), schizophrenia (Lett et al. 2014; Sheffield and Barch 2016; Whitfield-Gabrieli and Ford 2012), and autism spectrum disorder (Kercood et al. 2014; Uddin et al. 2013).

This chapter first outlines the cognitive and neural underpinnings of working memory. Then, the relationship between working memory and mind wandering is reviewed from both behavioral and neural perspectives. Next, the contribution of working memory to mind-reading or "theory of mind," which is one of the most basic abilities for social interactions, is discussed. Finally, anticipated avenues of future research into working memory in social cognition are suggested.

3.2 Cognitive and Neural Bases of Working Memory

Working memory refers to an emergent cognitive system generated from brain networks that provide temporary maintenance and manipulation of information necessary for complex cognitive tasks such as language comprehension, spatial manipulation, mental calculation, and reasoning (Baddeley 1986, 2007; Baddeley and Hitch 1974). Although short-term memory simply stores information for subsequent recall, working memory simultaneously and actively stores and processes information for subsequent cognitive activities. However, because short-term memory can be considered a function of working memory under conditions where there is no simultaneous cognitive load, some researchers have suggested that short-term memory is on the same continuum as working memory (Brouillet et al. 2011; Unsworth and Spillers 2010).

A variety of working memory models have been proposed (Miyake and Shah 1999) since Baddeley (1986) proposed the three-component model, which is composed of the central executive, phonological loop, and visuo-spatial sketchpad. Almost all multi-component models of working memory include an attentional control system and information stores. For instance, in Baddeley's (1986) original model, the central executive divides limited-capacity attention among several cognitive activities in an attentional control system, the phonological loop maintains and rehearses speech-based information, and the visuo-spatial sketchpad maintains and manipulates visual images. In his more recent models, Baddeley (2000, 2012) proposed a fourth component, the episodic buffer, which maintains binding information integrated from several sources (e.g., visual and verbal). Although there are differences among the models proposed by various researchers, it is generally assumed that working memory consists of a domain-general attentional control system and domain-specific information stores. This common view is compatible with neuropsychological and neuroimaging studies (Curtis and D'Esposito 2003; Smith and Jonides 1997; Wager and Smith 2003).

Neuroimaging studies investigating the neural bases of working memory consistently report activation in a lateral frontoparietal network (FPN) consisting of the dorsolateral prefrontal cortex (DLPFC; BA 9/46), the supplementary motor area (BA 6), and the intraparietal cortex (IPC; BA 7) (Curtis and D'Esposito 2003; Linden 2007; Wager and Smith 2003). These regions shown in Fig. 3.1 are commonly activated across several modalities (spatial, visual, verbal), and thus seem to correspond to modality-independent attentional control. Activity in these regions increases as the amount of information to be maintained in working memory increases (Braver et al. 1997; D'Esposito et al. 1995; Rypma et al. 1999). The n-back task, which is one of the most prevalent tasks in neuroimaging studies of working memory (Braver et al. 1997; Owen et al. 2005), requires participants to monitor a series of stimuli and to determine if a current stimulus is the same as or different from the one presented n stimuli before. Thus, this task relies heavily on participants' ability to simultaneously maintain and update information, and working memory demands increase as n increases. Using the n-back task, Braver et al.

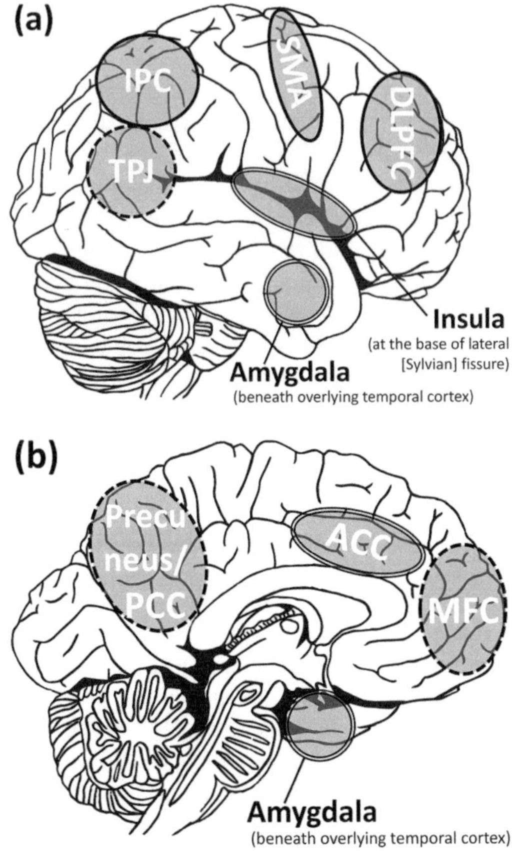

Fig. 3.1 (**a**) The lateral surface and (**b**) the medial surface of the brain. Solid-line ellipses represent the principal regions composing the frontoparietal network (FPN); broken-line ellipses represent the regions common to both the default mode network (DMN) and the social cognition network; double-line ellipses represent the major regions composing the salience network involved in switching between the FPN and DMN

(1997) showed that activation in DLPFC increased as *n* increased. In addition, individual differences in FPN activity correlate with working memory performance (Crone et al. 2006; Gray et al. 2003; Vogel and Machizawa 2004). For example, Gray et al. (2003) found that individual differences in DLPFC and IPC activation mediate the relation between general fluid intelligence (Raven's Advanced Progressive Matrices) and working memory performance (three-back task with a high level of interference).

In contrast to multimodal activation of the FPN, the maintenance of spatial, visual, and verbal materials frequently activates the superior parietal cortex (dorsal stream of visual processing), the inferior temporal cortex (ventral stream of visual processing), and the inferior frontal cortex (in the vicinity of Broca's area) respectively (Wager and Smith 2003). These regions seem to correspond to modality-specific information storage in multi-component working memory models. Maintenance of more social stimuli (e.g., human faces) in working memory should recruit different brain areas than those necessary for maintaining nonsocial stimuli. Using the *n*-back task with faces, Druzgal and D'Esposito (2001) found increased activation with increases in *n* in the fusiform face area (FFA), a brain region that shows selective activation for faces, and in the lateral prefrontal area. However, activation in the fusiform object area (FOA), a brain region that shows selective activation for nonface objects, did not increase as face *n*-back task demands increased. These findings indicate that the FFA, but not the FOA, plays a domain-specific role in maintaining faces in working memory. Another fMRI study also revealed that the amygdala and lateral orbitofrontal cortex, but not the FFA, exhibit elevated activity during maintenance of facial emotion, which indicates that there are domain-specific regions involved in storing emotional information in working memory (LoPresti et al. 2008).

3.3 Mind-Wandering and Working Memory

Mind-wandering refers to a spontaneous or unintended shift from ongoing tasks and events in the external environment to self-generated thoughts and feelings in the internal state (Smallwood and Schooler 2015). In daily life, we easily allow our mind to wander and engage in thoughts unrelated to the here and now. In fact, mind-wandering accounts for 50% of our waking time (Killingsworth and Gilbert 2010). Kane and colleagues (2007) investigated how individual differences in working memory capacity are related to mind-wandering in everyday life using experience-sampling methodology. Participants first completed several working memory span tasks to screen participants for high and low working memory capacity. In the experience-sampling study, the screened participants were told to carry a mobile data terminal with them for a week. The terminal signaled participants to answer a set of questions on the terminal eight times per day. The questions were related to on-task thoughts and mind-wandering episodes and were answered using a Likert scale. Participants with both high and low working memory capacity showed more mind-wandering when the current task required less concentration, and only participants with a low working memory capacity exhibited more mind-wandering when the current task was more challenging.

The findings of Kane et al. (2007) raise the possibility that there are at least two different types of mind-wandering: One occurs when the current task is easy enough to afford dividing attention to task-unrelated thoughts, and the other occurs in people with a low working memory capacity when the current task demands executive attention. The former type of mind-wandering is consistent with Mason et al. (2007), who found that participants reported more mind-wandering when engaged in a practiced compared with a novel working memory task because the practiced task did not require as much concentration. The latter type of mind wandering, which is caused by the interaction between working memory capacity and task difficulty, often has an obstructive influence on everyday cognition. For example, although it is well known that working memory capacity predicts reading comprehension (Daneman and Merikle 1996), it was recently shown that mind-wandering mediates the relationship between working memory capacity and reading comprehension (McVay and Kane 2012; Unsworth and McMillan 2013). In normal reading, working memory integrates and maintains external representations generated from lexical and propositional processing with internal representations retrieved from long-term memory to form a situational model, which provides a global context for interpreting the narrative events and helps readers link general knowledge to the specific episode in the sentences (Zwaan and Radvansky 1998). It is possible that mind-wandering prevents the formation of a situation model (Smallwood 2011). Individuals with a smaller working memory capacity are more vulnerable to mind-wandering while reading, which makes them pay more attention to the internal world (i.e., long-term memory representations) and less attention to the external world (i.e., lexical and propositional processing), such that they fail to construct a detailed situational model that supports reading comprehension.

During mind-wandering, the default mode network (DMN) exhibits increased neural activity (Allen et al. 2013; Christoff et al. 2009; Mason et al. 2007). The DMN is a distributed network of brain regions that show increased activation during passive control conditions, such as resting state, and relative deactivation during externally focused, attention-demanding, goal-directed tasks, such as working memory tasks (Raichle et al. 2001). Thus, the DMN and FPN show an inverse pattern of activation at rest and during working memory tasks (Greicius and Menon 2004; McKiernan et al. 2003; Raichle et al. 2001). Main cerebral cortices composing the DMN include the medial prefrontal cortex (MPFC; BA 9/10), the medial precuneus (BA 7) extending into the posterior cingulate cortex (PCC; BA 23/31), the temporo-parietal junction (TPJ; BA 39), the middle temporal gyrus (BA 21), and the temporal pole (BA 38) (Buckner et al. 2008; Gusnard and Raichle 2001; Yeo et al. 2011).

Because the DMN is associated with mind-wandering (Mason et al. 2007), one prediction is that the FPN might show increased deactivation during mind-wandering. However, an fMRI study suggests that the DMN and FPN sometimes co-activate during mind-wandering, and that neural recruitment in both networks is strongest when participants are unaware of their own mind wandering (Christoff et al. 2009). This may depend on the content of the self-generated thought during mind-wandering. The DMN and FPN show co-activation during autobiographical planning (Spreng et al. 2010), scene construction (Summerfield et al. 2010), simulation for personal goal pursuit (Gerlach et al. 2011), processing and maintenance of social information (Meyer et al. 2012), and generation of creative ideas (Ellamil et al. 2012). When engaging in these mental activities, people must concentrate on spontaneous, internally generated thoughts independently from immediate, externally triggered events, and transform and maintain representations of the self-generated thoughts to achieve the task goal. Hence, it is likely that when an internal stream of thought is generated and sustained in working memory, both the DMN and FPN are simultaneously recruited, even during mind-wandering (Smallwood et al. 2011). These results indicate that working memory is necessary for keeping available self-generated representations that cannot be directly perceived, such as the mental states of others.

Fig. 3.2 An illustration of the Sally-Anne false-belief task used to assess children's theory of mind (Pictures from Maehara 2014, Reproduced from Kyoto University Press)

3.4 Theory of Mind and Working Memory

We cannot directly look inside the minds of others. However, we can infer what others know and think to a certain extent. The ability to reason about mental states including beliefs, intentions, emotions, and knowledge is indispensable to our daily life. In cognitive science, this mind-reading ability is called "theory of mind" (ToM), and there are many studies on its typical and atypical development, and cognitive and neural mechanisms (Baron-Cohen et al. 2013). The false belief task is often used as a measure of whether a person can adequately use ToM (Wellman et al. 2001; Wimmer and Perner 1983). In this task, a person is given facts about a certain situation and then asked to reason about the mental states of another person who is not aware of the facts. In one version of the false belief task known as the Sally-Anne task (Fig. 3.2) (Baron-Cohen et al. 1985), a child listens to the following story and is asked to predict the behavior of the protagonist: "Sally places her toy in a red box and then goes outside to play. In her absence, another girl, Anne, moves the toy from the red box to a green box and then goes outside to play. Sally comes back and wants her toy. Where do you think Sally will look first?" Sally has a false belief about the toy's location, so the correct answer is the red box, not the green box that actually contains the toy. However, most 3-year-olds give the wrong answer (i.e., green box) even though the protagonist (Sally) does not know that fact, because they believe that what they know is equivalent to what others know.

Children with typical development generally pass false belief tasks by around 4 years of age (Perner 1991; Wellman et al. 2001). Development of working memory is considered an important factor for ToM in young children; indeed, developmental studies with preschoolers have confirmed that performance in the false belief tasks correlates with working memory capacity (Carlson et al. 2002; Davis and Pratt 1995; Keenan et al. 1998). However, studies with young children do not elucidate the specific cognitive mechanism by which working memory helps to solve the false belief task. This is because most of those studies adopt a correlational approach, and it is unclear whether working memory is needed for the acquisition or expression of ToM (Moses 2001). Thus, research with adults who have mature ToM is necessary to examine whether and how working memory contributes to ToM expression.

McKinnon and Moscovitch (2007) asked participants to read complex story sentences that included irony, humor, or white lies and then answer ToM questions regarding the mental states of the story's characters. In Experiment 1, elderly participants (mean age = 78 years) showed lower performance in advanced ToM problems than young participants (mean age = 20 years). In Experiment 4, young participants showed reduced performance in advanced ToM problems that was similar to performance observed in elderly participants in Experiment 1 when they were asked to simultaneously perform a two-back task that required working memory to be updated. This supports the notion that working memory contributes to the use of ToM in adults. Bull et al. (2008), however, revealed that a simultaneous two-back task while reading reduced performance for both ToM and non-ToM questions (e.g., physical causation). Thus, a secondary task that requires working memory generally interferes with the reading comprehension of complex sentences, not purely with reasoning about mental states. In the director task, which is a visual perspective-taking task, Lin et al. (2010) found that participants made more egocentric mistakes that neglected the partner's perspective when they were asked to maintain four two-digit numbers (high working memory load) than when they were asked to maintain one two-digit number (low working memory load). This research did not use complex sentences as materials, so that it could solve the problem that Bull et al. (2008) found; however, it did not ask what others "think" in the same way as standard ToM tasks. Furthermore, because cognitive load was weighted on the whole process of the task, it is not clear which processes (i.e., comprehension of task requirements or consideration of the partner's view) were affected by cognitive load.

Maehara and Saito (2011, 2013) proposed a working memory account of ToM, which assumes that first, more available information, including temporally new and/or perceptually salient information, is predominantly used for reasoning about the mental states of others, and second, the role of working memory is to keep activating necessary but less available information during mental-state reasoning. According to this account, the information that is most available in a false belief task while participants are reasoning about the mental states of a character is the fact that is temporally newest; therefore, working memory should activate the less available information that the character is ignorant of that fact. This account predicts that maintaining extra information in working memory would disrupt activation of the less available information and promote the use of more readily available information, because the amount of information that working memory can activate at a given time is limited (Cowan 2005). In Maehara and Saito (2011), adult participants read a simple ToM story (such as the Sally-Anne task), and were then asked to maintain a meaningless two-letter or seven-letter English string (i.e., low or high working memory load), and then estimate the probability of possible choices for a naive protagonist's behavior. This cognitive load manipulation did not interfere with reading comprehension for the ToM story. As a result, compared with participants with a low working memory load, participants with a high working memory load assigned a significantly higher probability to the choice indicating that the protagonist would behave as if she had known the fact. This suggests that cognitive load on working memory might have interfered with the use of information about the protagonist's mental state, which supports the notion that working memory keeps activating relevant but less accessible information.

3.5 Neural Perspectives on the Relationship Between ToM and Working Memory

Many neuroimaging and neuropsychological studies have sought to uncover the neural substrates of ToM (for influential reviews, see Amodio and Frith 2006; Carrington and Bailey 2009; Gallagher and Frith 2003; Mahy et al. 2014; Saxe et al. 2004). Meta-analyses of neuroimaging studies on social cognition commonly suggest that MPFC, TPJ, and PCC extending into the medial precuneus might be crucial regions for ToM that are activated when people are reasoning about mental states, irrespective of the task (Bzdok et al. 2012; Schurz et al. 2014; Van Overwalle 2009). MPFC is involved in a broad range of social cognitive processes, and, although incompletely specified, its function in ToM may be to represent knowledge about the self and others (Wagner et al. 2012). TPJ is thought to be involved in shifting attention between the self and the other's perspectives (Corbetta et al. 2008). PCC is thought to be involved in representing or simulating the perspective of others (Cavanna and Trimble 2006).

However, few studies have investigated the relationship between ToM and working memory from a neuroscience perspective. Stone et al. (1998) asked patients with frontal lobe lesions to complete false belief tasks with either high or low working memory load. Patients with DLPFC lesions had a lower performance in the high working memory load task; however, other patients with OFC (orbitofrontal cortex) lesions did not show such a performance deficit for the high working memory load task. Other studies of patients with traumatic brain injury also indicate that working memory plays a more critical role than inhibitory control in solving ToM tasks (Dennis et al. 2009; Honan et al. 2015). A study using repetitive transcranial magnetic stimulation (rTMS) with healthy adult participants (Costa et al. 2008) showed that rTMS over DLPFC delayed response times for false belief reasoning and reduced accuracy for the faux-pas task, which required participants to explain the mental states of story characters who unintentionally made a slip of the tongue. Therefore, neuropsychological research suggests that impaired DLPFC function, which is a core region of the FPN, might impair ToM expression. DLPFC may activate various representations necessary for reasoning with regard to others' mental states and keep them available. Although these observations support the importance of working memory in ToM, it should be noted that working memory is not necessarily a requirement for solving ToM problems. Indeed, in Stone et al. (1998), DLPFC patients did not show impaired performance on false belief tasks when there was no working memory load. Additionally, in Maehara and Saito (2011), participants with high working memory load did not attribute their knowledge about a fact to a story's character when the assumption that she would behave according to the fact was implausible. That is, even if working memory is impaired, ToM performance is not impaired in every situation. Working memory may not be required for ToM expression whenever people are reasoning with regard to the mental states of others, but may be necessary when there are many pieces of information that must be activated for reasoning.

The DMN offers another perspective on the relationship between working memory and ToM. There is a notable overlap between the DMN and brain regions involved in social cognition, including ToM; the overlap is specifically observed in MPFC, TPJ, and PCC extending into the medial precuneus shown in Fig. 3.1 (Hyatt et al. 2015; Mars et al. 2012; Reniers et al. 2012; Spreng and Grady 2010). Furthermore, DMN activity at rest may be a psychological baseline for social cognition, because in human evolution social cognitive skills have become so important for creating social bonds that we need to be always ready to use them (Schilbach et al. 2008). In an fMRI study using mentalizing and memory tasks, Hyatt et al. (2015) specified ten default mode subnetworks by spatial independent component analysis and then showed that three of them (the TPJ, PCC/precuneus, and MPFC subnetworks) were specifically engaged in mentalizing. The FPN and DMN often show an inverse pattern of activity (Greicius and Menon 2004; McKiernan et al. 2003; Raichle et al. 2001), so that cognitive load on working memory both activates the FPN and deactivates the DMN or regions related to ToM (i.e., MPFC, TPJ, PCC). Behavioral studies assume that excessive working memory load interferes with the use of information necessary for reasoning about the mental states of others (Lin et al. 2010; Maehara and Saito 2011; McKinnon and Moscovitch 2007). In contrast, neurofunctional findings suggest that overloaded working memory disrupts activation of ToM-related brain regions that play a crucial role in representing the mental states of others. Because the mental states of others are not physical stimuli that can be perceived by our senses, those representations are inherently self-generated. Therefore, an effort to maintain and manipulate external attention-demanding stimuli in working memory may lead to decreased DMN activity that prevents the production of internal, self-generated thoughts, including representations about the mental states of others. ToM performance may deteriorate when brain activity is excessively biased toward the FPN. Accordingly, when participants hold excessive to-be-maintained items in working memory in behavioral studies (Lin et al. 2010; Maehara and Saito 2011; McKinnon and Moscovitch 2007), the FPN may exhibit increased activity, and consequently, participants may fail to form clear representations about the mental states of others because of DMN deactivation.

Taken together, there is a distinct difference between psychological and neural perspectives on how working memory contributes to ToM. The psychological perspective argues that working memory plays a role in activating necessary but less available information for reasoning about the mental states of others. In contrast, the neural perspective assumes that when working memory is not occupied by a large amount of information, the DMN can generate representations about the mental states of others. Behavioral and neuroimaging studies need to cooperate and create an integrated account of the functions of working memory in social contexts.

3.6 Interaction Between Behavioral and Neuroscience Research to Explore Interactions Between Working Memory and Social Cognition

Psychiatric disorders called "social brain disorders," which are characterized by difficulties in social cognitive skills and interpersonal interactions, are also often described as "working memory disorders." These include major depression (Baune et al. 2014; Mulders et al. 2015), anxiety disorder (Koban and Pourtois 2014; Ouimet et al. 2009), schizophrenia (Lett et al. 2014; Sheffield and Barch 2016; Whitfield-Gabrieli and Ford 2012), and autism spectrum disorder (Kercood et al. 2014; Uddin et al. 2013). These psychiatric disorders may not be due to impairments in either the FPN or DMN, but rather impairments in a regulation system for both networks. The salience network that includes the anterior insula (BA 13), anterior cingulate cortex (ACC; BA 24), and amygdala is a candidate regulation system (Uddin 2015). It is activated not only when responding to salient stimuli, including interoceptive status (Seeley et al. 2007), but also when switching between the FPN and DMN (Goulden et al. 2014; Sidlauskaite et al. 2014; Sridharan et al. 2008). The salience network may balance and optimize the FPN and DMN (the principal regions of the three networks are illustrated in Fig. 3.1). Following the findings of network-based brain science, behavioral research should investigate whether the accuracy and efficiency of attentional shifting between external stimuli and internal states are associated with working memory capacity and social cognitive ability.

As reviewed in this article, working memory and social cognition are closely related, like two sides of the same coin. We will never fully understand working memory if we only explore its mechanisms in purely cognitive contexts. It is necessary to explore social, interpersonal contexts to further clarify the underlying cognitive mechanism. Although most studies to date have explored the influence of working memory on social cognition, future studies should pay more attention to the impact of social context on working memory. Towse et al. (2016) investigated on-line executive function performance under joint cognition instructions in which participants were asked to cooperate in a random number generation task that also required working memory. In the collaborative condition, paired participants were instructed to take turns generating numbers between 1 and 10 every 1.5 s, and to make the combined sequence as random as possible. Results revealed various cognitive costs and benefits of a social or interpersonal situation on executive functions. For example, the frequency of stereotypical response pairs was reliably higher (i.e., less random) in the collaborative condition than in the ignore condition, in which participants were instructed to ignore the partner's responses. From a brain network perspective, this may indicate that considering the partner's presence in the collaborative condition increased DMN activity and decreased FPN activity, which impaired the ability to monitor stereotypical responses maintained in working memory. Joint cognition is a promising approach for exploring the psychological constraints and models of working memory. At the same time, applying behavioral approaches used for joint cognition study to neuroimaging procedures such as fMRI and event-related potentials gives us many novel suggestions about the neural functions related to working memory in social or interpersonal situations.

References

Allen M, Smallwood J, Christensen J, Gramm D, Rasmussen B, Jensen CG, Roepstorff A, Lutz A (2013) The balanced mind: the variability of task-unrelated thoughts predicts error monitoring. Front Hum Neurosci 7:743

Amodio DM, Frith CD (2006) Meeting of minds: the medial frontal cortex and social cognition. Nat Rev Neurosci 7(4):268–277

Baddeley AD (1986) Working memory. Oxford University Press, Oxford

Baddeley AD (2000) The episodic buffer: a new component of working memory? Trends Cogn Sci 4(11):417–423

Baddeley AD (2007) Working memory, thought, and action. Oxford University Press, Oxford

Baddeley AD (2012) Working memory: theories, models, and controversies. Annu Rev Psychol 63:1–29

Baddeley AD, Hitch GJ (1974) Working memory. In: Bower GH (ed) The psychology of learning and motivation, vol 8. Academic Press, New York, pp 47–89

Baron-Cohen S, Leslie AM, Frith U (1985) Does the autistic child have a 'theory of mind'? Cognition 21(1):37–46

Baron-Cohen S, Tager-Flusberg H, Lombardo MV (eds) (2013) Understanding other minds: perspective from developmental social neuroscience. Oxford University Press, Oxford

Baune BT, Fuhr M, Air T, Hering C (2014) Neuropsychological functioning in adolescents and young adults with major depressive disorder: a review. Psychiatry Res 218(3):261–271

Braver TS, Cohen JD, Nystrom LE, Jonides J, Smith EE, Noll DC (1997) A parametric study of prefrontal cortex involvement in human working memory. NeuroImage 5(1):49–62

Brouillet P, Portrat S, Camos V (2011) On the law relating processing to storage in working memory. Psychol Rev 118(2):175–192
Buckner RL, Andrews-Hanna JR, Schacter DL (2008) The brain's default network: anatomy, function, and relevance to disease. Ann N Y Acad Sci 1124:1–38
Bull R, Phillips LH, Conway CA (2008) The role of control functions in mentalizing: dual-task studies of theory of mind and executive function. Cogn, 107(2):663–672
Bzdok D, Schilbach L, Vogeley K, Schneider K, Laird AR, Langner R, Eickhoff SB (2012) Parsing the neural correlates of moral cognition: ALE meta-analysis on morality, theory of mind, and empathy. Brain Struct Funct 217(4):783–796
Carlson SM, Moses LJ, Breton C (2002) How specific is the relation between executive function and theory of mind? Contributions of inhibitory control and working memory. Infant Child Dev 11(2):73–92
Carrington SJ, Bailey AJ (2009) Are there theory of mind regions in the brain? A review of the neuroimaging literature. Hum Brain Mapp 30(8):2313–2335
Cavanna AE, Trimble MR (2006) The precuneus: a review of its functional anatomy and behavioural correlates. Brain 129(3):564–583
Christoff K, Gordon AM, Smallwood J, Smith R, Schooler JW (2009) Experience sampling during fMRI reveals default network and executive system contributions to mind wandering. Proc Natl Acad Sci U S A 106(21):8719–8724
Corbetta M, Patel G, Shulman GL (2008) The reorienting system of the human brain: from environment to theory of mind. Neuron 58(3):306–324
Costa A, Torriero S, Oliveri M, Caltagirone C (2008) Prefrontal and temporo–parietal involvement in taking others' perspective: TMS evidence. Behav Neurol 19(1–2):71–74
Cowan N (2005) Working memory capacity (Essays in cognitive psychology). Psychology Press, New York
Crone EA, Wendelken C, Donohue S, van Leijenhorst L, Bunge SA (2006) Neurocognitive development of the ability to manipulate information in working memory. Proc Natl Acad Sci U S A 103(24):9315–9320
Curtis CE, D'Esposito M (2003) Persistent activity in the prefrontal cortex during working memory. Trends Cogn Sci 7(9):415–423
D'Esposito M, Detre JA, Alsop DC, Shin RK, Atlas S, Grossman M (1995) The neural basis of the central executive system of working memory. Nature 378(6554):279–281
Daneman M, Merikle PM (1996) Working memory and language comprehension: a meta–analysis. Psychon Bull Rev 3(4):422–433
Davis HL, Pratt C (1995) The development of children's theory of mind: the working memory explanation. Aust J Psychol 47(1):25–31
Dennis M, Agostino A, Roncadin C, Levin HS (2009) Theory of mind depends on domain–general executive functions of working memory and cognitive inhibition in children with traumatic brain injury. J Clin Exp Neuropsychol 31(7):835–847
Druzgal TJ, D'Esposito M (2001) Activity in fusiform face area modulated as a function of working memory load. Cogn Brain Res 10(3):355–364
Ellamil M, Dobson C, Beeman M, Christoff K (2012) Evaluative and generative modes of thought during the creative process. NeuroImage 59(2):1783–1794
Gallagher HL, Frith CD (2003) Functional imaging of 'theory of mind'. Trends Cogn Sci 7(2):77–83
Gerlach KD, Spreng RN, Gilmore AW, Schacter DL (2011) Solving future problems: Default network and executive activity associated with goal-directed mental simulations. NeuroImage 55(4):1816–1824
Goulden N, Khusnulina A, Davis N, Bracewell RM, Bokde AL, McNulty JP, Mullins PG (2014) The salience network is responsible for switching between the default mode network and the central executive network: replication from DCM. NeuroImage 99:180–190
Gray JR, Chabris CF, Braver TS (2003) Neural mechanisms of general fluid intelligence. Nat Neurosci 6(3):316–322

Greicius MD, Menon V (2004) Default–mode activity during a passive sensory task: uncoupled from deactivation but impacting activation. J Cogn Neurosci 16(9):1484–1492

Gusnard DA, Raichle ME (2001) Searching for a baseline: functional imaging and the resting human brain. Nat Rev Neurosci 2(10):685–694

Honan CA, McDonald S, Gowland A, Fisher A, Randall RK (2015) Deficits in comprehension of speech acts after TBI: the role of theory of mind and executive function. Brain Lang 150(1):69–79

Hyatt CJ, Calhoun VD, Pearlson GD, Assaf M (2015) Specific default mode subnetworks support mentalizing as revealed through opposing network recruitment by social and semantic fMRI tasks. Hum Brain Mapp 36(8):3047–3063

Kane MJ, Brown LH, McVay JC, Silvia PJ, Myin-Germeys I, Kwapil TR (2007) For whom the mind wanders, and when. Psychol Sci 18(7):614–621

Keenan T, Olson DR, Marini Z (1998) Working memory and children's developing understanding of mind. Aust J Psychol 50(1):76–82

Kercood S, Grskovic JA, Banda D, Begeske J (2014) Working memory and autism: a review of literature. Res Autism Spect Dis 8(10):1316–1332

Killingsworth MA, Gilbert DT (2010) A wandering mind is an unhappy mind. Science 330(6006):932

Koban L, Pourtois G (2014) Brain systems underlying the affective and social monitoring of actions: an integrated review. Neurosci Biobehav Rev 46:71–84

Lett TA, Voineskos AN, Kennedy JL, Levine B, Daskalakis ZJ (2014) Treating working memory deficits in schizophrenia: a review of the neurobiology. Biol Psychiatry 75(5):361–370

Lin S, Keysar B, Epley N (2010) Reflexively mindblind: using theory of mind to interpret behavior requires effortful attention. J Exp Soc Psychol 46(3):551–556

Linden DE (2007) The working memory networks of the human brain. Neuroscientist 13(3):257–267

LoPresti ML, Schon K, Tricarico MD, Swisher JD, Celone KA, Stern CE (2008) Working memory for social cues recruits orbitofrontal cortex and amygdala: a functional magnetic resonance imaging study of delayed matching to sample for emotional expressions. J Neurosci 28(14):3718–3728

Maehara Y (2014) Mind over–reading: Psychological research on working memory in theory of mind. Kyoto University Press, Kyoto (in Japanese)

Maehara Y, Saito S (2011) I see into your mind too well: working memory adjusts the probability judgment of others' mental states. Acta Psychol 138(3):367–376

Maehara Y, Saito S (2013) Cognitive load on working memory both encourages and discourages reasoning bias regarding the mental states of others. Aust J Psychol 65(3):163–171

Mahy CEV, Moses LJ, Pfeifer JH (2014) How and where: theory–of–mind in the brain. Dev Cogn Neurosci 9(1):68–81

Mars RB, Neubert FX, Noonan MP, Sallet J, Toni I, Rushworth MF (2012) On the relationship between the "default mode network" and the "social brain". Front Hum Neurosci 6:189

Mason MF, Norton MI, Van Horn JD, Wegner DM, Grafton ST, Macrae CN (2007) Wandering minds: the default network and stimulus-independent thought. Science 315(5810):393–395

McKiernan KA, Kaufman JN, Kucera-Thompson J, Binder JR (2003) A parametric manipulation of factors affecting task-induced deactivation in functional neuroimaging. J Cogn Neurosci 15(3):394–408

McKinnon MC, Moscovitch M (2007) Domain-general contributions to social reasoning: theory of mind and deontic reasoning re-explored. Cognition 102(2):179–218

McVay JC, Kane MJ (2012) Why does working memory capacity predict variation in reading comprehension? On the influence of mind wandering and executive attention. J Exp Psychol Gen 141(2):302–320

Meyer ML, Spunt RP, Berkman ET, Taylor SE, Lieberman MD (2012) Evidence for social working memory from a parametric functional MRI study. Proc Natl Acad Sci U S A 109(6):1883–1888

Miyake A, Shah P (eds) (1999) Models of working memory: mechanisms of active maintenance and executive control. Cambridge University Press, New York
Moses LJ (2001) Executive accounts of theory-of-mind development. Child Dev 72(3):688–690
Mulders PC, van Eijndhoven PF, Schene AH, Beckmann CF, Tendolkar I (2015) Resting-state functional connectivity in major depressive disorder: a review. Neurosci Biobehav Rev 56:330–344
Ouimet AJ, Gawronski B, Dozois DJA (2009) Cognitive vulnerability to anxiety: a review and an integrative model. Clin Psychol Rev 29(6):459–470
Owen AM, McMillan KM, Laird AR, Bullmore E (2005) N-back working memory paradigm: a meta-analysis of normative functional neuroimaging studies. Hum Brain Mapp 25(1):46–59
Perner J (1991) Understanding the representational mind. MIT Press, Cambridge, MA
Raichle ME, MacLeod AM, Snyder AZ, Powers WJ, Gusnard DA, Shulman GL (2001) A default mode of brain function. Proc Natl Acad Sci U S A 98(2):676–682
Reniers RL, Corcoran R, Vollm BA, Mashru A, Howard R, Liddle PF (2012) Moral decision-making, ToM, empathy and the default mode network. Biol Psychol 90(3):202–210
Rypma B, Prabhakaran V, Desmond JE, Glover GH, Gabrieli JD (1999) Load-dependent roles of frontal brain regions in the maintenance of working memory. NeuroImage 9(2):216–226
Saxe R, Carey S, Kanwisher N (2004) Understanding other minds: linking developmental psychology and functional imaging. Annu Rev Psychol 55:87–124
Schilbach L, Eickhoff SB, Rotarska-Jagiela A, Fink GR, Vogeley K (2008) Minds at rest? Social cognition as the default mode of cognizing and its putative relationship to the "default system" of the brain. Conscious Cogn 17(2):457–467
Schurz M, Radua J, Aichhorn M, Richlan F, Perner J (2014) Fractionating theory of mind: a meta-analysis of functional brain imaging studies. Neurosci Biobehav Rev 42:9–34
Seeley WW, Menon V, Schatzberg AF, Keller J, Glover GH, Kenna H, Kenna H, Reiss AL, Greicius MD (2007) Dissociable intrinsic connectivity networks for salience processing and executive control. J Neurosci 27(9):2349–2356
Sheffield JM, Barch DM (2016) Cognition and resting-state functional connectivity in schizophrenia. Neurosci Biobehav Rev 61:108–120
Sidlauskaite J, Wiersema JR, Roeyers H, Krebs RM, Vassena E, Fias W, Brass B, Achten E, Sonuga-Barke E (2014) Anticipatory processes in brain state switching: Evidence from a novel cued-switching task implicating default mode and salience networks. NeuroImage 98:359–365
Smallwood J (2011) Mind-wandering while reading: attentional decoupling, mindless reading and the cascade model of inattention. Lang Ling Compass 5(2):63–77
Smallwood J, Schooler JW (2015) The science of mind wandering: empirically navigating the stream of consciousness. Annu Rev Psychol 66:487–518
Smallwood J, Brown K, Baird B, Schooler JW (2011) Cooperation between the default mode network and the frontal-parietal network in the production of an internal train of thought. Brain Res 1428:60–70
Smith EE, Jonides J (1997) Working memory: a view from neuroimaging. Cogn Psychol 33(1):5–42
Spreng RN, Grady C (2010) Patterns of brain activity supporting autobiographical memory, prospection and theory-of-mind and their relationship to the default mode network. J Cogn Neurosci 22(6):1112–1123
Spreng R, Stevens WD, Chamberlain JP, Gilmore AW, Schacter DL (2010) Default network activity, coupled with the frontoparietal control network, supports goal-directed cognition. NeuroImage 53(1):303–317
Sridharan D, Levitin DJ, Menon V (2008) A critical role for the right fronto-insular cortex in switching between central-executive and default-mode networks. Proc Natl Acad Sci U S A 105(34):12569–12574
Stone VE, Baron-Cohen S, Knight RT (1998) Frontal lobe contributions to theory of mind. J Cogn Neurosci 10(5):640–656
Summerfield JJ, Hassabis D, Maguire EA (2010) Differential engagement of brain regions within a "core" network during scene construction. Neuropsychologia 48(5):1501–1509

Towse JN, Towse AS, Saito S, Maehara Y, Miyake A (2016) Joint cognition: thought contagion and the consequences of cooperation when sharing the task of random sequence generation. PLoS One 11(3):e0151306

Uddin LQ (2015) Salience processing and insular cortical function and dysfunction. Nat Rev Neurosci 16(1):55–61

Uddin LQ, Supekar K, Menon V (2013) Reconceptualizing functional brain connectivity in autism from a developmental perspective. Front Hum Neurosci 7:458

Unsworth N, McMillan BD (2013) Mind wandering and reading comprehension: examining the roles of working memory capacity, interest, motivation, and topic experience. J Exp Psychol Learn Mem Cogn 39(3):832–842

Unsworth N, Spillers GJ (2010) Working memory capacity: attention control, secondary memory, or both? A direct test of the dual-component model. J Mem Lang 62(4):392–406

Van Overwalle F (2009) Social cognition and the brain: a meta-analysis. Hum Brain Mapp 30(3):829–858

Vogel EK, Machizawa MG (2004) Neural activity predicts individual differences in visual working memory capacity. Nature 428(6984):748–751

Wager TD, Smith EE (2003) Neuroimaging studies of working memory: a meta-analysis. Cogn Affect Behav Neurosci 3(4):255–274

Wagner DD, Haxby JV, Heatherton TF (2012) The representation of self and person knowledge in the medial prefrontal cortex. WIREs Cogn Sci 3(4):451–470

Wellman HM, Cross D, Watson J (2001) Meta-analysis of theory-of-mind development: the truth about false belief. Child Dev 72(3):655–684

Whitfield-Gabrieli S, Ford JM (2012) Default mode network activity and connectivity in psychopathology. Annu Rev Clin Psychol 8:49–76

Wimmer H, Perner J (1983) Beliefs about beliefs: representation and constraining function of wrong beliefs in young children's understanding of deception. Cognition 13(1):103–128

Yeo BTT, Krienen FM, Sepulcre J, Sabuncu MR, Lashkari D, Hollinshead M, Roffman JL, Smoller JW, Zöllei L, Polimeni JR, Fischl B, Liu H, Buckner RL (2011) The organization of the human cerebral cortex estimated by intrinsic functional connectivity. J Neurophysiol 106(3):1125–1165

Zwaan RA, Radvansky GA (1998) Situation models in language comprehension and memory. Psychol Bull 123(2):162–185

Chapter 4
Neural Mechanisms Underlying the Effects of Emotional Arousal on Memory

Nilgun Turkileri and Michiko Sakaki

Abstract Emotional events (e.g., seeing a snake while hiking) typically stay in the memory longer and in more detail than neutral ones (e.g., seeing a bird while hiking). This emotion-induced memory enhancement has been attributed to the amygdala's modulation on other brain regions, such as the medial temporal lobe and visual cortices. In line with this amygdala modulation hypothesis, previous brain imaging research revealed enhanced amygdala activity when encoding and retrieving emotional stimuli compared with non-emotional stimuli. However, emotion does not always enhance memory. Indeed, researchers have long been aware that emotion sometimes enhances and sometimes impairs memory. Yet, the underlying mechanisms of the opposing effects of emotion are relatively unknown. For example, if the amygdala's modulation on other brain regions is critical in the emotion-memory interaction, how does it result in the enhancement effects for some aspects, while impairing other aspects of memory? In this chapter, we argue an alternative possibility: that norepinephrine released by the locus coeruleus under arousal plays a critical role in the complex effects of emotion on memory.

Keywords Emotion • Arousal • Amygdala • Norepinephrine • Glutamate • Priority • Arousal-biased competition

4.1 Introduction

It has been widely accepted that emotional events stay in the memory longer than neutral events (e.g., Bradley et al. 1992; Schaefer and Philippot 2005). However, emotion does not always enhance memory (Easterbrook 1959; Levine and Edelstein 2009; Mather and Sutherland 2011). For example, a witness to a crime tends to remember the weapon a perpetrator was holding but forgets the perpetrator's face

N. Turkileri • M. Sakaki (✉)
School of Psychology and Clinical Language Sciences, University of Reading, Harry Pitt Building, Earley gate, Whitenights road, Reading RG6 7BE, UK
e-mail: m.sakaki@reading.ac.uk

T. Tsukiura, S. Umeda (eds.), *Memory in a Social Context*,
https://doi.org/10.1007/978-4-431-56591-8_4

(Steblay 1992). Encountering emotionally arousing events also induces retrograde amnesia and impairs memory for an event that has occurred before the arousing event (Loftus and Burns 1982). In this chapter, we review past functional imaging research on the emotion and memory interaction. We then provide an account for why emotion sometimes enhances and sometimes impairs memory.

4.2 Mechanisms by Which Emotionally Arousing Events Are Remembered Well

Ample research has demonstrated that people remember emotionally salient stimuli better than non-emotional stimuli (for review, see LaBar and Cabeza 2006). The emotion-induced enhancement effects were observed in both negative and positive stimuli (e.g., Hamann et al. 1999; Sakaki et al. 2012) and irrespective of stimulus types, including pictures (e.g., Bradley et al. 1992; Sakaki et al. 2012), words (e.g., Kensinger and Corkin 2003), and autobiographical memories (e.g., St. Jacques and Levine 2007). Thus, it appears that increased arousal induced by emotionally salient stimuli facilitates memory of that stimuli, irrespective of their valence and stimulus type. How then does the brain enable this arousal-induced memory enhancement effect?

Functional MRI research reveals that successful encoding of emotional material is associated with enhanced amygdala activation and enhanced interactions between the amygdala and other regions critical for memory (for a review, see Murty et al. 2010), such as the prefrontal cortex (PFC; Kensinger and Corkin 2004; Kumfor et al. 2013) and the medial temporal cortex (MTL; Dolcos et al. 2004a; Kensinger and Corkin 2004; Richardson et al. 2004). The amygdala's connectivity with the MTL and the PFC was predictive of memory performance a week later to a greater degree than memory performance after a short delay (20 min; Ritchey et al. 2008), suggesting that the amygdala's interactions with these regions might be critical for the consolidation of emotional stimuli.

Animal research also supports the importance of the amygdala in emotional memory formation. In particular, previous research demonstrated that when encountering emotionally salient events, it results in norepinephrine (NE) being released in the basolateral nucleus of the amygdala (BLA) which modulates memory processing in the hippocampus (for review, see McIntyre et al. 2012). NE is released from the locus coeruleus (LC) when exposed to emotionally salient stimuli, irrespective of whether they are positive or negative (Chen and Sara 2007; Grant et al. 1988). The amygdala has strong reciprocal connections with the LC (Bouret et al. 2003; Sears et al. 2013; van Bockstaele et al. 1998) and the MTL (Pitkanen et al. 2000). Thus, the amygdala is in an ideal position to modulate memory processing through the noradrenergic mechanisms. Indeed, administration of NE into the BLA in rats immediately after the training of a water maze task enhanced retention for the location of a platform in the maze (Hatfield and McGaugh 1999). Research also shows that BLA stimulation helps a transient early long-term potentiation (LTP) transform into a lasting late LTP, but this modulation effect of the BLA on LTP was diminished by β-adrenoceptor antagonists (e.g., Frey et al. 2001).

In addition to modulating the hippocampal function, the amygdala and NE also modulate the visual processing pathway (Markovic et al. 2014; Pourtois et al. 2013). In fact, the amygdala has strong connections with the visual cortex (Amaral et al. 2003), which enables its modulation of visual processing. For example, the presentation of fearful faces activates the fusiform face area (an area critical for face processing) more strongly than the presentation of neutral faces, but this enhanced activation of the fusiform face area was diminished in patients with the amygdala lesion (Vuilleumier et al. 2004). Administration of the selective NE re-uptake inhibitor also resulted in enhanced activation for fearful faces not only in the amygdala, but also in sensory areas including the fusiform gyrus and higher order visual cortical areas (Onur et al. 2009).

The amygdala's modulation of the sensory cortex, together with its modulation of the hippocampus, suggests that emotional events are perceived and remembered differently from non-emotional memories (for a review, see Phelps and Sharot 2008). In fact, behavioral studies show that emotional events are recollected more vividly than non-emotional ones (e.g., Kensinger and Corkin 2003; Ochsner 2000). This subjective feeling of vividness has been associated with enhanced activations in the amygdala and the visual cortex. For example, in one study, participants were shown emotional or neutral scenes that were overlaid with visual noise; thus, the images were perceptually unclear owing to random noises (Todd et al. 2012). Despite the constant level of noise overlaid on images across valence, participants reported less noisiness for emotional than for non-emotional images, indicating higher subjective feelings of perceptual vividness for emotional stimuli. Furthermore, the amygdala showed stronger connectivity with the sensory area when presented with emotional rather than non-emotional images and stronger activation of the sensory area was predictive of stronger perceptual vividness ratings. These results suggest that the amygdala's modulation of the visual area might be the key to forming vivid memories for emotional events (see also Kensinger et al. 2011).

On the one hand, strong subjective feelings of vividness do not necessarily imply high memory accuracy for emotional events (Rimmele et al. 2011; Talarico and Rubin 2003). However, there is cumulative evidence that people are in fact able to remember perceptual details of emotional items more accurately than those of non-emotional items (e.g., Kensinger et al. 2006). The enhanced memory for the details of emotional stimuli was associated with enhanced activity in the amygdala (Kensinger et al. 2011; Kensinger and Schacter 2007) and sensory cortex (Mickley and Kensinger 2008; Mickley Steinmetz and Kensinger 2009). The amygdala also showed stronger connectivity with the sensory cortex when learning emotionally arousing stimuli (Mickley Steinmetz et al. 2010; Talmi et al. 2008). Recent behavioral research further demonstrates that emotional arousal can increase the precision of sensory representations stored in the memory (Spachtholz et al. 2014).

Another line of research examines the effects of emotion on pattern separation and provides further evidence against the claim that emotional events result in vivid but inaccurate representations. Pattern separation refers to a process by which similar representations are stored into distinctive memories to minimize interferences (Yassa and Stark 2011). For example, imagine that you are in a parking lot of a shopping mall and need to distinguish your car from another car with the same make and model. To avoid recognizing someone else's car as yours, you must combat interference from the similar features. Recent research indicates that arousal enhances pattern separation processing. For example, the dentate gyrus, which is critical for pattern separation (Bakker et al. 2008) showed stronger activity when learning similar but different emotional lures than learning neutral lures (Leal et al. 2014). Increased levels of salivary alpha amylase (which is correlated with central NE release) lead to facilitated pattern separation as well (Segal et al. 2012). Taken together, past findings suggest that emotional arousal enhances memory accuracy for perceptual details and these detailed visual representations may contribute to an increased subjective feeling of vividness when remembering those events.

In summary, previous research suggests that when exposed to emotionally salient information, enhanced activity of the amygdala may occur, which in turn stimulates LC, leading to NE release across regions including the amygdala. Existing evidence suggests that the amygdala might also modulate the MTL and sensory cortices to form perceptually rich memories for the behaviorally relevant information via the noradrenergic mechanisms.

4.3 Mechanisms That Support Retrieval of Emotionally Arousing Events

In addition to the encoding processing of emotional events, past studies have also addressed the mechanisms underlying retrieval of emotional memories. These studies suggest that regions important for encoding of emotional stimuli (e.g., the amygdala, hippocampus, sensory cortex and PFC) might also be involved in remembering emotional stimuli, irrespective of whether they are images, words or autobiographical memories (e.g., Daselaar et al. 2008; Dolan et al. 2000; Ford et al. 2014; Greenberg et al. 2005; Kensinger and Schacter 2005a; Maratos et al. 2001; Taylor et al. 1998).

For instance, Dolcos et al. (2005) examined brain activation when participants saw emotional and neutral stimuli that were learned 1 year ago. They found that emotionally arousing stimuli were more vividly remembered than neutral stimuli. In addition, the amygdala and the hippocampus were more strongly activated when participants saw old emotional images that were remembered vividly than old neutral pictures that were remembered vividly. Other researchers observed enhanced activity in the amygdala when retrieving memories for emotional items relative to neutral items (Keightley et al. 2011; Kensinger and Schacter 2005a, 2007; Sharot et al. 2004). In addition, retrieval of emotional memories is accompanied by enhanced activation in the sensory cortex as observed for encoding of emotional memories (Keightley et al. 2011; Taylor et al. 1998). In fact, there is accumulating evidence for the similarity between processes involved in encoding emotional

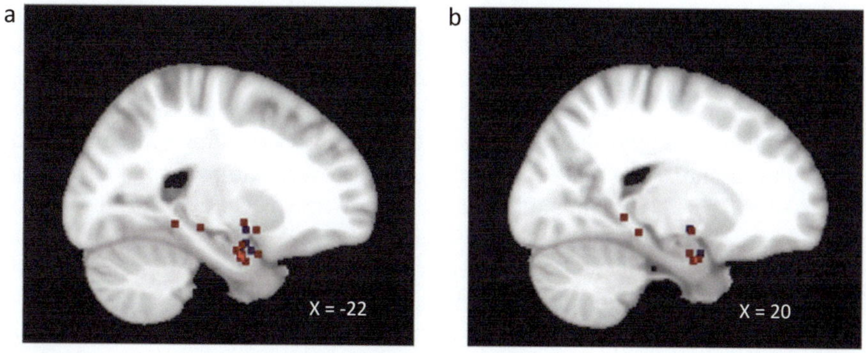

Fig. 4.1 Overlap in (**a**) left and (**b**) *right* medial temporal lobe activity during encoding (red) and retrieval (blue) of emotional stimuli compared with neutral stimuli (Studies used to create this figure are shown in References with an asterisk (*))

memories versus those involved in remembering emotional memories (Dew et al. 2014; Fenker et al. 2005; Hofstetter et al. 2012; Kark and Kensinger 2015). Thus, it appears that brain areas involved in encoding emotional memories are also critical for the retrieval of emotional memories (Fig. 4.1).

However, it is not entirely clear from these studies whether the activity in the amygdala, MTL, and the sensory cortex is induced by retrieved memory representations or by stimuli presented during memory tests. For example, in a typical recognition memory test, participants are presented with images shown in a previous encoding phase in addition to new images and asked to indicate whether or not they have seen each image. In these experiments, brain activity induced by old emotional images may simply reflect encoding processes for those stimuli presented during the recognition test, rather than retrieval processes per se.

To address this issue, other researchers have developed experimental paradigms to tease apart the processes induced by stimuli presented during memory tests from the processes involved in retrieving mental representations (e.g., Shafer and Dolcos 2014; Smith et al. 2004, 2005). For example, Smith et al. (2006) asked participants to learn neutral stimuli while manipulating the emotionality of the context, followed by a memory test where participants were shown neutral stimuli and asked to remember the associated context. They found that retrieval of associated emotional memories, relative to neutral memories, was linked with stronger connectivity between the amygdala and hippocampus and stronger connectivity between the hippocampus and visual areas. Retrieval of emotional memories also induced greater involvement of the medial prefrontal cortex, the amygdala and the hippocampus. These results support the notion that the areas implicated in emotional memory encoding are also activated during retrieval.

It should be noted that the similar brain activities between encoding and retrieval are not unique to emotional memories: memory retrieval is typically accompanied by brain activities similar to those in encoding, even when individuals remember non-emotional memories. In fact, episodic memory retrieval processes have been considered to involve reinstatement of processes engaged when an event is encoded (Norman and O'Reilly 2003; Rugg et al. 2008; Rugg and Vilberg 2013): retrieval of episodic memories relies on the *recapitulation* of the brain activity involved when forming memories for the original (encoded) experience. This idea is supported by a number of neuroimaging findings showing that successful memory retrieval is associated with activations in brain areas that overlap with those activated during encoding (Johnson et al. 2009; Nyberg et al. 2000; Ritchey et al. 2012; Wheeler et al. 2000). Thus, according to this account, activation in the amygdala, MTL, and sensory cortices during the retrieval of emotional memories observed in past studies may reflect the reinstatement of processes involved when those experiences are learned.

Kark and Kensinger (2015) provide evidence consistent with this idea. In this study, participants were asked to study low-resolution line drawings of emotional or neutral images followed by the complete images. In a subsequent retrieval test, participants saw the line drawings as retrieval cues and were asked to indicate whether the drawing had been studied before. The results indicated that occipital and temporal cortices activated during the encoding of emotional images overlapped with those regions activated during the retrieval of emotional images. Critically, stronger activity of these regions in retrieval was associated with better memory performance. In addition, the bilateral fusiform gyrus, which exhibited strong connectivity with the amygdala during encoding of negative images, showed enhanced activity during successful retrieval of negative images, suggesting that the amygdala's modulation of this area during encoding might help reactivation of the original representations of that memory at the time of recollection.

4.4 Emotion Does Not Always Enhance Memory

So far, we have reviewed past evidence that the amygdala plays a key role in the enhancement effects of emotion on memory. One naïve prediction suggested by this amygdala modulation hypothesis is that emotional arousal should always enhance memory; if the amygdala's modulation is the primary route for emotion to modulate memory, NE released in the amygdala under arousal should enhance memory for any stimuli processed at the same time in the brain.

However, people do not show good memory performance for *all* aspects of emotionally arousing events (cf. Yonelinas and Ritchey 2015). For example, emotional arousal impairs memory for information that is not relevant to the goals of the observers (Levine and Edelstein 2009). It is also known that the constituent features of emotionally arousing stimuli (e.g., the color of a gun held by the perpetrator) typically stay in the memory of individuals longer, whereas memory for between-item associations for arousing stimuli (e.g., associations between the gun and the perpetrator to identify who is holding the weapon) is impaired (Mather 2007). Furthermore, although emotionally salient aspects are remembered well, memory for nearby neutral information is often impaired (Kensinger et al. 2007a; Strange et al. 2003). To complicate the story further, exposure to emotionally arousing information sometimes enhances, rather than impairs, memory for nearby neutral information (Anderson et al. 2006; Kensinger et al. 2007a).

To explain the complex effects of arousal, in their arousal-biased competition (ABC) theory, Mather and Sutherland (2011) argue that the effects of emotional arousal on memory depend on the priority of stimuli at the time of the onset of arousal. According to the ABC theory, arousal enhances perception and memory of high-priority stimuli, but impairs perception and memory of low-priority stimuli. Priority is determined by either bottom–up salience (e.g., brightness, contrast) or by relevance to individuals' top–down goals. Consistent with this theory, participants remembered more high-priority salient letters than low-priority salient ones when the letters were preceded by negative rather than neutral sounds (Sutherland and Mather 2012), suggesting that arousal might have different effects on short-term memory, depending on the stimuli's bottom–up saliency. Recent fMRI research has also shown that emotional arousal increases brain activation in an area responsible for perceptually salient stimuli, whereas it decreases brain activation in an area responsible for nonsalient stimuli (Lee et al. 2014).

Another study examined the effects of top–down goals (Sakaki et al. 2014) and further confirmed the prediction from the ABC theory. In this study, participants were shown image sequences, each of which included several neutral objects with no frame and one oddball image highlighted by a black frame. Oddball images were sometimes emotionally arousing and sometimes emotionally neutral. Participants' top–down goal was manipulated by telling them to focus on oddball images (i.e., "thumb" in Fig. 4.2) or objects that came right before the oddball images (i.e., "banana" in Fig. 4.2). The results indicated that emotional oddball images enhanced memory for preceding objects when participants focused on the objects, whereas emotional oddball images impaired memory for preceding objects when participants focused on the oddball images.

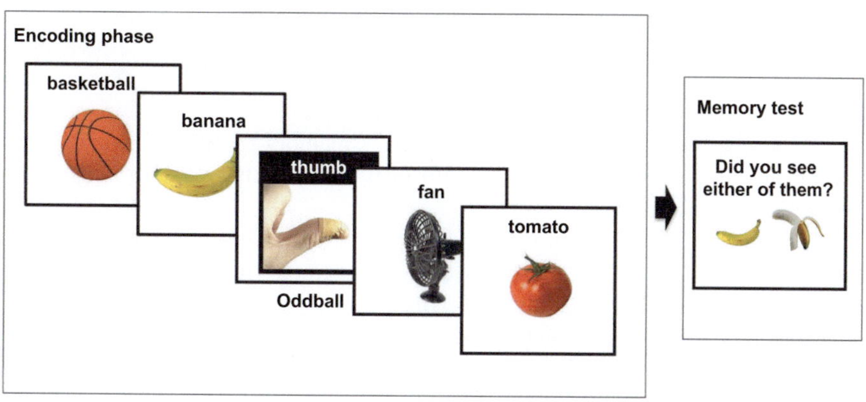

Fig. 4.2 Schematic representation of a trial in the encoding phase and the memory test in Sakaki et al. (2014)

How then does the brain allow for these opposing effects of arousal depending on priority? The glutamate amplifies noradrenergic effects (GANE) model provides an answer to this question (Mather et al. 2016). As in the amygdala modulation model explained above, the GANE model focuses on the role of the noradrenergic system and posits that when something arousing happens, the LC releases NE throughout the brain. However, unlike other models (e.g., Markovic et al. 2014; McGaugh 2013; McIntyre et al. 2012), the GANE model proposes that NE concentration levels might be modulated locally depending on levels of glutamate (i.e., the primary excitatory neurotransmitter of the brain). In addition, it posits that NE should produce different effects depending on its concentration levels; the different NE concentration levels should lead to engagement of different sub-types of noradrenaline receptors, resulting in different NE effects across regions.

According to the GANE model, neurons transmitting high-priority information release more glutamate than those transmitting low-priority information; thus, priority is signaled by the amount of glutamate released by glutamate neurons. The phasic NE release under arousal interacts with glutamate receptors, resulting in greater glutamate release, which in turn leads to greater NE release. This positive feedback loop between NE and glutamate creates local NE hotspots near the neurons that transmit high-priority information. Higher NE levels in the NE hotspots in turn enable low-affinity β-adrenoceptors to be engaged, resulting in LTP and memory enhancement effects. In contrast, lower levels of NE outside hotspots results in the engagement of high-affinity α1-receptors that induce long-term depression (LTD), thus resulting in memory impairment.

The GANE model was initially developed to explain how emotional arousal sometimes enhances and sometimes impairs memory for other nearby information, but this model can also explain why emotionally arousing events stay in the memory longer (as reviewed earlier in this chapter). Emotional information tends to have higher priority over non-emotional information because of its relevance to an individual's wellbeing and goals. Therefore, encountering emotionally arousing events should not only induce phasic arousal and NE release, but should also result in a greater glutamate release in neurons that process the emotional events. The co-occurrence of the enhanced glutamate release and NE release should help the neurons that process the arousing stimuli to win the competition for limited resources. Thus, the GANE model explains why encountering emotionally arousing information sometimes enhances and sometimes impairs memory for nearby neutral information, and why emotionally arousing information stays in the memory longer than non-emotional information.

Although the GANE model still needs further investigations and more tests, initial evidence provides support for its use. For example, in one study on young women, enhanced NE release by squeezing a ball as hard as possible selectively enhanced memory for emotionally negative stimuli over positive stimuli (Nielsen et al. 2015). As negative stimuli tend to have high priority (Baumeister et al. 2001), these results support the prediction of the GANE model: that NE enhances memory for high-priority information.

4.5 Conclusion

Decades of research have addressed how and why emotion influences memory. As we reviewed in this chapter, substantial research demonstrates that emotional information stays in the memory longer than non-emotional information. Given the limited resources in the brain and given that emotional arousal is typically evoked by an important factor for an individual's survival and reproduction, it is adaptive to remember emotional information. However, the emotion–memory interaction does not always lead to the adaptive effects. The fact that emotional arousal impairs memory for low-priority information suggests that a person should be cautious when listening to memories from a witness who has experienced a stressful event. Studies reviewed in this chapter help us to understand which aspects of memories from those witnesses are accurate and reliable. In addition, research on emotion and memory has clinical implications for post-traumatic stress disorder (PTSD), a condition characterized by intrusive memories of the traumatic event. Further research on the mechanisms by which emotion modulates memory may provide important insights into pharmacological, behavioral, and/or brain stimulation interventions to help PTSD patients' symptoms and help to increase their overall well-being.

References

Amaral DG, Behniea H, Kelly JL (2003) Topographic organization of projections from the amygdala to the visual cortex in the macaque monkey. Neuroscience 118(4):1099–1120

Anderson AK, Wais PE, Gabrieli JDE (2006) Emotion enhances remembrance of neutral events past. Proc Natl Acad Sci U S A 103(5):1599–1604

Bakker A, Kirwan CB, Miller M, Stark CEL (2008) Pattern separation in the human hippocampal CA3 and dentate gyrus. Science 319(5870):1640–1642

Baumeister RF, Bratslavsky E, Finkenauer C, Vohs KD (2001) Bad is stronger than good. Rev Gen Psychol 5(4):323–370

Bellace M, Williams JM, Mohamed FB, Faro SH (2012) An fMRI study of the activation of the hippocampus by emotional memory. Int J Neurosci 123(2):121–127*

Bouret S, Duvel A, Onat S, Sara SJ (2003) Phasic activation of locus ceruleus neurons by the central nucleus of the amygdala. J Neurosci 23(8):3491–3497

Bradley MM, Greenwald MK, Petry MC, Lang PJ (1992) Remembering pictures: pleasure and arousal in memory. J Exp Psychol Learn Mem Cogn 18(2):379–390

Cahill L, Uncapher M, Kilpatrick L, Alkire MT, Turner J (2004) Sex-related hemispheric lateralization of amygdala function in emotionally influenced memory: an fMRI investigation. Learn Mem 11(3):261–266*

Chen FJ, Sara SJ (2007) Locus coeruleus activation by foot shock or electrical stimulation inhibits amygdala neurons. Neuroscience 144(2):472–481

Daselaar SM, Rice HJ, Greenberg DL, Cabeza R, LaBar KS, Rubin DC (2008) The spatiotemporal dynamics of autobiographical memory: neural correlates of recall, emotional intensity, and reliving. Cereb Cortex 18(1):217–229

Dew ITZ, Ritchey M, LaBar KS, Cabeza R (2014) Prior perceptual processing enhances the effect of emotional arousal on the neural correlates of memory retrieval. Neurobiol Learn Mem 112:104–113

Dolan RJ, Lane R, Chua P, Fletcher P (2000) Dissociable temporal lobe activations during emotional episodic memory retrieval. NeuroImage 11(3):203–209
Dolcos F, LaBar KS, Cabeza R (2004a) Interaction between the amygdala and the medial temporal lobe memory system predicts better memory for emotional events. Neuron 42(5):855–863
Dolcos F, LaBar KS, Cabeza R (2004b) Dissociable effects of arousal and valence on prefrontal activity indexing emotional evaluation and subsequent memory: an event-related fMRI study. NeuroImage 23(1):64–74*
Dolcos F, LaBar KS, Cabeza R (2005) Remembering one year later: role of the amygdala and the medial temporal lobe memory system in retrieving emotional memories. Proc Natl Acad Sci U S A 102(7):2626–2631
Dolcos F, Iordan AD, Kragel J, Stokes J, Campbell R, McCarthy G, Cabeza R (2013) Neural correlates of opposing effects of emotional distraction on working memory and episodic memory: an event-related fMRI investigation. Front Psychol 4:293*
Dougal S, Phelps EA, Davachi L (2007) The role of medial temporal lobe in item recognition and source recollection of emotional stimuli. Cogn Affect Behav Neurosci 7(3):233–242*
Easterbrook JA (1959) The effect of emotion on cue utilization and the organization of behavior. Psychol Rev 66(3):183–201
Fenker DB, Schott BH, Richardson-Klavehn A, Heinze H-J, Düzel E (2005) Recapitulating emotional context: activity of amygdala, hippocampus and fusiform cortex during recollection and familiarity. Eur J Neurosci 21(7):1993–1999*
Ford JH, Morris JA, Kensinger EA (2014) Effects of emotion and emotional valence on the neural correlates of episodic memory search and elaboration. J Cogn Neurosci 26(4):825–839*
Frey S, Bergado-Rosado J, Seidenbecher T, Pape H-C, Frey JU (2001) Reinforcement of early long-term potentiation (Early-LTP) in dentate gyrus by stimulation of the basolateral amygdala: heterosynaptic induction mechanisms of late-LTP. J Neurosci 21(10):3697–3703
Grant SJ, Aston-Jones G, Redmond DE Jr (1988) Responses of primate locus coeruleus neurons to simple and complex sensory stimuli. Brain Res Bull 21(3):401–410
Greenberg DL, Rice HJ, Cooper JJ, Cabeza R, Rubin DC, LaBar KS (2005) Co-activation of the amygdala, hippocampus and inferior frontal gyrus during autobiographical memory retrieval. Neuropsychologia 43(5):659–674
Hamann S, Ely TD, Grafton ST, Kilts CD (1999) Amygdala activity related to enhanced memory for pleasant and aversive stimuli. Nat Neurosci 2(3):289–293
Harvey PO, Fossati P, Lepage M (2007) Modulation of memory formation by stimulus content: specific role of the medial prefrontal cortex in the successful encoding of social pictures. J Cogn Neurosci 19(2):351–362*
Hatfield T, McGaugh JL (1999) Norepinephrine infused into the basolateral amygdala posttraining enhances retention in a spatial water maze task. Neurobiol Learn Mem 71(2):232–239
Hofstetter C, Achaibou A, Vuilleumier P (2012) Reactivation of visual cortex during memory retrieval: content specificity and emotional modulation. NeuroImage 60(3):1734–1745*
Johnson JD, McDuff SGR, Rugg MD, Norman KA (2009) Recollection, familiarity, and cortical reinstatement: a multivoxel pattern analysis. Neuron 63(5):697–708
Kark SM, Kensinger EA (2015) Effect of emotional valence on retrieval-related recapitulation of encoding activity in the ventral visual stream. Neuropsychologia 78:221–230
Keightley ML, Chiew KS, Anderson JAE, Grady CL (2011) Neural correlates of recognition memory for emotional faces and scenes. Soc Cogn Affect Neurosci 6(1):24–37
Kensinger EA, Corkin S (2003) Memory enhancement for emotional words: are emotional words more vividly remembered than neutral words? Mem Cogn 31(8):1169–1180
Kensinger EA, Corkin S (2004) Two routes to emotional memory: distinct neural processes for valence and arousal. Proc Natl Acad Sci U S A 101(9):3310–3315
Kensinger EA, Schacter DL (2005a) Retrieving accurate and distorted memories: neuroimaging evidence for effects of emotion. NeuroImage 27(1):167–177
Kensinger EA, Schacter DL (2005b) Emotional content and reality-monitoring ability: fMRI evidence for the influences of encoding processes. Neuropsychologia 43(10):1429–1443*

Kensinger EA, Schacter DL (2006) Amygdala activity is associated with the successful encoding of item, but not source, information for positive and negative stimuli. J Neurosci 26(9):2564–2570*
Kensinger EA, Schacter DL (2007) Remembering the specific visual details of presented objects: neuroimaging evidence for effects of emotion. Neuropsychologia 45(13):2951–2962*
Kensinger EA, Garoff-Eaton RJ, Schacter DL (2006) Memory for specific visual details can be enhanced by negative arousing content. J Mem Lang 54(1):99–112
Kensinger EA, Garoff-Eaton RJ, Schacter DL (2007a) Effects of emotion on memory specificity:memory trade-offs elicited by negative visually arousing stimuli. J Mem Lang 56(4):575–591
Kensinger EA, Garoff-Eaton RJ, Schacter DL (2007b) How negative emotion enhances the visual specificity of a memory. J Cogn Neurosci 19(11):1872–1887*
Kensinger EA, Addis DR, Atapattu RK (2011) Amygdala activity at encoding corresponds with memory vividness and with memory for select episodic details. Neuropsychologia 49(4):663–673
Kumfor F, Irish M, Hodges JR, Piguet O (2013) The orbitofrontal cortex is involved in emotional enhancement of memory: evidence from the dementias. Brain 136(10):2992–3003
LaBar KS, Cabeza R (2006) Cognitive neuroscience of emotional memory. Nat Rev Neurosci 7(1):54–64
Leal SL, Tighe SK, Jones CK, Yassa MA (2014) Pattern separation of emotional information in hippocampal dentate and CA3. Hippocampus 24(9):1146–1155
Lee T-H, Sakaki M, Cheng R, Velasco R, Mather M (2014) Emotional arousal amplifies the effects of biased competition in the brain. Soc Cogn Affect Neurosci 9(12):2067–2077
Levine LJ, Edelstein RS (2009) Emotion and memory narrowing: a review and goal-relevance approach. Cognit Emot 23(5):833–875
Loftus E, Burns T (1982) Mental shock can produce retrograde amnesia. Mem Cogn 10(4):318–323
Maratos EJ, Dolan RJ, Morris JS, Henson RNA, Rugg MD (2001) Neural activity associated with episodic memory for emotional context. Neuropsychologia 39(9):910–920
Markovic J, Anderson AK, Todd RM (2014) Tuning to the significant: neural and genetic processes underlying affective enhancement of visual perception and memory. Behav Brain Res 259:229–241
Mather M (2007) Emotional arousal and memory binding: an object-based framework. Perspect Psychol Sci 2(1):33–52
Mather M, Sutherland M (2011) Arousal-biased competition in perception and memory. Perspect Psychol Sci 6(2):114–133
Mather M, Clewett D, Sakaki M, Harley CW (2016) Norepinephrine ignites local hot spots of neuronal excitation: how arousal amplifies selectivity in perception and memory. Behav Brain Sci 39:e200
McGaugh JL (2013) Making lasting memories: remembering the significant. Proc Natl Acad Sci U S A 110(S2):10402–10407
McIntyre CK, McGaugh JL, Williams CL (2012) Interacting brain systems modulate memory consolidation. Neurosci Biobehav Rev 36(7):1750–1762
Medford N, Phillips ML, Brierley B, Brammer M, Bullmore ET, David AS (2005) Emotional memory: separating content and context. Psychiatry Res 138(3):247–258*
Mickley KR, Kensinger EA (2008) Emotional valence influences the neural correlates associated with remembering and knowing. Cogn affect Behav Neurosci 8(2):143–152
Mickley Steinmetz KR, Kensinger EA (2009) The effects of valence and arousal on the neural activity leading to subsequent memory. Psychophysiology 46(6):1190–1199*
Mickley Steinmetz KR, Addis DR, Kensinger EA (2010) The effect of arousal on the emotional memory network depends on valence. NeuroImage 53(1):318–324
Mickley Steinmetz KR, Schmidt K, Zucker HR, Kensinger EA (2012) The effect of emotional arousal and retention delay on subsequent-memory effects. Cogn Neurosci 3(3-4):150–159*

Murty VP, Ritchey M, Adcock RA, LaBar KS (2010) fMRI studies of successful emotional memory encoding: a quantitative meta-analysis. Neuropsychologia 48(12):3459–3469
Nielsen SE, Barber SJ, Chai A, Clewett DV, Mather M (2015) Sympathetic arousal increases a negative memory bias in young women with low sex hormone levels. Psychoneuroendocrinology 62:96–106
Norman KA, O'Reilly RC (2003) Modeling hippocampal and neocortical contributions to recognition memory: a complementary-learning-systems approach. Psychol Rev 110(4):611–646
Nyberg L, Persson J, Habib R, Tulving E, McIntosh AR, Cabeza R, Houle S (2000) Large scale neurocognitive networks underlying episodic memory. J Cogn Neurosci 12(1):163–173
Ochsner KN (2000) Are affective events richly recollected or simply familiar? The experience and process of recognizing feelings past. J Exp Psychol Gen 129(2):242–261
Onur OA, Walter H, Schlaepfer TE, Rehme AK, Schmidt C, Keysers C, Maier W, Hurlemann R (2009) Noradrenergic enhancement of amygdala responses to fear. Soc Cogn Affect Neurosci 4(2):119–126
Phelps EA, Sharot T (2008) How (and why) emotion enhances the subjective sense of recollection. Curr Dir Psychol Sci 17(2):147–152
Pitkanen A, Pikkarainen M, Nurminen N, Ylinen A (2000) Reciprocal connections between the amygdala and the hippocampal formation, perirhinal cortex, and postrhinal cortex in rat: a review. Ann N Y Acad Sci 911(1):369–391
Pourtois G, Schettino A, Vuilleumier P (2013) Brain mechanisms for emotional influences on perception and attention: what is magic and what is not. Biol Psychol 92(3):492–512
Richardson MP, Strange BA, Dolan RJ (2004) Encoding of emotional memories depends on amygdala and hippocampus and their interactions. Nat Neurosci 7(3):278–285
Rimmele U, Davachi L, Petrov R, Dougal S, Phelps EA (2011) Emotion enhances the subjective feeling of remembering, despite lower accuracy for contextual details. Emotion 11(3):553–562
Ritchey M, Dolcos F, Cabeza R (2008) Role of amygdala connectivity in the persistence of emotional memories over time: an event-related FMRI investigation. Cereb Cortex 18(11):2494–2504*
Ritchey M, LaBar KS, Cabeza R (2011) Level of processing modulates the neural correlates of emotional memory formation. J Cogn Neurosci 23(4):757–771*
Ritchey M, Wing EA, LaBar KS, Cabeza R (2012) Neural similarity between encoding and retrieval is related to memory via hippocampal interactions. Cereb Cortex 23(12):2818–2828
Rugg MD, Vilberg KL (2013) Brain networks underlying episodic memory retrieval. Curr Opin Neurobiol 23(2):255–260
Rugg MD, Johnson JD, Park H, Uncapher MR (2008) Encoding-retrieval overlap in human episodic memory: a functional neuroimaging perspective. In: Sossin WS, Lacaille JC, Castellucci VF, Belleville S (eds) Progress in brain research, vol 169. Elsevier, Amsterdam, pp 339–352
Sakaki M, Niki K, Mather M (2012) Beyond arousal and valence: the importance of the biological versus social relevance of emotional stimuli. Cogn affect Behav Neurosci 12(1):115–139
Sakaki M, Fryer K, Mather M (2014) Emotion strengthens high-priority memory traces but weakens low-priority memory traces. Psychol Sci 25(2):387–395
Schaefer A, Philippot P (2005) Selective effects of emotion on the phenomenal characteristics of autobiographical memories. Memory 13(2):148–160
Schwarze U, Bingel U, Sommer T (2012) Event-related nociceptive arousal enhances memory consolidation for neutral scenes. J Neurosci 32(4):1481–1487*
Sears RM, Fink AE, Wigestrand MB, Farb CR, de Lecea L, LeDoux JE (2013) Orexin/hypocretin system modulates amygdala-dependent threat learning through the locus coeruleus. Proc Natl Acad Sci U S A 110(50):20260–20265
Sergerie K, Lepage M, Armony JL (2006) A process-specific functional dissociation of the amygdala in emotional memory. J Cogn Neurosci 18(8):1359–1367*
Segal SK, Stark SM, Kattan D, Stark CE, Yassa MA (2012) Norepinephrine-mediated emotional arousal facilitates subsequent pattern separation. Neurobiol Learn Mem 97(4):465–469
Shafer AT, Dolcos F (2012) Neural correlates of opposing effects of emotional distraction on perception and episodic memory: an event-related FMRI investigation. Front Integr Neurosci 6:70*

Shafer AT, Dolcos F (2014) Dissociating retrieval success from incidental encoding activity during emotional memory retrieval, in the medial temporal lobe. Front Behav Neurosci 8:177

Sharot T, Delgado MR, Phelps EA (2004) How emotion enhances the feeling of remembering. Nat Neurosci 7(12):1376–1380

Smith APR, Henson RNA, Dolan RJ, Rugg MD (2004) fMRI correlates of the episodic retrieval of emotional contexts. NeuroImage 22(2):868–878*

Smith APR, Henson RNA, Rugg MD, Dolan RJ (2005) Modulation of retrieva processing reflects accuracy of emotional source memory. Learn Mem 12(5):472–479*

Smith APR, Stephan KE, Rugg MD, Dolan RJ (2006) Task and content modulate amygdalahippocampal connectivity in emotional retrieval. Neuron 49(4):631–638*

Spachtholz P, Kuhbandner C, Pekrun R (2014) Negative affect improves the quality of memories: trading capacity for precision in sensory and working memory. J Exp Psychol Gen 143(4):1450–1456

St. Jacques PL, Levine B (2007) Ageing and autobiographical memory for emotional and neutral events. Memory 15(2):129–144

Steblay NM (1992) A meta-analytic review of the weapon focus effect. Law Hum Behav 16(4):413–424

Sterpenich V, D'Argembeau A, Desseilles M, Balteau E, Albouy G, Vandewalle G, Degueldre C, Luxen A, Collette F, Maquet P (2006) The locus ceruleus is involved in the successful retrieval of emotional memories in humans. J Neurosci 26(28):7416–7423*

Sterpenich V, Albouy G, Darsaud A, Schmidt C, Vandewalle G, TTD V, Desseilles M, Phillips C, Degueldre C, Balteau E, Collette F (2009) Sleep promotes the neural reorganization of remote emotional memory. J Neurosci 29(16):5143–5152*

Strange BA, Hurlemann R, Dolan RJ (2003) An emotion-induced retrograde amnesia in humans is amygdala- and beta-adrenergic-dependent. Proc Natl Acad Sci U S A 100(23):13626–13631*

Sutherland MR, Mather M (2012) Negative arousal amplifies the effects of saliency in short-term memory. Emotion 12(6):1367–1372

Talarico JM, Rubin DC (2003) Confidence, not consistency, characterizes flashbulb memories. Psychol Sci 14(5):455–461

Talmi D, Anderson AK, Riggs L, Caplan JB, Moscovitch M (2008) Immediate memory consequences of the effect of emotion on attention to pictures. Learn Mem 15(3):172–182*

Taylor SF, Liberzon I, Fig LM, Decker LR, Minoshima S, Koeppe RA (1998) The effect of emotional content on visual recognition memory: a PET activation study. NeuroImage 8(2):188–197

Todd RM, Talmi D, Schmitz TW, Susskind J, Anderson AK (2012) Psychophysical and neural evidence for emotion-enhanced perceptual vividness. J Neurosci 32(33):11201–11212

van Bockstaele EJ, Colago EEO, Valentino RJ (1998) Amygdaloid corticotropin-releasing factor targets locus coeruleus dendrites: substrate for the co-ordination of emotional and cognitive limbs of the stress response. J Neuroendocrinol 10(10):743–758

Vuilleumier P, Richardson MP, Armony JL, Driver J, Dolan RJ (2004) Distant influences of amygdala lesion on visual cortical activation during emotional face processing. Nat Neurosci 7(11):1271–1278

Waring JD, Kensinger EA (2011) How emotion leads to selective memory: neuroimaging evidence. Neuropsychologia 49(7):1831–1842*

Wheeler ME, Petersen SE, Buckner RL (2000) Memory's echo: vivid remembering reactivates sensory-specific cortex. Proc Natl Acad Sci U S A 97(20):11125–11129

Yassa MA, Stark CEL (2011) Pattern separation in the hippocampus. Trends Neurosci 34(10):515–525

Yonelinas AP, Ritchey M (2015) The slow forgetting of emotional episodic memories: an emotional binding account. Trends Cogn Sci 19(5):259–267

Chapter 5
Current Issues and Emerging Directions in the Impact of Emotion on Memory: A Review of Evidence from Brain Imaging Investigations

Florin Dolcos, Yuta Katsumi, Ekaterina Denkova, Mathias Weymar, and Sanda Dolcos

Abstract What are the neural markers of encoding and retrieving memories of emotional events with increased efficacy? In recent years, this question has captured the attention of neuroscientists, who have been fervently engaged in addressing it using a multitude of approaches. The present chapter emphasizes evidence from brain imaging investigations regarding three emerging research directions in the field: the role of social information in emotional memory, the role of emotion regulation in the impact of emotion on memory, and the impact of emotion on associative or relational memory. Overall, this evidence provides insights into the brain mechanisms that make emotional memories special, points to possible alterations that may explain negative affective biases in encoding and retrieving emotional

F. Dolcos (✉)
Social, Cognitive, Personality, and Emotional (SCoPE) Neuroscience Laboratory, University of Illinois at Urbana-Champaign, Urbana, IL, USA

Psychology Department, University of Illinois at Urbana-Champaign, Champaign, IL, USA

Neuroscience Program, University of Illinois at Urbana-Champaign, Urbana, IL, USA

Beckman Institute for Advanced Science & Technology, University of Illinois at Urbana-Champaign, Urbana, IL, USA
e-mail: fdolcos@illinois.edu; http://dolcoslab.beckman.illinois.edu/

Y. Katsumi • S. Dolcos (✉)
Social, Cognitive, Personality, and Emotional (SCoPE) Neuroscience Laboratory, University of Illinois at Urbana-Champaign, Urbana, IL, USA

Psychology Department, University of Illinois at Urbana-Champaign, Champaign, IL, USA

Beckman Institute for Advanced Science & Technology, University of Illinois at Urbana-Champaign, Urbana, IL, USA
e-mail: sdolcos@illinois.edu

E. Denkova
Department of Psychology, University of Miami, Coral Gables, FL, USA

M. Weymar
Department of Psychology, University of Potsdam, Potsdam, Germany

T. Tsukiura, S. Umeda (eds.), *Memory in a Social Context*,
https://doi.org/10.1007/978-4-431-56591-8_5

memories observed in affective disorders, and highlights specific aspects to be clarified in future research.

Keywords Emotion–cognition interactions • Modulation hypothesis • Social cognition • Emotion regulation • Associative/relational memory • Neuroimaging • Amygdala • Hippocampus • Prefrontal cortex • Large-scale functional networks

5.1 Introduction

The memory-enhancing effect of emotion has long been documented, but elucidation of the neural mechanisms of this phenomenon has only relatively recently been the focus of intense research in humans, using functional neuroimaging tools. The present chapter builds upon our recent reviews (Dolcos et al. 2011, 2012), by incorporating new evidence regarding three emerging research directions in the field[1]: the role of social information in emotional memory, the role of emotion regulation in the impact of emotion on memory, and the impact of emotion on associative or relational memory. The emphasis is on evidence from functional neuroimaging studies in neurologically intact humans investigating the role of the amygdala (AMY) and its interaction with memory-related medial temporal lobe (MTL) brain regions, along with the role of other brain regions (e.g., prefrontal cortex, PFC) in the effect of emotion during both the encoding and retrieval of episodic memories. Following a brief introduction to basic animal and human research regarding the neural mechanisms of emotional memory, along with an introduction to the important methodological developments that facilitated such findings, we discuss in detail new evidence from the literature circumscribed by the three emerging topics mentioned above. The chapter ends with concluding remarks and a brief presentation of open issues and future directions.

[1] It is important to note that besides the emerging aspects discussed here, aspects emphasizing the importance of understanding individual differences in the research of emotional memory are also important. However, because of space limitations, these aspects are not the focus of the present discussion (but see Dolcos et al. 2012, 2017; Katsumi et al. 2017). It should be mentioned, however, that investigation of how the neural mechanisms involved in the memory-enhancing effects of emotion vary among individuals allows a better understanding of the underlying mechanisms in both normal functioning and in clinical conditions, such as affective disorders, which are typically characterized by dysfunctional emotional memory. Thus, it is crucial to understand the factors that influence individual variation in the engagement of and interactions among brain regions during the encoding and retrieval of emotional memories. Such factors could be linked to personality, sex, and age-related differences, among others, in basic emotional responses and in emotion regulation (reviewed in Dolcos et al. 2017).

5.1.1 Basic Evidence from Animal Research

Historically, considerable evidence for the involvement of the AMY in emotional memory emerged from animal models, which provided the foundation for the investigation of neural mechanisms underlying emotional learning and memory in humans (Delgado et al. 2006; LeDoux 2000; McGaugh 2000, 2004; Phelps and LeDoux 2005). Animal research has primarily emphasized the role of the AMY in emotional learning and memory, but its specific role continues to be a matter of current debate. The two most influential hypotheses concerning its role are the *modulation* and the *plasticity* hypotheses, which posit different forms of involvement of the AMY in emotional memory (Dolcos and Denkova 2008). Specifically, the modulation hypothesis suggests a *modulatory* role of the AMY on the encoding and consolidation of memory processes occurring in other brain regions (McGaugh 2000, 2004), whereas the plasticity hypothesis proposes that the AMY itself is a *site of plasticity*[2] underlying learning and memory of fear conditioning, as well as reconsolidation of the memory trace following retrieval (LeDoux 2000, 2007). It should be noted that, although these hypotheses could be seen as conflicting (e.g., Cahill et al. 1999; Vazdarjanova and McGaugh 1998), in reality they are complementary (Fanselow and LeDoux 1999; Phelps and LeDoux 2005). The apparent contradiction seems to result from the fact that they tended to be tested with slightly different behavioral paradigms (i.e., inhibitory avoidance vs Pavlovian fear conditioning, respectively) that emphasize different aspects of the role of the AMY in mediating the impact of emotion on learning and memory (relatively more *explicit/declarative* vs *more implicit/procedural* forms of memory). However, given that real-life situations typically involve both declarative and nondeclarative aspects of behavior, and that emotion may influence memory not only during the initial stages, but also during the actual retrieval, it is more reasonable to consider both accounts when investigating the neural correlates of the memory-enhancing effect of emotion, rather than considering either one or the other of the mechanisms suggested by the two hypotheses. Evidence from functional neuroimaging studies in humans is actually consistent with this idea (Dolcos and Denkova 2008; Dolcos et al. 2004b, 2005, 2011, 2012).

[2]As previously noted, this nomenclature does not refer to the notion of neuronal plasticity *sensu stricto*, as is generally accepted in neuroscience linked to memory processes (see Martin et al. 2000), but rather to the place where plasticity is expected in relation to processing, leading to increased emotional memories, as predicted by the two views. Both the *modulation* and the *plasticity* hypotheses predict neural plasticity, but although the former view predicts that the AMY influences plasticity occurring in other brain regions, the latter view posits that the AMY itself is a main site of plasticity that contributes to the memory-enhancing effect of emotion.

5.1.2 Important Conceptual and Methodological Implementations That Facilitated Elucidation of the Neural Correlates of Emotional Memory in Human Research

Consistent with animal research, human research involving lesions (Adolphs et al. 1997, 2000; Cahill et al. 1995; LaBar et al. 1998; Phelps et al. 1998) and pharmacological methods (Cahill et al. 1994; Strange and Dolan 2004; Strange et al. 2003) highlighted the role of the AMY in emotional memory. More recently, human research has also greatly benefited from the advent of brain imaging methods, and from conceptual and methodological implementations that allowed identification with increased specificity of changes in brain activity linked to episodic memory processing and to dissociable effects of basic affective properties on emotional memory, in neurologically intact human brains.

Event-Related Designs A significant conceptual and methodological contribution to our understanding of the neural mechanisms of memory has been made by the introduction of *event-related designs* as opposed to blocked designs (D'Esposito et al. 1999; Donaldson and Buckner 2001; Rosen et al. 1998). Different from blocked designs, constrained to averaging brain activity to several trials of the same type over time, *event-related designs* allow for analysis of brain imaging data on a stimulus-by-stimulus basis, so that the individual responses to single events can be specifically identified. An important advantage of the event-related data over the blocked data is that the former can be analyzed and categorized *post-hoc* according to the participant's task performance. This advantage has proven particularly important in investigating the neural correlates of memory processes, as it allows comparison of brain activity for items that are subsequently remembered versus forgotten in a memory test, thereby allowing the possibility of establishing a direct link between brain activity and memory performance (Paller and Wagner 2002).

The Subsequent Memory Paradigm One such event-related experimental paradigm that allows comparison of brain activity according to participants' memory performance is the *subsequent memory paradigm (SMP)*. For instance, as illustrated in Fig. 5.1, activity can be compared for items that are remembered versus those that are forgotten in a memory test, thereby establishing a direct link between brain activity and successful memory performance in individual participants. Comparing brain activity for remembered versus forgotten items can be done both during the encoding (learning) phase and during the retrieval (test) phase. By sorting brain activity recorded during encoding based on whether stimuli processed during the study phase are subsequently remembered (R) or forgotten (F) in a later memory test, the so-called *difference in memory* or *Dm effect* – i.e., greater encoding activity for remembered than for forgotten items – can be calculated (e.g., Paller et al. 1987; Paller and Wagner 2002). Brain regions showing a positive Dm effect (R > F) are assumed to mediate processes that lead to successful memory encoding, and are

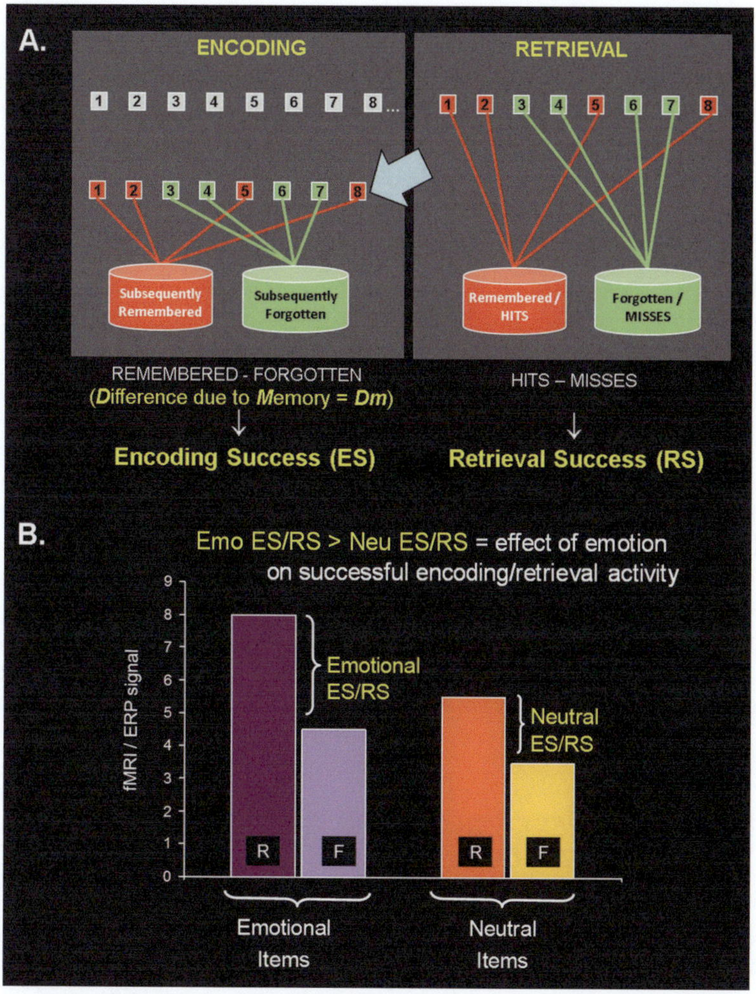

Fig. 5.1 The subsequent memory paradigm (SMP) – measuring the impact of emotion on encoding success (ES) and retrieval success (RS) activity. (**a**) General procedure involved in the SMP. (**b**) Diagram of the comparisons that allow identification of the brain regions susceptible to emotional ES and RS. *R* subsequently remembered items, *F* subsequently forgotten items, *fMRI* functional magnetic resonance imaging, *ERP* event-related potential

associated with *encoding success* (ES). In the case of retrieval, the contrast "remembered-minus-forgotten" identifies regions showing greater activity for items in which retrieval was successful (hits) than for items in which retrieval failed (misses). The activity difference between remembered and forgotten items during retrieval is known as *retrieval success* (RS) (Prince et al. 2005; Weis et al. 2004).

In the context of investigating the impact of emotion on memory, one of the main advantages of Dm/ES and RS analyses is that they can eliminate the confounding effect of general emotion processing and specifically identify brain activity linked to memory processes and to the memory-enhancing effect of emotion (Shafer et al. 2011). By contrasting the emotional ES/RS with the neutral ES/RS, the brain regions whose memory-related activity is susceptible to emotional modulation during encoding and/or retrieval can be identified. Therefore, neuroimaging studies using event-related designs and the SMP have been highly influential in revealing brain regions that show an interaction between emotion and memory at different stages of memory processing; basic findings from such studies are reviewed below.

Manipulating Basic Affective Properties Another important conceptual and methodological implementation concerns consideration of basic affective properties in experimental manipulations investigating the neural correlates of emotional memories. Emotional *arousal* and *valence*[3] are such affective dimensions (Lang et al. 1993; Russell 1980). Before the implementation of the proper manipulation of these dimensions (as illustrated in Shafer et al. 2011), the literature on emotion processing in general and on emotional memory in particular was dominated by the notion that negative stimuli produce stronger effects on memory than positive ones, which was linked to a negative bias in the involvement of basic emotion processing brain regions such as the AMY in these effects. However, subsequent studies directly comparing the effect of equally arousing positive and negative stimuli revealed similar memory-enhancing effects of such stimuli, regardless of their valence, and identified the sensitivity of AMY responses to processing positively valenced arousing stimuli (reviewed in Dolcos et al. 2012).

5.1.3 Basic Findings Regarding the Neural Correlates of Emotional Memory

Based on animal research, early human studies primarily emphasized the role of the AMY and its interaction with memory-related MTL regions in mediating the memory-enhancing effect of emotion, during encoding (e.g., Dolcos et al. 2004b; Kensinger and Corkin 2004; Kensinger and Schacter 2006a; Ritchey et al. 2008; Sergerie et al. 2006), consolidation (Ritchey et al. 2008), and retrieval (e.g., Dolcos et al. 2005; Kensinger and Schacter 2005; Sergerie et al. 2006) of emotional memories. Given its multiple connections with core memory structures, and with numerous cortical and subcortical regions involved in various aspects of information processing, such as perception, emotion, elaborative cognitive processes, and social cognition (see reviews by Dolcos and Denkova 2008; Dolcos et al. 2006b, 2012; Hamann 2001; LaBar and Cabeza 2006; LeDoux 2000; McGaugh 2004; Phelps 2004), the AMY together with the memory-related MTL regions and the PFC play

[3]*Arousal* refers to a continuum that varies from calm to excitement, whereas *valence* refers to a continuum that varies from pleasant to unpleasant, with neutral as an intermediate value (for methods of assessing these dimensions, see Bradley and Lang 1994).

a critical role in emotional memory. Most of these studies revealed that activity in the AMY, the MTL, and their interactions are primarily driven by the arousal and not the valence of the information, and that their interactions with the PFC contribute to valence-related effects in emotional memory (Dolcos et al. 2012).

Emotional Memory Encoding Early studies suggested a link between AMY activity at encoding and later retrieval of emotionally arousing material (e.g., Cahill et al. 1996; Hamann et al. 1999). However, as they used blocked designs, these studies could not distinguish between brain activity for successfully and unsuccessfully encoded stimuli within participants, and thus could not specifically assess the role of the AMY in the *successful* encoding of emotional stimuli, as opposed to linking it to general emotion processing. By contrast, studies involving event-related designs in conjunction with the SMP have allowed identification of the role of the AMY and its interactions with the memory-related MTL brain regions (e.g., hippocampus, HC) in memory enhancement by emotion, in neurologically intact human brain (e.g., Canli et al. 2000; Dolcos et al. 2004b; Hamann et al. 1999; Kensinger and Corkin 2004; Kensinger and Schacter 2006a; Kilpatrick and Cahill 2003; Richardson et al. 2004; Ritchey et al. 2008; Sergerie et al. 2006), and that AMY–MTL interactions contribute to the persistence of emotional memories over time (Ritchey et al. 2008). More recently, using a combination of blocked and event-related designs, encoding of emotionally arousing stimuli over an extended period (>20 min) was found to induce a sustained arousal-related brain state (reflected in increased AMY–HC functional connectivity), which overall contributed to greater recollection of unrelated neutral items encoded minutes following the initial encoding of emotional items (Tambini et al. 2017). This suggests that transient exposure to emotional arousal can create persistent "carry-over" effects, resulting in similar memory-enhancing effects for subsequently encountered neutral items.

Although the main focus of the earlier neuroimaging studies investigating the facilitating effect of emotion on memory was on the AMY and its interaction with the memory-related MTL regions, increasing emphasis has been placed on the influence of emotion on memory through cognitive processes supported by other brain regions, such as the PFC (Dolcos et al. 2011). It has been suggested that, whereas the AMY and the MTL are part of basic/direct neurohormonal mechanisms underlying the memory enhancement effect of emotion, the PFC has an indirect/mediated involvement in the formation of emotional memories (e.g., by enhancing strategic, semantic, and working memory processes) (Dolcos and Denkova 2008; LaBar and Cabeza 2006). Moreover, whereas AMY–MTL mechanisms are modulated primarily by arousal, the involvement of the PFC also seems to be influenced by valence (Dolcos et al. 2004a; Kensinger 2004; Kensinger and Schacter 2006b), probably reflecting the PFC involvement in higher-order emotion processing associated with evaluation of emotional valence and/or cognitive control strategies (see Sect. 5.2.2 below). Finally, available evidence also shows that the interactions of the AMY and the MTL memory regions are linked to the valence of emotional information (Ritchey et al. 2011). Ritchey et al. (2011) showed that the intrinsic AMY–MTL interaction is stronger for encoding of negative stimuli, whereas the extrinsic interactions between the MTL and the PFC are stronger for encoding positive stimuli (see also Mickley Steinmetz et al. 2010).

Further supporting the idea of valence-related differences linked to differential connectivity, there is evidence that successful encoding of positive items involves a fronto-parietal network, whereas successful encoding of negative items involves a temporo-occipital network (Kensinger and Schacter 2008; Mickley Steinmetz and Kensinger 2009). These findings are consistent with those of other studies, revealing that encoding of positive information is associated with activity in specific PFC subregions (Botzung et al. 2010a; Dolcos et al. 2004a), probably because of more elaborative processing demanding cognitive resources, whereas encoding of negative information is associated with temporo-occipital regions (Mickley and Kensinger 2008), probably because of increased sensory processing. These findings suggest that successful encoding of positive and negative information might also depend on slightly different processing, with positive items involving more elaborate processing and engaging the PFC regions associated with semantic processing (Kapur et al. 1996; Poldrack et al. 1999; Shallice et al. 1994), working memory operations (D'Esposito et al. 2000; Owen et al. 1999; Petrides 1995), memory control (Anderson et al. 2004), and negative items involving perceptual processing and engaging posterior areas associated with the visual processing of emotional information (Vuilleumier et al. 2004).

Emotional Memory Retrieval Although, as discussed above, involvement of the AMY in the formation of emotional memories has been well documented (for reviews see Dolcos and Denkova 2008; Dolcos et al. 2006b, 2012; LaBar and Cabeza 2006; Phelps 2004), until relatively recently it has been more difficult to demonstrate the involvement of the AMY in the retrieval of emotional memories (for review see Buchanan 2007). This may be partly due to memory testing methodology, using different kinds of events – e.g., *laboratory micro-events* versus *autobiographical real-life* events. Laboratory events are simpler experimenter-generated stimuli, such as lists of words or sets of pictures that are encoded in laboratory settings and retrieved at different intervals following encoding (e.g., minutes, hours, days, weeks, months). By contrast, autobiographical events are events from an individual's own history that are encoded in real-world settings and may be retrieved after much longer intervals of years or even decades. Hence, retrieval of laboratory events is dissimilar to retrieval of autobiographical events in at least two aspects: time elapsed between encoding and retrieval, and self-relevance. However, testing memory with both types of events presents advantages: autobiographical events have the advantage of being close to personal real-life events and to cover temporally dispersed past episodes, whereas laboratory events have the advantage of being well controlled experimentally for neuroimaging applications (see also Cabeza et al. 2004).

Similar to the early studies of encoding, the early studies of emotional memory retrieval share many of the same limitations, as most of them either used blocked designs that did not allow assessment of the functional neuroimaging data on a trial-by-trial basis (e.g., Dolan et al. 2000; Kosslyn et al. 1996; Taylor et al. 1998), or they did not compare activity associated with successfully versus unsuccessfully retrieved items to distinguish brain activity specifically associated with RS (e.g., Fossati et al. 2004; Maratos et al. 2001). On the other hand, studies involving event-related designs in conjunction with the SMP have also proved influential in elucidating the role of the AMY during *successful retrieval* of emotional memory (e.g., Dolcos et al. 2005; Kensinger and Schacter 2005; Sergerie et al. 2006). These studies provided strong evidence that successful retrieval of emotional memories involves AMY engagement similar to that identified during successful encoding of emotional memories. Furthermore, refined examinations of AMY activity linked to a differential impact of emotion on recollection- versus familiarity-based retrieval[4] pointed to a specific role of the AMY in the enhancement of recollection rather than familiarity in retrieving emotional memories (Dolcos et al. 2005; Sharot et al. 2004; Talarico et al. 2004).

More recent neuroimaging evidence also suggests that the AMY might play a role in the successful retrieval of personally relevant autobiographical memories (AMs), following shorter retention intervals. By investigating basketball fans' memories of specific basketball games, encoded a few days before the scanning session, Botzung et al. (2010b) showed that AMY activity was modulated by the emotional intensity of the recollected events, and was greater for extremely high- vs. low-intensity events. Hence, this study also points to the involvement of the AMY in the retrieval of memories for emotional and personally relevant events (see also Muscatell et al. 2010; Sharot et al. 2007a). Taken together, these studies provide evidence for the involvement of the AMY in the recollection of highly arousing and personally relevant recent events.

Unlike the involvement of the AMY in the retrieval of laboratory or relatively recent personally relevant events, which has been well documented, the engagement of the AMY during retrieval of remote autobiographical episodes temporally dispersed in an individual's history was, until relatively recently, unclear (Buchanan 2007). Whereas some studies found increased activity in the AMY for emotional autobiographical events (Markowitsch et al. 2000, 2003), others did not observe such an effect (Oddo et al. 2008; Vandekerckhove et al. 2005). In the same vein, although some studies reported that AMY activity may be modulated by emotional intensity during the early stages of memory retrieval (Daselaar et al. 2008), others failed to observe a link between AMY activity and emotional intensity (Addis et al. 2004; Maguire and Frith 2003). One of the factors that may account for these inconsistencies in AMY engagement during retrieval of AMs (Denkova et al. 2006; Greenberg et al. 2005) may be related to differences in the engagement of effortful processing, which is increased when remembering temporally dispersed past events, and thus could divert the *attentional resources* from the emotional value of recollections (Phan et al. 2002). Hence, engagement of the AMY seems to depend on the task instructions (Smith et al. 2006), as discussed below.

[4] Recollection-based retrieval refers to remembering specific contextual details of an event (e.g., about the time and place of its occurrence), whereas familiarity-based retrieval refers to only knowing that certain events occurred, without retrieving specific contextual details (Tulving 1985).

The link between task instructions and AMY response was recently investigated by directing the participants' *attention focus* either to emotional (*emotion* condition) or to other contextual (*context* condition) details during recollection of positive and negative AMs (Denkova et al. 2013). As expected, emotion compared with the context focus condition yielded increased activity in the left AMY for both positive and negative memories. Moreover, greater engagement of the AMY when focusing on emotional compared to the contextual details was associated with greater subjective re-experience of emotion of the recollected AMs. This finding is consistent with the emotion research, suggesting that the engagement of the AMY could be modulated by attention, current goals, and task demands (Lieberman et al. 2007; Shafer et al. 2012), and extends the available evidence by revealing that this effect also applies to the retrieval of both positive and negative AMs. In contrast to the left AMY, right AMY activity was not modulated by the current retrieval goals in the case of negative AMs, thus suggesting a hemispheric dissociation in the AMY as a function of retrieval focus and valence during AM recollection. One potential explanation for this AMY lateralization could be linked to Glascher and Adolph's model (Glascher and Adolphs 2003), which suggests that the right AMY is involved in an initial, automatic detection of emotions, whereas the left AMY is involved in a more elaborate, cognitive representation of emotions (Morris et al. 1999; Phelps et al. 2001). Overall, these findings shed light on the involvement of the AMY during recollection of AMs as a function of retrieval focus and emotional valence.

In addition to the evidence highlighting the engagement of the AMY during emotional memory retrieval, further evidence showed increased interactions between an AMY-based emotion processing system and the memory-related MTL regions during retrieval (Dolcos et al. 2005; Greenberg et al. 2005). These findings are consistent with the idea that the AMY and the MTL memory system are part of a synergistic mechanism in which emotion enhances recollection and recollection enhances emotion. Emotion may enhance recollection because reinstating the affective context of the original episode is likely to facilitate the recovery of contextual details, such as where, when, and how the original events happened. In turn, recollection of the context surrounding an emotional effect is likely to augment the emotional arousal elicited by the event during retrieval (Dolcos et al. 2005).

There is also evidence that the interaction between the AMY and the MTL memory-related regions may be modulated by the involvement of the PFC. For instance, a study by Smith et al. (2006) used dynamic causal modeling to examine the effective connectivity among the AMY, HC, and PFC during retrieval, reporting increased bidirectional connectivity between the AMY and the HC, and increased engagement of the medial PFC. Moreover, the latter influenced activity in the AMY and the HC during a task requiring explicit recollection of the emotional information compared with a task not requiring explicit focus on emotional information. Consistent with the evidence discussed above, this finding suggests that the involvement of the AMY could be modulated by an individual's goals (Cunningham et al. 2008), and points to the involvement of top–down processes through the medial PFC, during explicit retrieval of emotional information (see also Denkova et al. 2015). Finally, similar to evidence from encoding, studies of memory retrieval for emotional laboratory (Erk et al. 2005;

Maratos et al. 2001; Smith et al. 2004b, 2006) and autobiographical (Botzung et al. 2010b; Markowitsch et al. 2003; Piefke et al. 2003) events also point to the involvement of the PFC linked to emotional valence. For instance, increased activity in the medial and/or the orbital PFC was associated with retrieval of positive contextual information (e.g., Erk et al. 2005) and of positive personal experiences (Markowitsch et al. 2003; Piefke et al. 2003). The medial orbital PFC has been associated with general emotion processing (Davidson and Irwin 1999; Phan et al. 2002), reward-related (Dolan 2007; O'Doherty 2004) processing, and self-referential processing (for review see Heinzel and Northoff 2009; Ochsner et al. 2004). Therefore, it is possible that the engagement of the medial orbital PFC during retrieval of positive emotional information is linked to the elaboration of self-relevant rewarding experience. Related to this, increased activity in the orbital PFC, together with its increased connectivity with the HC, was associated with the processing of socially rewarding stimuli (Tsukiura and Cabeza 2008) (see also Sect. 5.2.1 below).

In summary, the evidence discussed above supports the notion that the memory-enhancing effect of emotion is associated with increased activity and interaction between an emotion-based system involving the AMY and the memory-based MTL structures, during both encoding and retrieval. Whereas the MTL-based mechanism seems to be involved in arousal-dependent effects, valence effects could be revealed in the connectivity of these regions within and outside the MTL. It should also be noted that, although overall these findings provide strong evidence for the modulation hypothesis, the AMY findings, showing greater encoding and RS for emotional than for neutral stimuli, are also consistent with the plasticity hypothesis. Regarding the role of the PFC, the available evidence suggests that its involvement might be sensitive to the processing of valence and might reflect higher order cognitive processing (e.g., semantic memory, working memory, cognitive control, and self-referential processing), during both encoding and retrieval of emotional information.

5.2 Current Issues and Emerging Directions in the Impact of Emotion on Memory

This section discusses new evidence regarding three emerging research directions, as follows: the role of social information in emotional memory, the role of emotion regulation in the impact of emotion on memory, and the impact of emotion on associative or relational memory.

5.2.1 The Role of Social Information in Emotional Memory

Recent evidence suggests that the memory-enhancing effect of emotion might involve more complex interplays among the AMY, MTL, PFC, and other brain regions, when besides the typical basic affective dimensions (arousal and valence), other more abstract psychological dimensions, such as *social relevance* are considered. Evidence supporting these ideas is discussed in this section.

Elucidation of the complex interplay between emotion and social cognition and its impact on memory is important, as it may shed light on our understanding of emotional memory for more ecologically valid situations, involving interactions with other people and having direct relevance to our own social behavior. Knowledge acquired through previous social interactions is critical for our ability to navigate successfully the highly complex social world, guiding our adaptive behavior in a variety of social situations (Ciaramelli et al. 2013; Spreng and Andrews-Hanna 2015; Spreng and Mar 2012; Tsukiura 2012), which is influenced by personality traits affecting social behavior. For instance, trait empathy has been positively associated with recognition memory performance in healthy individuals (Wagner et al. 2015), and was significantly reduced in patients with hippocampal amnesia showing declarative memory impairments (Beadle et al. 2013), thus suggesting an important link between memory processes and healthy social behavior (see also Laurita and Spreng 2017). Moreover, recent neuroimaging evidence points to the existence of a network of brain regions subserving memory processes for stimuli with *social relevance*, such as other people's faces and impressions formed based on them (e.g., Gilron and Gutchess 2012; Tsukiura 2012).

Given that social information may carry emotional significance, at a basic level, it should probably influence memory through similar mechanisms to those engaged by emotion. However, because of its complexity, it is also expected to engage additional brain systems. Consistent with this idea, there is evidence that social relevance requires more elaborative processing (e.g., interpretation of the stimulus meaning depending on the context and individual differences), and seems to enhance memory only when sufficient cognitive resources are available (Sakaki et al. 2012). Along these lines, Sakaki et al. (2012) suggested that, although basic processing of biologically emotional stimuli might be automatic (mediated by increased activity in and connectivity between the AMY and the visual cortex), processing of socially emotional stimuli might also depend on more elaborative processes involving enhanced activity in and connectivity between the AMY and the medial PFC. In support of this idea, neuroimaging studies have also demonstrated the involvement of the AMY and the medial/orbital PFC in complex social cognitive functions, such as detecting social cues in the environment, interpreting and monitoring affective reactions, engaging in self-referential processes, or processing reward and punishment (e.g., Botzung et al. 2010a; Gilron and Gutchess 2012; Harvey et al. 2007; Somerville et al. 2006; Tsukiura 2012; Yaoi et al. 2015). The subsections below discuss evidence regarding the neural mechanisms of encoding and retrieving socially relevant information.

Encoding of Socio-emotional Memories Recent studies have suggested that the role of the AMY might extend beyond successful encoding of general emotional stimuli, to include successful encoding of socially (personally) relevant stimuli (Botzung et al. 2010a; Harvey et al. 2007; Kleinhans et al. 2007; Tsukiura 2012) and, more broadly, processing of stimuli with motivational significance (Adolphs

2010; Cunningham and Brosch 2012). For instance, by using personally relevant social stimuli (e.g., portions of basketball games), Botzung et al. (2010a, b) reported that AMY activity was preferentially sensitive to highly emotional memories, especially those regarding positively valenced plays. In this context, it is interesting to note that increased AMY activity was also observed when people imagined positive future events relative to negative ones (Sharot 2011; Sharot et al. 2007b). It is possible that basketball fans are more apt to consider positive (versus negative) plays as more personally significant, and thus this greater self-relevance could lead to increased AMY involvement in the encoding of positive personal episodes (but see Northoff et al. 2009). Overall, these findings suggest that the involvement of the AMY in emotional memory encoding might be influenced by personal involvement during encoding of positive episodes, which identifies the social belonging of the individual.

There is also evidence pointing to the role of the AMY in encoding both positive and negative information about social others (e.g., faces, traits) (Said et al. 2009; Schiller et al. 2009; Vrtička et al. 2012). It is possible that, owing to its intrinsic motivational value and importance for survival, social relevance in stimuli is detected by the AMY independently of the valence and arousal dimensions (Harvey et al. 2007; Vrtička et al. 2012). These findings are consistent with the involvement of the AMY in social cognition and behavior, and with a more general role of this region in tracking subjective significance or relevance in the environment based on the current goals (Adolphs 2010; Cunningham and Brosch 2012).

Besides the AMY, extant evidence also suggests the involvement of medial/orbital PFC regions in encoding socially relevant stimuli (Gilron and Gutchess 2012; Gutchess et al. 2015; Harvey et al. 2007; Tsukiura 2012; Vrtička et al. 2012; Yaoi et al. 2015). For instance, increased activity in the medial PFC has been associated with both enhanced memory for the information encoded with reference to oneself – i.e., the self-reference effect (Greenwald and Banaji 1989; Gutchess et al. 2015; Macrae et al. 2004; Yaoi et al. 2015) – and with encoding of impressions of other people, based on face–behavior associations (Cassidy et al. 2013; Gilron and Gutchess 2012; Mitchell et al. 2004). In addition, activity in the medial orbitofrontal cortex (OFC) was associated with encoding of faces signaling positive social cues (Tsukiura and Cabeza 2008, 2011a, b) and scenes with a social content in general (Harvey et al. 2007; Vrtička et al. 2012). These findings are consistent with the role of the medial PFC regions in the representations of an individual's own and of others' minds (Northoff et al. 2006; Wagner et al. 2012), in addition to the encoding and integration of the subjective value of the stimuli (Delgado et al. 2016).

Moreover, greater activity in the medial OFC, together with enhanced functional connectivity with the HC, have been observed during successful encoding of socially rewarding stimuli (i.e., smiling and attractive faces) (Tsukiura and Cabeza 2008, 2011a). These findings emphasize the interplay between the reward-related brain regions (medial PFC/OFC) and the memory-related regions (HC) during the formation of memory for positive social stimuli. In contrast, the enhancing effect of socially negative signals (e.g., untrustworthy or unattractive faces) on memory was mediated by increased activity in and connectivity between the insula and the HC (see also Botzung et al. 2010a; Tsukiura and Cabeza 2011b; Tsukiura et al. 2013). Finally, activity in the anterior temporal lobe (ATL) structures along with connectivity with the AMY, MTL, and medial PFC regions have also been implicated in socio-emotional memory encoding, particularly with regard to its role in representing and storing personal identity information (Collins and Olson 2014; Olson et al. 2013; Spreng and Andrews-Hanna 2015; Tsukiura et al. 2010).

Retrieval of Socio-emotional Memories The involvement of the AMY and the memory-related MTL regions, together with the engagement of the PFC, have been revealed during the retrieval of socio-emotional memories. First, increased AMY activity was linked to retrieval of faces that had previously been encoded with emotional descriptions of behaviors in general (Somerville et al. 2006), and with social fairness learned in the context of an economic game (Singer et al. 2004). Second, increased activity in, and interactions between the AMY and the HC have been identified for socially induced memory errors during retrieval (Edelson et al. 2011), thus suggesting that social interaction could have long-lasting effects on memory through AMY–HC mechanisms. Regarding PFC regions, the right medial PFC appears to mediate retrieval of social contexts, whereas the left medial PFC seems to underlie retrieval of self-generated contexts (Mano et al. 2011), thus revealing that dissociable regions within the PFC mediate social and self-referential processing during episodic memory retrieval. Finally, increased activity in the ATL along with its enhanced connectivity with the HC have been associated with the retrieval of personal identity information (Collins and Olson 2014; Tsukiura et al. 2008, 2011).

Functional Systems and Networks in Socio-emotional Memories As discussed above, there is ample evidence supporting the idea that the encoding and retrieval of information with social relevance are subserved by the interaction of a host of regions broadly involved in emotion processing, social cognition, and memory processes. In integrating the differential contributions of brain regions to socioemotional memory, Tsukiura (2012) has posited that the impact of affective information on memory for faces may be mediated by the interaction between the two neural systems: *affective* (AMY, medial OFC, and insula) and *memory* (HC, fusiform face area [FFA]) systems (Fig. 5.2). In this context, the AMY may detect general emotional intensity in faces and work in concert with the medial OFC and the insula to process positive and negative signals from the faces respectively. These emotion processing regions may then modulate activity in the HC and the FFA, which have been implicated in encoding and retrieval of neutral faces (Prince et al. 2009), in enhancing memory for faces with affective information (Tsukiura 2012). This model is consistent with evidence that the memory-enhancing effect of emotion in general involves AMY-MTL interactions with top–down modulatory influences from the PFC regions (Dolcos et al. 2012), and with the emerging view of the AMY as a functional hub in the large-scale functional networks subserving various facets of social cognition and behavior in humans (Bickart et al. 2014). Moreover, evidence from the recent studies examining the dynamics of large-scale functional networks of the brain suggests that the complex processes involved in socio-emotional memory may be subserved by a network of regions whose activity repre-

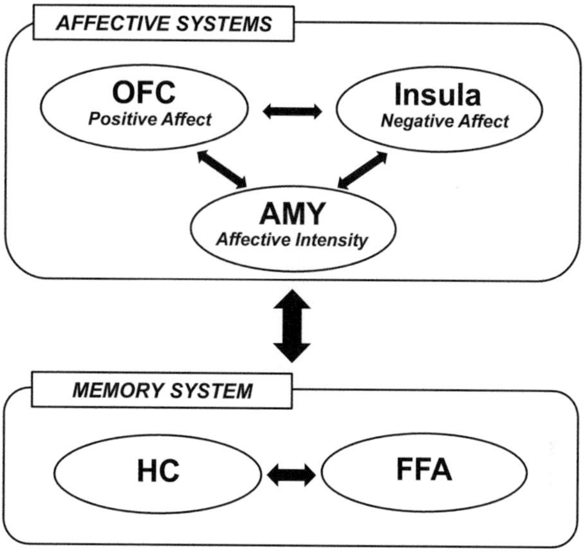

Fig. 5.2 A hypothetical model of the neural mechanisms underlying the effect of face-based affective signals on memory for faces. *AMY* amygdala, *OFC* orbitofrontal cortex, *INS* insular cortex, *HIP* hippocampus, *FFA* fusiform face area (Reproduced from Tsukiura 2012, with permission)

sents a "default state" of the brain. These studies identified considerable overlap between the networks of regions typically activated in tasks involving emotion processing, social cognition, and episodic memory, and brain regions consistently activated when individuals are *at rest* – i.e., the *default mode network* (DMN) (Li et al. 2014; Schilbach et al. 2012; Spreng and Mar 2012). For instance, a recent meta-analysis which employed conjunction analyses to identify the common set of regions implicated consistently in studies of emotion processing, social cognition, and resting state confirmed the common involvement of a dorsal subregion of the medial PFC (dmPFC) along with the precuneus, suggesting that introspective processes may be a common denominator across the three processes (Schilbach et al. 2012). The dmPFC is considered part of the dorsomedial subsystem of the DMN (Spreng and Andrews-Hanna 2015), and more importantly, this region has been implicated in a broad range of social cognitive processes, including successful memory encoding of first impressions (Gilron and Gutchess 2012; Mitchell et al. 2004) and temporary storage and maintenance of information about others' traits (Meyer and Lieberman 2012; Meyer et al. 2015).

Interestingly, previous studies assessing resting-state connectivity within the DMN have shown that greater intrinsic functional connectivity among regions within this network is linked to more frequent retrieval of AMs in a social context (Yang et al. 2013) and higher scores in emotional intelligence (Takeuchi et al. 2013). Moreover, in support of these findings, and further extending the traditional view of the DMN, emerging evidence suggests that the AMY, through its high intrinsic and task-evoked connectivity with a subset of the DMN regions, might constitute a functional network involved in both maintenance of the brain's default state and various socio-affective processes (i.e., the *extended social-affective default network*) (Amft et al. 2015; see also Bickart et al. 2014; Schilbach et al. 2008). Taken together, these findings identify the critical role of the DMN and associated regions in socio-emotional memory processes, and further demonstrate that variability in intrinsic connectivity among the DMN regions is linked to behavioral indices of emotional, social, and memory processes. Future research directly examining the relationships between task-evoked and intrinsic activity/connectivity in the (extended) DMN regions, in addition to behavioral performance, would be helpful in elucidating the role of large-scale networks in memory encoding and retrieval of information with social relevance.

In summary, recent neuroimaging studies examining the impact of social relevance on the emotional enhancing effect on memory emphasize the involvement of the AMY, HC, PFC, along with other regions involved in emotion processing and social cognition (e.g., insula, ATL), during both encoding and retrieval of socio-emotional memories. The engagement of the AMY/HC and the medial PFC was linked to memory for positive stimuli with increased personal or social relevance, whereas the HC and the insula were linked to memory for negative social stimuli, such as untrustworthy and unattractive faces. In addition, the AMY, medial PFC, and ATL were also associated with encoding and retrieval of social information, regardless of their valence, and these findings are consistent with the involvement of these regions in social cognition and behavior. Emerging evidence from the investigations of large-scale functional networks has also begun to reveal the link between the brain's intrinsic functional architecture and various social-affective and cognitive processes subserving socio-emotional memory, with the dmPFC possibly serving as a "hub" of these processes. Therefore, memory encoding and retrieval of information with social relevance seem to involve complex interactions among both distinct and overlapping neural networks subserving basic emotion processing, social cognition, and memory processes, which may collectively allow us to integrate information from internal and external sources to adaptively navigate through the social landscape.

5.2.2 The Role of Emotion Regulation in the Impact of Emotion on Memory

The topic of emotion regulation (ER) has gained considerable interest, as the ability to cope adaptively with emotionally challenging situations is vital for physical and mental health (Gross 2008, 2015). Important progress has been made in understanding the neural correlates of ER associated with the immediate effects of engaging specific strategies, which are typically reflected in reduced emotional experiences, if down-regulated (e.g., Lieberman et al. 2011; Ray et al. 2010). However, less is known about the long-term effects of ER on memory for emotional events. These issues are discussed in the present section.

The two most widely studied ER strategies in brain imaging research are *cognitive reappraisal,* which involves attempts to change the meaning of stimuli/situations (by thinking for instance that the situation is not real), and *expressive suppression,* which involves attempts to decrease emotionally expressive behavior

(Gross 2008). Extensive research suggests that reappraisal might have an advantage over suppression in reducing emotional experiences (Eippert et al. 2007; Kalokerinos et al. 2015; Olatunji et al. 2015), and a differential engagement of brain regions by these ER strategies (Goldin et al. 2008; Hermann et al. 2014). More recently, ER research has also considered the impact of *attentional deployment strategies*, such as *focused attention (FA)*, which involves shifts in attention to or away from the emotional aspects of emotion eliciting stimuli or events (Gross 2008; Sheppes et al. 2014). The effectiveness of attentional deployment ER strategies has been confirmed by a recent meta-analysis (Webb et al. 2012), and the underlying neural mechanisms have also been investigated in some recent functional magnetic resonance imaging (fMRI) studies (Dorfel et al. 2014; Kanske et al. 2011; McRae et al. 2010). Recent studies also started clarifying how the engagement of various ER strategies affects the impact of emotion on memory (Ahn et al. 2015; Dillon et al. 2007; Richards and Gross 2000), and the underlying neural circuitry (Binder et al. 2012; Erk et al. 2010; Hayes et al. 2010). Better understanding of the impact of ER on emotional memories has important implications for understanding and treating affective disorders, which are characterized by an excessive focus on negative memories (Rubin et al. 2011; Williams and Moulds 2010) and emotional dysregulation (Gotlib and Joormann 2010; Mayberg 1997; Sheppes et al. 2015). Evidence emerging from the literature regarding the impact of ER on emotional memory is discussed below.

ER and Emotional Memory Encoding Overall, available research reveals that engaging cognitive reappraisal during memory encoding enhances subsequent memory for the reappraised information (Dillon et al. 2007; Liu et al. 2015; Richards and Gross 2000), even after longer intervals (Ahn et al. 2015; Kim and Hamann 2012), whereas suppression tends to impair memory for the suppressed items (Richards and Gross 2000). One potential explanation that has been put forward to explain the beneficial effects of reappraisal on memory is linked to semantic elaboration processes involved during reappraisal, which may lead to a deeper level of encoding of the reappraised items (Dillon et al. 2007). Interestingly, the memory advantages of reappraisal have also been linked to increased bias toward enhancing positive memories (Levine et al. 2012). Regarding the neural correlates of reappraisal and suppression of emotional memory, recent neuroimaging studies point to a differential neural engagement of the HC and PFC for successfully encoded emotional items that were reappraised or suppressed (Binder et al. 2012; Hayes et al. 2010). Specifically, Hayes et al. (2010) showed that the memory enhancement by reappraisal was linked to increased engagement and co-activation of the HC and the left lateral PFC (Fig. 5.3), whereas Binder et al. (2012) revealed that the memory impairment by suppression was associated with decreased engagement and co-activation of the HC and the lateral PFC. These findings suggest that the core memory region (HC) and a higher-level cognitive processing region (PFC), along with their interplay during memory formation of "regulated" emotional items, are differently affected by the engagement of these ER strategies, which results in different effects on the subsequent emotional memory.

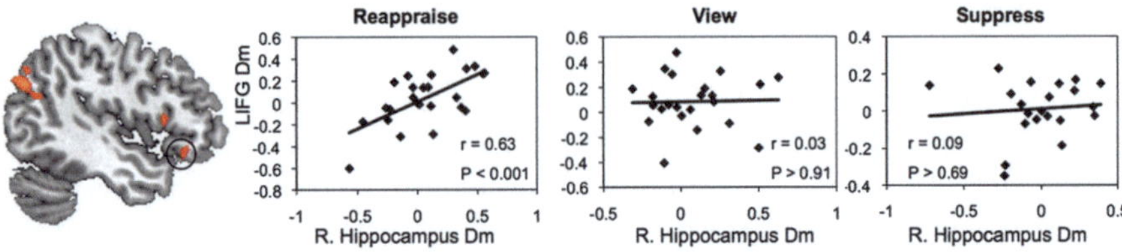

Fig. 5.3 Stronger positive correlation between the left inferior frontal gyrus (*IFG*) and hippocampus (*HC*) when using reappraisal. Compared with passive viewing and suppression, reappraisal was associated with increased IFG–HC correlation for memory-related activity (Dm effect) as a function of the emotion regulation (ER) strategies used during encoding. *Dm* difference due to memory (Reproduced from Hayes et al. 2010, with permission)

ER and Emotional Memory Retrieval *Laboratory events.* One of the most frequently studied strategies for inhibiting retrieval of laboratory-based episodic memories is *thought suppression* (Benoit et al. 2015; Depue et al. 2006, 2007; Kupper et al. 2014). It is referred to as an attentional deployment strategy (Sheppes and Gross 2012) and is typically studied using the think/no-think paradigm, during which participants first learn cue–target pair associations. They are then presented with the cues and instructed to suppress retrieval of the associated targets. In general, suppressing retrieval of unwanted emotionally neutral memories has been linked to top–down influences of the lateral PFC on the HC (Benoit et al. 2015; Gagnepain et al. 2014), and this effect was even stronger in participants who are better at suppressing memories (Benoit and Anderson 2012). More specifically, suppressing retrieval of *emotional* information has been linked to two-stage neural mechanisms, involving initial inhibition by cognitive control regions (inferior frontal gyrus) on visual regions (fusiform gyrus), followed by a subsequent inhibition by cognitive control regions (middle frontal gyrus) on emotional memory regions (AMY and HC) (Depue et al. 2007). This finding highlights the importance of considering the emotional content of unwanted memory and suggests that, although common mechanisms involving the PFC and the HC might play a role in suppressing both emotional and non-emotional memory retrieval, specific patterns of activity might be linked to suppressing the retrieval of emotional information.

Autobiographical Events. Investigation of the neural correlates underlying the impact of ER during autobiographical recollection is of particular relevance not only for everyday activities in healthy populations, but also for affective disorders. The very few studies investigating ER in the context of emotional AM retrieval focused particularly on reappraisal (Fabiansson et al. 2012; Holland and Kensinger 2013; Kross et al. 2009), and reported that it can lead to a decrease in emotional experience accompanying the retrieval of AMs. However, evidence regarding the associated neural correlates is not conclusive, particularly concerning the role of emotion-related brain regions. For instance, Holland and Kensinger (2013) reported increased activity in the lateral and the medial PFC, when participants down-regulated their emotional reactions using reappraisal during the (re)construction of AMs, but this was also associated with increased, rather than decreased, activity in

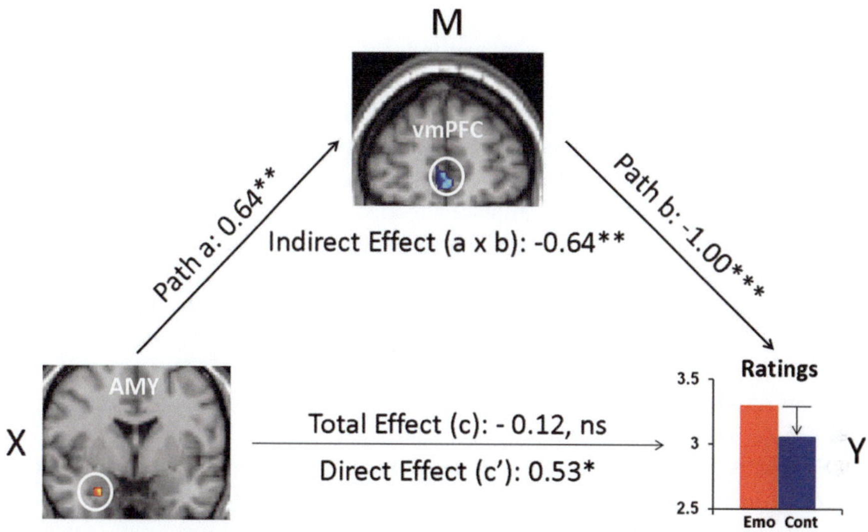

Fig. 5.4 Ventromedial PFC mediates the link between AMY and emotional ratings, when focusing away from the emotional aspects of autobiographical recollections. Mediation analysis identified a significant ($p = 0.009$) negative mediation effect of the ventromedial PFC (vmPFC) on the relation between the AMY and emotional ratings, while focusing on non-emotional aspects of personal memories, and a significant ($p = 0.03$) positive direct effect (path c', X to Y controlling for M) between AMY and emotional ratings, when controlling for the influence of the vmPFC. Standardized coefficients and significance noted with *asterisks* are reported for each path. $^{*}p < 0.05$; $^{**}p < 0.01$; $^{***}p < 0.001$ (two-tailed); *ns* not significant (Reproduced from Denkova et al. 2015, with permission)

emotion-related regions. The other two studies did not report specific activations related to reappraisal, which could be due to the use of only a few memories repeatedly presented across different conditions and/or to delayed instructions to regulate (i.e., after the engagement of memory retrieval) (Fabiansson et al. 2012; Holland and Kensinger 2013; Kross et al. 2009).

A recent study identified the neural underpinnings of *FA*, as an effective *attentional deployment* ER strategy during emotional autobiographical recollection (Denkova et al. 2015). This study revealed that focusing on the non-emotional contextual aspects (time, location, other persons present, etc.), and away from the emotional aspects of highly emotional personal memories led to decreased self-reported ratings of emotional experiences associated with those memories. This behavioral effect was accompanied by increased activity in the ventromedial PFC (vmPFC) and decreased activity in the AMY. Moreover, mediation analysis suggested that the vmPFC might play the role of a functional hub, integrating affective signals from the AMY and mediating their impact on the subjective re-experiencing of emotion according to the current retrieval focus (Fig. 5.4). Importantly, the finding regarding the role of the AMY (see also Dolcos et al. 2006a) challenges the view promoted by most of the functional neuroimaging studies of ER, which mainly emphasize top–

down influences on the response of the AMY from the PFC regions involved in cognitive control, rather than reciprocal influences between these regions and the AMY.

It is important to note that the manipulation used here involves simply switching the retrieval focus, whereas the main task remains the same (i.e., recollection of personal memories). This remarkably subtle manipulation is consistent with ER studies showing that the manipulation of attention focus either on emotional or non-emotional aspects during working memory tasks can alter the emotional responses (Thiruchselvam et al. 2011) and performance (Iordan 2016). Interestingly, this manipulation can also be linked to emerging evidence in the AM literature highlighting the beneficial impact of *episodic specificity induction*, which involves remembering events in great detail (Madore and Schacter 2016). This recent line of research revealed that focusing on very specific episodic details related to places and people may be beneficial to a range of cognitive tasks (Madore et al. 2015; McFarland et al. 2017), and to psychological well-being (Jing et al. 2016). These emerging lines of evidence are also consistent with previous studies showing that training people to be more concrete and specific in recollecting memories may have beneficial effects on depression (Raes et al. 2009; Watkins et al. 2009). The findings from the study by Denkova et al. (2015) extend such investigations to *emotional* AMs, by revealing that focusing on non-emotional aspects of AMs can influence the emotional (re)experience of such memories. These findings provide support for cognitive behavioral therapies involving ER training to "distract" from emotional aspects of personal memories, by focusing and elaborating on non-emotional contextual aspects of retrieved AMs, which in turn leads to reduced self-experienced emotions.

In summary, emerging evidence regarding the impact of ER on emotional memory suggests that the use of ER strategies influences both formation and retrieval of emotional memories at behavioral and neural levels. In particular, using different ER strategies while encoding or retrieving emotional information seems to result in differential effects on subsequent memory (enhanced versus impaired). More specifically, available evidence suggests that using cognitive reappraisal during encoding leads to enhanced emotional memory, whereas using suppression leads to impaired memory, and that reappraisal tends to favor positive memories. Using attentional deployment strategies, such FA, can have beneficial effects on the retrieval of emotional personal memories, because it can enhance the emotional impact of positive memories and reduce the impact of distressing ones. These behavioral effects are coupled with differential engagement of, and connectivity between, specific lateral and medial PFC regions, associated with cognitive control, and the MTL regions, associated with emotion (AMY) and memory (HC). Notably, the patterns of activity and connectivity linked to the engagement of ER are also consistent with the idea discussed in Sect. 5.1 that the PFC plays an indirect/mediating role in the formation of emotional memories, and further suggests that ER strategies might affect the basic MTL-related mechanisms underlying emotional memories through the involvement of the indirect PFC-related mechanisms.

5.2.3 *The Impact of Emotion on Associative or Relational Memory*

As highlighted in the previous section, there is abundant evidence that emotions play a critical modulatory role in episodic memory (Dolcos et al. 2012; Weymar and Hamm 2013). For instance, when emotional and neutral items (e.g., words, faces or scenes) are presented in isolation and memory is tested at a later time, emotional items are better remembered than neutral ones (e.g., Bradley et al. 1992; Dolcos et al. 2005; Weymar et al., 2009). This memory-enhancing effect of emotion has been attributed to the involvement of the AMY and its interaction with the memory-related MTL regions (HC, perirhinal and parahippocampal cortices) (Dolcos et al. 2012), which can also be influenced by social context and cognitive control (see also Liu et al. 2016). An important aspect of human memory, however, is the binding of contextual information (e.g., time, place, or associative cues), which constitutes many disparate features of a unified event (Davachi 2006; Ranganath 2010). This section discusses evidence regarding the impact of emotion on such associations, and the underlying neural mechanisms.

When experiencing a traffic accident, associated cues, such as the make of the car, the color of the surrounding vehicles, a radio song or a brake noise, that were part of such experience, are also able to reactivate strong emotional memories when exposed to them later. Understanding such fundamental associative mechanisms is crucial for everyday life, and is also relevant for clinical conditions. For instance, events associated with traumatizing contexts leading to post-traumatic stress disorder (PTSD) can involuntarily trigger vivid distressing memories in the form of intrusive thoughts, flashbacks, or nightmares (e.g., Flor and Nees 2014; Wilker et al. 2014), in which binding mechanisms seem to be compromised. Recent research has, therefore, focused on the effects of emotion on memory for items as a part of, or in relation to, other items (reviewed by Chiu et al. 2013). Both enhancing and impairing effects of emotion on memory binding have been observed and, as discussed below, these opposing effects have been explained in the context of different (but not mutually exclusive) views (Bisby et al. 2016; Chiu et al. 2013; Christianson 1992; Kensinger 2009; Mather and Sutherland 2011).

Behavioral Findings It has long been known from the attention literature (Easterbrook 1959) that emotional arousal narrows attention to central cues, at the cost of peripheral, irrelevant cues. This prioritization of resources toward central aspects of information can result in better memory, compared with neutral peripheral or contextual information, which tends to be attended less (Buchanan and Adolphs 2002; Burke et al. 1992; Christianson 1992). For instance, memory is better for central aspects in a scenario (e.g., the weapon itself), but worse for other surrounding contextual details (e.g., about the perpetrator holding the weapon). Thus, emotional salience can weaken the integration of central aspects with peripheral contextual information into a unified memory representation (Mather 2007). This phenomenon is also referred to as *emotion-induced memory trade-off*

(Kensinger 2009), in which memory is enhanced for emotional items but impaired for the associated neutral backgrounds.

Among the various accounts for these opposing effects of emotion on memory (e.g., Chiu et al. 2013; Kensinger 2009; Mather 2007), one suggests that emotional arousal has differential effects on memory binding depending on certain aspects related to an event. Mather (2007) proposes that, on the one hand, emotion can boost memory for within-binding features, but on the other hand, it impairs memory for between-object features, which do not take advantage of emotional arousal (see also Kensinger 2009, for relationship to the intrinsic versus extrinsic distinction). These assumptions were derived from a number of studies using either within-item or between-item binding tasks (for overview see Chiu et al. 2013). For instance, there is evidence that specific intrinsic features, such as color (e.g., D'Argembeau and Van der Linden 2004; Doerksen and Shimamura 2001) or location (Mather and Nesmith 2008; Nashiro and Mather 2011) of emotional stimuli, are better remembered than the same features associated with neutral events. By contrast, memory-imparing effects are observed for extrinsic aspects and contextual details of emotional stimuli, when emotional objects are accompanied by neutral scenes (Kensinger et al. 2007), when neutral stimuli (e.g., objects) are overlaid on emotional scenes (Touryan et al. 2007b), or when emotional scenes are associated with a colored frame (Rimmele et al. 2011).

Mather's view regarding these effects has been recently postulated in the so-called *arousal-biased competition (ABC) theory* (Mather and Sutherland 2011), which highlights another important factor contributing to memory binding, namely, *attentional priority*. According to this model, emotional arousal enhances processing of whatever has the highest priority (because of bottom–up salience or top–down relevance), and impairs processing of information with lower priority (Mather and Sutherland 2011). Thus, emotional arousal may enhance associative memory for features of high-priority items (e.g., color or location of an item) and impair memory for neutral items, when presented at the same time (or almost the same time, see Sakaki et al. 2014) as emotional items. For instance, presentation of emotional images can enhance memory for preceding neutral objects when these objects receive high priority, but can impair memory when these objects are not prioritized (Sakaki et al. 2014).

This theory also addresses previous behavioral findings, in which better memory was found for neutral words presented in the context of emotional words (Guillet and Arndt 2009) or when neutral objects were placed on emotional scenes (e.g., Smith et al. 2004b, 2005; Ventura-Bort et al. 2016b), which should result in worse memory for the neutral event following the attention-narrowing (Easterbrook 1959) or between-binding concept (Mather 2007). In these studies, participants were instructed to learn word–word associations (Guillet and Arndt 2009) or to mentally integrate or connect objects and pictures at encoding (e.g., Smith et al. 2004b), which gives such associations high "attentional priority" that may facilitate binding of emotional and neutral information and diminish competition.

To summarize, behavioral findings point to both enhancing and impairing effects of emotion on memory binding, and these opposing effects can be explained by differences in attentional deployment toward, and prioritization of, emotional items and associated features (Easterbrook 1959; Kensinger 2009; Mather 2007; Mather and Sutherland 2011). The next subsections discuss evidence concerning the neural mechanisms of these effects.

Neural Findings: Encoding At the neural level, the effect of emotion on associative memory remains largely unexplored. Available evidence from emotional memory research reviewed in the previous sections suggests that the effects of emotion on binding are mediated by the AMY (in interaction with regions supporting sensory and perceptual processing, e.g., Lang and Bradley 2010), which may modulate regions in the MTL (e.g., HC) that are important for relational representations, such the binding of item and context information (Davachi 2006; Ranganath 2010). Empirical evidence, however, has not been conclusive so far, partly because of differences in experimental designs and aspects tested (e.g., within versus between features) related to an event. In a study investigating the *trade-off* in memory, enhanced encoding activity was observed in emotional memory regions, such as the AMY, HC, and PFC regions (Waring and Kensinger 2011), when participants remembered emotional items better than neutral items presented in neutral scenes. Similarly, increased AMY activity was reported for emotional words that were subsequently remembered, compared with neutral words (Kensinger and Schacter 2006a). This is also in line with Dougal et al. (2007), who found AMY and MTL activation showing covariation with subsequent item memory for emotional words. However, such memory-enhancing effects in the AMY (and the MTL) were not observed for emotional events when associated contextual details (i.e., semantic judgment or color) were tested (Dougal et al. 2007; Kensinger and Schacter 2006a). This finding is consistent with evidence regarding the role of the AMY in emotional gist memory (e.g., Adolphs et al. 2005). Hippocampal activity, on the other hand, was related to correct memory of contextual details (Ranganath 2010) for both emotional and neutral events.

Together, these findings support the view that the AMY is specifically involved in memory-enhancing effects for aspects that are intrinsically linked to the emotional event itself (see also Thoresen et al. 2012), but not to extrinsic aspects. However, this prioritized processing of emotion-associated information, which could lead to better memory storage, is not restricted to intrinsic aspects of the emotional event. When participants are instructed to intentionally connect, for instance, objects and their locations with emotional or neutral background scenes, better memory is observed for the location of the objects when presented with emotional scenes, and this was associated with enhanced AMY activity during encoding (Luck et al. 2014). It remains to be clarified, however, whether or not such an emotion-enhancing effect is related to preferential processing due to high attentional prioritization, as suggested by Mather and Sutherland (2011). Extant evidence also points to a network consisting of the AMY and visual processing areas involved during encoding, depending on whether or not information is prioritized (Lee et al. 2014). This indicates that emotional arousal can in fact amplify visual processing of high-priority events, which may influence subsequent MTL-based binding into a long-lasting memory representation. This idea is consistent with event-related potential (ERP) findings of early enhanced perceptual and elaborative processing in conditioning studies using aversive electrical shocks (for review see Miskovic and Keil 2012) or emotional background pictures (Ventura-Bort et al. 2016a), as unconditioned stimuli, which provide support for the finding by Lee et al. (2014).

Neural Findings: Retrieval Neural evidence for emotion effects on associative binding also comes from studies investigating memory retrieval. When measuring neural activity during a recognition memory test for words that had been encoded in the context of emotional and neutral sentences, enhanced activation was found for words from the emotional context in the AMY, HC, and PFC (Maratos et al. 2001), consistent with previous retrieval findings testing emotional item memory (e.g., Dolcos et al. 2005). Increased AMY activation during retrieval was also observed for recognized unpleasant scenes compared with forgotten ones, when cued by either an unpleasant or a neutral associated scene (Bisby et al. 2016). Moreover, increased retrieval activity was identified in the AMY, HC, and PFC when objects (and source information) from emotional background scenes were remembered (e.g., Smith et al. 2004b, 2005). Notably, both the AMY and the HC showed enhanced connectivity (Smith et al. 2006), during retrieval of objects that had been associated with emotional scenes during encoding, suggesting that the AMY and the HC support the retrieval of emotion-associated information from episodic memories when successfully integrated (see encoding instructions by Smith et al. 2004b, 2005, 2006; but see also Takashima et al. 2016 for HC involvement when emotional context memory is inaccurate).

The retrieval data from imaging studies, showing a memory-enhancing effect of emotion on associative binding, are also supported by a recent ERP study (Ventura-Bort et al. 2016b). Using a similar paradigm to the one used by Smith and colleagues (2004a, b, 2005, 2006), this study found larger parietal positivity during retrieval of neutral objects previously encoded with emotional, but not neutral, background scenes (Fig. 5.5a) (Smith et al. 2005; Ventura-Bort et al. 2016b). The observed parietal old/new effect (>400 ms after stimulus onset) has been related to recollection-based remembering (Rugg and Curran 2007; Weymar and Hamm 2013), suggesting that objects from emotional contexts were better recollected than those from neutral contexts. A similar ERP retrieval signature has been found for stimuli encoded under the threat of shock (Weymar et al. 2013, 2014). Specifically, when a painful electric shock was merely anticipated, signaled by the color of a stimulus (word; within-object binding), compared with a safety (no-shock) condition, words from the anticipatory threat condition were better recollected than words from the safety condition (Fig. 5.5b). Interestingly, the observed context effects on memory were most reliable for emotional events (Weymar et al. 2013, 2014), which fits with the ABC theory positing that arousal (e.g., threat of shock) during encoding may later facilitate the recollection of prioritized information (e.g., emotional salient words). These studies also show that a context that is bound to an event is not restricted to the actual experience, but also to its mere anticipation (see for example, cognitive contexts: Maren et al. 2013).

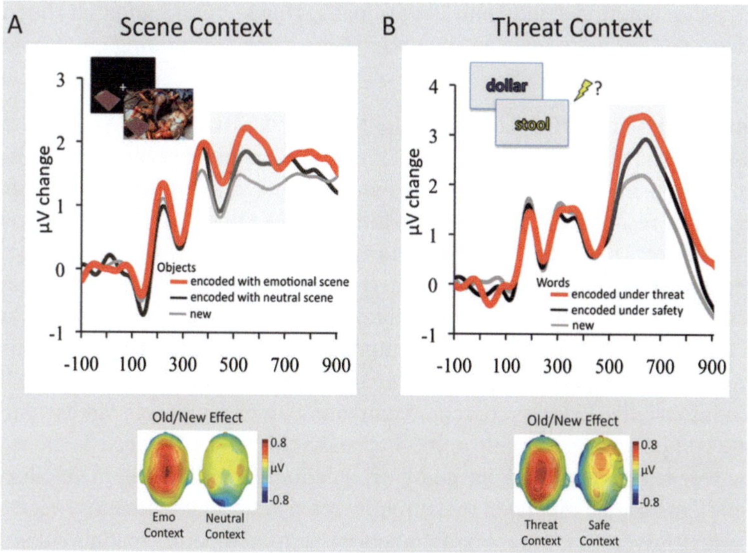

Fig. 5.5 Retrieval-related ERP signature (old/new effect) of stimuli previously associated with emotional and neutral contexts. (**a**) Grand average ERP waveforms at a representative centro-parietal sensor cluster for correctly recognized objects that had been encoded in the context of an emotional background scene (red line) or neutral background scene (black line) and correctly classified new objects (grey line). the encoding sequence of the experiment is displayed upper left. In this experiment 144 objects were presented in the context of 144 background scenes (48 pleasant, 48 neutral, 48 unpleasant). Objects were presented first followed by the background scene, to avoid direct competition between emotional backgrounds and neutral objects. To facilitate memory binding, participants were instructed to imagine that the object is a part of the scene. The graph below illustrates the scalp topographies of the ERP difference (old minus new; 400–700 ms) separately for objects originally paired with emotional or neutral scenes (Adapted from Ventura-Bort et al. 2016b, with permission). (**b**) Grand average ERP waveforms at a representative centro-parietal sensor cluster for correctly remembered words encoded in a font color (see encoding sequence *upper left*) that signaled threat of shock (red line) or safety (black line) and correctly classified new words (gray line). The graph below illustrates the scalp topographies of the ERP difference (old minus new; 500–700 ms) separately for emotional words originally encoded under threat or safety (Adapted from Weymar et al. 2013, with permission)

The study by Ventura-Bort et al. (2016b) points to an important factor that may have an impact on memory binding – the *retention interval*. Specifically, enhanced ERP old/new effects were observed for objects from emotional contexts when tested *1 week* after initial encoding, but not in previous studies using immediate or 24-h delayed testing (e.g., Jaeger et al. 2009; Jaeger and Rugg 2012; Smith et al. 2004a). A number of studies have demonstrated that longer retention periods facilitate consolidation processes, resulting in memory-enhancing effects for highly arousing emotional events, compared with less arousing neutral ones (e.g., Dolcos et al. 2005; LaBar and Cabeza 2006; Ritchey et al. 2008; Weymar and Hamm 2013; Yonelinas and Ritchey 2015). Thus, emotion may facilitate associative binding after longer delays, which is also supported by recent findings from Pierce and Kensinger

(2011), using emotional and neutral word pairs. Hence, future research should also consider retention interval as an important factor when examining emotion effects on associative binding.

An Emerging Alternative Account: The Role of "Unitization" Another interesting explanation for the opposing effects of emotion on memory binding has recently been proposed by Chiu et al. (2013), who suggest that enhancing or impairing effects might result from the way in which items are represented in the memory (*unitized* or not). This view was derived from a growing body of literature pointing to distinct MTL regions that support item (perirhinal cortex) and relational (parahippocampal cortex/HC) memory representations (e.g., Cohen et al. 1999; Davachi 2006; Ranganath 2010). Recent fMRI studies (Haskins et al. 2008; Staresina and Davachi 2010) revealed that the perirhinal cortex may also contribute to simpler forms of associative learning, based on unitization, which involves representation of separate components as a single unit (Graf and Schacter 1989), such as the association between an object and its color or location. Therefore, item memory and unitized items can be mediated by similar mechanisms, unlike memory representations that involve more complex associations (e.g., temporal, spatial, situational), which rely on HC-dependent mechanisms (Konkel and Cohen 2009). This finding is also supported by recent ERP data showing differential ERP old/new effects related to familiarity and recollection for highly unitized and less unitized associations (Diana et al. 2011).

Based on the extant evidence, Chiu et al. (2013) suggest that emotion might lead to enhancement of "item-only" or unitized memory representations, but it impairs more complex HC-dependent relational representations. Therefore, emotion leads to memory enhancement in tasks where the nature of the item–source association is more intrinsic (e.g., color or location) and allows a single representation, but it produces impairments when complex HC-dependent associations are required, for instance, in tasks using object–background associations (Kensinger et al. 2007) or item pairs (Mather and Knight 2008). However, when separate objects or features receive equivalent attention, for instance, by instructions to mentally integrate or connect certain items (e.g., Smith et al. 2004a, b; Ventura-Bort et al. 2016b), this can facilitate unitized processing and later remembering. A recent study by Murray and Kensinger (2014) using emotional and neutral word pairs found that such integrative mental imagery relies more on amygdalar and parietal processing, and less on frontal and hippocampal processing, indicating that unitized emotional associations may be less frequently mediated by HC-dependent mechanisms (see also Ventura-Bort et al. 2016b for a discussion on the involvement of familiarity-based recognition and valence). However, more research is required to substantiate this view, as to our knowledge it has not been empirically tested using systematic manipulations of unitization.

To summarize, both enhancing and impairing effects of emotion on memory binding have been observed that can be explained by differences in attentional deployment toward emotional items and associated features during encoding (Easterbrook 1959; Kensinger 2009; Mather 2007; Mather and Sutherland 2011) that are processed in a unitized or complex manner (Chiu et al. 2013), and by differences in consolidation (retention retrieval) (Pierce and Kensinger 2011; Ventura-Bort et al. 2016b). The extant literature (e.g., Kensinger et al. 2011; Luck et al. 2014; Thoresen et al. 2012) shows that emotion may facilitate memory for contextual details (e.g., color or location) or other surrounding stimuli when viewed as intrinsic or united to the emotional event, via involvement of the AMY, which interacts with perceptual regions (see also Mather et al. 2015 for the possible role of the locus coeruleus) to promote binding in the MTL regions (Davachi 2006; Ranganath 2010). When context information is successfully bound to the emotional event, enhanced AMY and MTL activation (Smith et al. 2006), along with enhanced parietal electrophysiological processing (Ventura-Bort et al. 2016b) have been observed during memory retrieval of emotion-associated information.

5.3 Conclusions and Future Directions

The overarching goal of the present review was to discuss findings from brain imaging studies investigating the neural correlates of encoding and retrieving emotional memories, and how they are modulated by factors linked to social information, ER, and associative/relational memory processes. The available evidence points to the involvement of and interaction between *direct* MTL-based (i.e., AMY–MTL) and *indirect* (i.e., PFC, parietal) mechanisms for the memory-enhancing effect of emotion (Fig. 5.6), which seem to be differentially modulated by basic affective properties (arousal and valence) and by the three factors discussed here.

Regarding the role of *social information*, available evidence suggests that when such complex dimensions are taken into consideration, a spatially distributed yet functionally interconnected network of brain regions is engaged in concert with the bottom–up AMY–MTL mechanisms involved in emotional memory. The medial PFC seems to play an important role, possibly because of the need to integrate social and affective information from both external and internal sources. The involvement of other brain regions, such as the orbital PFC and the insula, has also been implicated in the enhancement of memory for socially positive and negative information respectively. Furthermore, more recently, emerging evidence linking these social–affective regions with the brain's intrinsic functional architecture has started to paint a more comprehensive picture of the possible role of large-scale functional networks in socio-emotional memory.

Regarding the role of *ER*, emerging evidence identified the effect of specific ER strategies that can have an impact on the encoding and retrieval of emotional memories, for both laboratory micro-events and real-life autobiographical episodes. Overall, the evidence reviewed here suggests that the use of different ER strategies could lead to different patterns of emotional experience and memory (increased versus decreased), which are linked to differential interactions between the MTL and the PFC regions. For instance, reappraisal during encoding was associated with enhanced, whereas suppression was associated with impaired, subsequent emo-

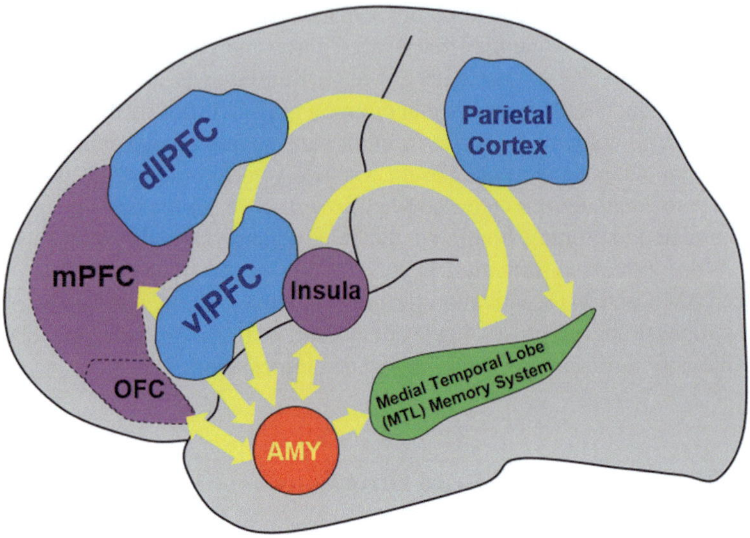

Fig. 5.6 Diagram summarizing the neural correlates of the memory-enhancing effect of emotion, as resulted from brain imaging studies. Two main basic mechanisms involved in the memory-enhancing effect of emotion were identified: one based in the medial temporal lobe (MTL;(AMY and MTL memory system = HC and associated parahippocampal cortices) and the other also involving non-MTL regions, such as the medial and dorso-/ventrolateral prefrontal cortex (mPFC and dlPFC/vlPFC respectively), among others (e.g., parietal cortex). The AMY and the MTL memory regions interact through direct/automatic neurohormonal mechanisms that contribute to the memory enhancement effect of emotion (bottom–up mechanism), whereas PFC is part of a mechanism that has an indirect/mediated involvement in emotional memories, by enhancing strategic, semantic, working memory, and attentional processes (top–down mechanism). Moreover, investigation of emotional memory for social aspects identified valence-specific engagement of other brain regions that contribute to enhanced emotional memories in social contexts – i.e., memory for socially relevant information involves activity in and interactions between the medial OFC and the MTL in the case of items with positive connotations and between the insula and the MTL for items with negative connotations. Finally, investigation of the impact of ER on emotional memory identified bidirectional relations between the MTL and the PFC regions associated with specific emotion regulation strategies, involving the lateral/medial PFC in top–down modulation of the AMY–MTL mechanisms in emotional memory encoding and retrieval, and the AMY signaling the medial PFC the need to exert control over emotional stimuli, resulting in reduced emotional experience overall, during autobiographical retrieval (Adapted from Dolcos et al. 2011, 2012 and Denkova et al. 2015, with permission)

tional memory. These opposing effects are linked to differential activity in and connectivity between the HC and the lateral PFC (increased versus decreased respectively). Emerging evidence also highlights the efficacy of thought suppression and attentional deployment strategies in inhibiting recollection of unwanted memories and in modulating emotional re-experiencing of emotional personal events respectively. The successful suppression of unwanted emotional memories has been linked to top–down modulatory effects from the PFC on the AMY and the HC, whereas effectively switching attentional focus away from emotions during AM recollections has been linked to the central role of the vmPFC in integrating affective signals from the AMY and reducing the subjective re-experiencing of emotion.

Finally, available evidence regarding *associative/relational memories* points to opposing effects of emotion, which may be explained by differential effects of prioritization and to a possible role of operations allowing unitization of information. Emotion may enhance or impair memory for contextual details depending on whether such information is encoded as *intrinsic/united* or *extrinsic* to the emotional event, respectively (Chiu et al. 2013; Kensinger 2009; Murray and Kensinger 2013), possibly because of prioritized attentional processing for both item and contextual details (Mather and Sutherland 2011). Neuroimaging evidence regarding the enhancing effects of emotion in associative/relational memory points to the role of the AMY in influencing the activity of sensory/perceptual regions, the locus coeruleus, and memory-related MTL regions in facilitating memory-binding during encoding. ERP evidence has also identified enhanced parietal positivity during retrieval of emotion-associated information, which might point to the involvement of parietal brain regions (Vilberg & Rugg 2009; Weymar et al. 2011). Consolidation also seems to facilitate associative binding, as demonstrated by evidence showing enhanced ERP effects associated with recognition memory after longer versus shorter consolidation periods/retention intervals. Finally, opposing effects of emotion on memory binding may also be explained by the complex mechanisms involving the *enhancement* of item-based memory representations versus *impairment* of HC-dependent relational memory representations (Chiu et al. 2013). Overall, these findings highlight the important role of the AMY and the MTL regions and point to electrophysiological evidence indexing memory-enhancing effects of emotion-associated information.

Despite a rapidly growing body of literature elucidating the mechanisms underlying emotional memory in the context of these emerging directions, a number of issues are still unclear. Regarding the role of *social information*, one issue to be clarified in future investigations concerns the link with the other two factors emphasized in the present review. First, the mechanisms underlying the impact of ER on memory processes in a social context remain unclear, despite evidence from previous studies showing that the engagement of various cognitive control strategies modulates the impact of emotion on memory (e.g., Ahn et al. 2015; Dillon et al. 2007; Richards and Gross 2000) and the associated neural correlates (Binder et al. 2012; Erk et al. 2010; Hayes et al. 2010) (see Sect. 5.2.2). This is surprising, given the evidence that the ability to control and experience emotions appropriately is essential in successful social interaction, and has been associated with measures of social support and personal likeability (Gross 2002). A few behavioral studies have investigated the impact of ER on memory in a social context. For instance, Richards et al. (2003) showed that reappraisal and suppression were associated with enhanced memory for conversation content versus emotional reactions in the context of discussing relationship conflicts respectively. Also, Pasupathi (2003) found that talking about negative AMs was associated with a decreased negative emotional experience, particularly when the listeners agreed with the participants' view of what had happened. More recently, it has been shown that social regulation (e.g., holding someone's hand during an emotional experience) can lead to decreased memory for negative images (Flores and Berenbaum 2017). Complementing these findings, recent neuroimaging evidence suggests that the regulation of emotions experienced through interactions with others may be subserved by dissociable neural networks, compared with those involved in regulating emotions experienced individually (for a review see Grecucci et al. 2015). Thus, clarification of the mechanisms involved in regulating how we experience and interpret our own emotions and those of others, and how these processes influence our remembrance of social experience, may allow us to better understand the complex nature of human social interaction.

Second, it also remains unclear how *association and integration* of complex social information from different sources may influence the neural mechanisms underlying emotional memory for social context. Although most studies on social cognition to date have focused on examining the associated neural mechanisms using mainly facial stimuli (Adolphs 2010; Mende-Siedlecki et al. 2013; Tsukiura 2012), it is not clear whether similar or dissociable mechanisms are engaged when processing information about others from their faces only versus in the context of other information available (e.g., body language, social setting). When interacting with others in real-life situations, we rarely encounter their faces (or bodies) in isolation, but rather as an integrated whole (de Gelder et al. 2015). Therefore, to fully elucidate the mechanisms associated with memory processes in a social context, it is of particular importance to investigate these in the context of real-life social interactions.

A few behavioral studies investigating the impact of social contextual information on subsequent memory for faces have provided evidence for the role of social/affective context in face memory. For instance, Van den Stock and de Gelder (2012) presented compound stimuli consisting of faces with either whole-body expressions or scenes (categorized as fearful, happy, or neutral), and asked participants to remember only faces, regardless of contextual information. The authors identified poorer subsequent memory for faces that had been encoded with either emotional bodily expressions or scenes than those encoded with neutral ones, suggesting that social/emotional context hampers subsequent recognition of faces (Van den Stock and de Gelder 2012). On the other hand, in the Mattarozzi et al. (2015) study, participants were first presented with faces embedded in a newspaper layout, with a headline illustrating an action (negative, positive, or neutral) performed by the person depicted, and were later shown a series of faces to judge whether or not they had previously seen the faces. The authors found that faces encoded in emotional contexts were more likely to be remembered with episodic details, whereas those encoded in neutral context tended to be judged only as familiar (Mattarozzi et al. 2015). Although seemingly inconsistent findings from these two studies may, in part, be driven by the differences in task designs (e.g., instructed versus incidental encoding, nonverbal versus verbal context), these findings provide evidence that contextual information can have an impact on subsequent memory for faces. Future neuroimaging research addressing these issues would be useful to clarify the role of social context in memory encoding and retrieval of others' faces.

Regarding the role of *ER* on emotional memory, given that this research is still in its infancy, several issues remain poorly understood. For instance, further investigations are needed to clarify the impact of various ER strategies that can be used to cope with emotional memories. Given that emerging research highlights the effectiveness of attentional deployment strategies (distraction, FA) (Denkova et al. 2015; Depue et al. 2007), future investigations of the impact of ER on memory should include more elaborate practices and training that rely on attentional control. *Mindfulness*, which involves attention to the present moment without emotional reactivity and elaboration, is such a technique that could be used as a training method to reduce the impact of distressing emotional memories. This line of future research seems promising, as there is preliminary evidence showing that mindfulness training, which has been associated with increased well-being, can have an impact on the neural correlates of emotional processing (Allen et al. 2012; Desbordes et al. 2012) and increase the specificity of AMs in healthy (Heeren et al. 2009) and formerly depressed (Williams et al. 2000) individuals. Second, because most studies focused on the impact of instructed ER, little is known about the role of *habitual, spontaneous, or uninstructed* use of ER strategies in emotional memories and the associated neural correlates (for a review, see Katsumi et al. 2017). Available evidence points to possible sex differences – e.g., engaging suppression as a habitual coping strategy in women may be inefficient and may come at a cost, as it is associated with overall increased retrieval of negative AMs and post-retrieval negative affect (Denkova et al. 2012). However, the neural mechanisms of these effects are not clear. Thus, further research is needed to advance our understanding of the neural correlates underlying the impact of the spontaneous use of ER strategies on emotional memory.

Regarding *associative/relational emotional memory*, the neural mechanisms of associative binding of emotional and neutral information are still not well understood. The following four issues are particularly unclear.

First, the engagement of and interaction between the MTL (AMY, HC) and the PFC regions when emotional and neutral information are successfully integrated into a unified representation or not (e.g., Murray and Kensinger 2014; Okada et al. 2011) needs further clarification. Recent data indicate that the AMY is involved not only during facilitated but also during disruptive encoding processes (Okada et al. 2011), coupled with reduced PFC and MTL activation (Murray and Kensinger 2014), but the mechanisms dissociating these opposing effects are not clear.

Second, the neural basis of ABC has not been sufficiently described for memory-binding processes, which is, in fact, difficult to test in an experimental environment. A novel approach that may help to overcome this issue is to use steady-state visually-evoked potentials, which allow time–frequency analyses (by frequency-tagging), when different stimuli are presented at the same time (Wieser and Keil 2014). This approach provides a continuous measure of the visual resource allocation to specific stimuli amid competing stimuli and thus makes it suitable to compare emotional and neutral information with regard to competition and memory binding.

Third, although arousal may influence associative memory to a greater extent, unpleasant valence can also influence the retention of associative information (Pierce and Kensinger 2011), which supports some previous findings showing a memory advantage for unpleasant stimuli over pleasant ones (Weymar et al. 2011). Valence may, therefore, be considered an important factor when examining the effect of emotion on associative processes.

Fourth, related to the third issue, the role of stress in neural memory binding has been less frequently explored. Recent lines of research (e.g., Hermans et al. 2014; Joels et al. 2011; van Ast et al. 2013; Vogel et al. 2016) emphasize two different corticosteroid actions critical for memory processes. On the one hand, to promote survival, shortly after stress, nongenomic corticosteroids interact with noradrenaline to enable immediate fight–flight responses and to focus attention on gist-based information at the cost of remembering contextual details. On the other hand, some hours after stress, slower, long-lasting gene-mediated corticosteroid actions are thought to facilitate restorative processes, involving enhanced executive control (Hermans et al. 2014) and remembering events in a more contextual manner (van Ast et al. 2013). Future studies investigating opposing effects of emotion on associative binding may consider these time-dependent stress actions (see also Dolcos 2014).

Finally, aside from these topic-related open questions, an important issue concerns the role of individual differences in emotional memory. As briefly discussed in the footnotes, investigation of how the neural mechanisms involved in memory-enhancing effects of emotion may vary between individuals (e.g., linked to personality, sex, and age; see also Dolcos et al., 2017) allows a better understanding of the underlying mechanism in both normal functioning and in clinical conditions, such as affective disorders, which are typically characterized by dysfunctional emotional memory. In this context, it is important to emphasize the need for future studies considering how individual differences in cognitive abilities, in addition to differences influencing emotion processing, modulate the impact of emotion on memory and the underlying neural mechanisms. Although investigation of individual differences in emotion-related dimensions seems promising to explain differences in the heightened sensitivity to emotional information and hence, enhanced encoding of emotional information (e.g., Haas and Canli 2008), little is known about the impact of individual differences in the cognitive domain. The latter may be of particular importance regarding the implementation and impact of ER strategies during memory formation. For instance, there is evidence that higher working memory capacity is associated with better suppression of negative emotions (Schmeichel et al. 2008) and that working memory training may lead to better emotion control through enhanced efficiency of the frontoparietal network (Schweizer et al. 2013). Hence, this seems a promising research avenue.

In sum, the evidence reviewed here shed light on the mechanisms through which emotion influences how we remember or forget certain information in different contexts. Emphasis is on evidence from brain imaging investigations regarding three emerging research directions: The role of social information in emotional memory; The role of emotion regulation in the impact of emotion on memory; and The impact of emotion on associative or relational memory. This emerging evidence also opens up new avenues for future research, including intervention studies aimed at facilitating adaptive emotional memory encoding and retrieval, both in healthy and pathological functioning.

References

Addis DR, Moscovitch M, Crawley AP, McAndrews MP (2004) Recollective qualities modulate hippocampal activation during autobiographical memory retrieval. Hippocampus 14(6):752–762

Adolphs R (2010) What does the amygdala contribute to social cognition? Ann N Y Acad Sci 1191(1):42–61

Adolphs R, Cahill L, Schul R, Babinsky R (1997) Impaired declarative memory for emotional material following bilateral amygdala damage in humans. Learn Mem 4(3):291–300

Adolphs R, Tranel D, Denburg N (2000) Impaired emotional declarative memory following unilateral amygdala damage. Learn Mem 7(3):180–186

Adolphs R, Tranel D, Buchanan TW (2005) Amygdala damage impairs emotional memory for gist but not details of complex stimuli. Nat Neurosci 8(4):512–518

Ahn HM, Kim SA, Hwang IJ, Jeong JW, Kim HT, Hamann S, Kim SH (2015) The effect of cognitive reappraisal on long-term emotional experience and emotional memory. J Neuropsychol 9(1):64–76

Allen M, Dietz M, Blair KS, van Beek M, Rees G, Vestergaard-Poulsen P, Lutz A, Roepstorff A (2012) Cognitive-affective neural plasticity following active-controlled mindfulness intervention. J Neurosci 32(44):15601–15610

Amft M, Bzdok D, Laird AR, Fox PT, Schilbach L, Eickhoff SB (2015) Definition and characterization of an extended social-affective default network. Brain Struct Funct 220(2):1031–1049

Anderson MC, Ochsner KN, Kuhl B, Cooper J, Robertson E, Gabrieli SW, Glover GH, Gabrieli JDE (2004) Neural systems underlying the suppression of unwanted memories. Science 303(5655):232–235

Beadle JN, Tranel D, Cohen NJ, Duff M (2013) Empathy in hippocampal amnesia. Front Psychol 4:69

Benoit RG, Anderson MC (2012) Opposing mechanisms support the voluntary forgetting of unwanted memories. Neuron 76(2):450–460

Benoit RG, Hulbert JC, Huddleston E, Anderson MC (2015) Adaptive top-down suppression of hippocampal activity and the purging of intrusive memories from consciousness. J Cogn Neurosci 27(1):96–111

Bickart KC, Dickerson BC, Barrett LF (2014) The amygdala as a hub in brain networks that support social life. Neuropsychologia 63:235–248

Binder J, de Quervain DJ, Friese M, Luechinger R, Boesiger P, Rasch B (2012) Emotion suppression reduces hippocampal activity during successful memory encoding. NeuroImage 63(1):525–532

Bisby JA, Horner AJ, Horlyck LD, Burgess N (2016) Opposing effects of negative emotion on amygdalar and hippocampal memory for items and associations. Soc Cogn Affect Neurosci 11(6):981–990

Botzung A, LaBar KS, Kragel P, Miles A, Rubin DC (2010a) Component neural systems for the creation of emotional memories during free viewing of a complex, real-world event. Front Hum Neurosci 4:34

Botzung A, Rubin DC, Miles A, Cabeza R, Labar KS (2010b) Mental hoop diaries: emotional memories of a college basketball game in rival fans. J Neurosci 30(6):2130–2137

Bradley MM, Lang PJ (1994) Measuring emotion: i self-assessment manikin and the semantic differential. J Behav Ther Exp Psychiatry 25(1):49–59

Bradley MM, Greenwald MK, Petry MC, Lang PJ (1992) Remembering pictures: pleasure and arousal in memory. J Exp Psychol Learn Mem Cogn 18(2):379–390

Buchanan TW (2007) Retrieval of emotional memories. Psychol Bull 133(5):761–779

Buchanan TW, Adolphs R (2002) The role of the human amygdala in emotional modulation of long-term declarative memory. In: Moore SC, Oaksford M (eds) Emotional cognition: from brain to behaviour, Advances in consciousness research, vol. 44. John Benjamins Publishing Company, Amsterdam, pp 9–34

Burke A, Heuer F, Reisberg D (1992) Remembering emotional events. Mem Cogn 20(3):277–290
Cabeza R, Prince SE, Daselaar SM, Greenberg DL, Budde M, Dolcos F, LaBar KS, Rubin DC (2004) Brain activity during episodic retrieval of autobiographical and laboratory events: aN fMRI study using a novel photo paradigm. J Cogn Neurosci 16(9):1583–1594
Cahill L, Prins B, Weber M, McGaugh JL (1994) Beta-adrenergic activation and memory for emotional events. Nature 371(6499):702–704
Cahill L, Babinsky R, Markowitsch HJ, McGaugh JL (1995) The amygdala and emotional memory. Nature 377(6547):295–296
Cahill L, Haier RJ, Fallon J, Alkire MT, Tang C, Keator D, Wu J, McGaugh JL (1996) Amygdala activity at encoding correlated with long-term, free recall of emotional information. Proc Natl Acad Sci 93(15):8016–8021
Cahill L, Weinberger NM, Roozendaal B, McGaugh JL (1999) Is the amygdala a locus of "conditioned fear"? Some questions and caveats. Neuron 23(2):227–228
Canli T, Zhao Z, Brewer J, Gabrieli JD, Cahill L (2000) Event-related activation in the human amygdala associates with later memory for individual emotional experience. J Neurosci 20(19):RC99
Cassidy BS, Leshikar ED, Shih JY, Aizenman A, Gutchess AH (2013) Valence-based age differences in medial prefrontal activity during impression formation. Soc Neurosci 8(5):462–473
Chiu Y-C, Dolcos F, Gonsalves BD, Cohen NJ (2013) On opposing effects of emotion on contextual or relational memory. Front Psychol 4:103
Christianson SA (1992) Emotional stress and eyewitness memory: a critical review. Psychol Bull 112(2):284–309
Ciaramelli E, Bernardi F, Moscovitch M (2013) Individualized Theory of Mind (iToM): when memory modulates empathy. Front Psychol 4:4
Cohen NJ, Ryan J, Hunt C, Romine L, Wszalek T, Nash C (1999) Hippocampal system and declarative (relational) memory: summarizing the data from functional neuroimaging studies. Hippocampus 9(1):83–98
Collins JA, Olson IR (2014) Beyond the FIthe role of the ventral anterior temporal lobes in face processing. Neuropsychologia 61:65–79
Cunningham WA, Brosch T (2012) Motivational salience: amygdala tuning from traits, needs, values, and goals. Curr Dir Psychol Sci 21(1):54–59
Cunningham WA, Van Bavel JJ, Johnsen IR (2008) Affective flexibility: evaluative processing goals shape amygdala activity. Psychol Sci 19(2):152–160
D'Argembeau A, Van der Linden M (2004) Influence of affective meaning on memory for contextual information. Emotion 4(2):173–188
D'Esposito M, Zarahn E, Aguirre GK (1999) Event-related functional MRI: implications for cognitive psychology. Psychol Bull 125(1):155–164
D'Esposito M, Postle BR, Rypma B (2000) Prefrontal cortical contributions to working memory: evidence from event-related fMRI studies. Exp Brain Res 133(1):3–11
Daselaar SM, Rice HJ, Greenberg DL, Cabeza R, LaBar KS, Rubin DC (2008) The spatiotemporal dynamics of autobiographical memory: neural correlates of recall, emotional intensity, and reliving. Cereb Cortex 18(1):217–229
Davachi L (2006) Item, context and relational episodic encoding in humans. Curr Opin Neurobiol 16(6):693–700
Davidson RJ, Irwin W (1999) The functional neuroanatomy of emotion and affective style. Trend Cogn Sci 3(1):11–21
De Gelder B, de Borst AW, Watson R (2015) The perception of emotion in body expressions. Wiley Interdiscip Rev Cogn Sci 6(2):149–158
Delgado MR, Olsson A, Phelps EA (2006) Extending animal models of fear conditioning to humans. Biol Psychol 73(1):39–48
Delgado MR, Beer JS, Fellows LK, Huettel SA, Platt ML, Quirk GJ, Schiller D (2016) Viewpoints: dialogues on the functional role of the ventromedial prefrontal cortex. Nat Neurosci 19(12):1545–1552

Denkova E, Botzung A, Scheiber C, Manning L (2006) Implicit emotion during recollection of past events: a nonverbal fMRI study. Brain Res 1078(1):143–150
Denkova E, Dolcos S, Dolcos F (2012) Reliving emotional personal memories: affective biases linked to personality and sex-related differences. Emotion 12(3):515–528
Denkova E, Dolcos S, Dolcos F (2013) The effect of retrieval focus and emotional valence on the medial temporal lobe activity during autobiographical recollection. Front Behav Neurosci 7:109
Denkova E, Dolcos S, Dolcos F (2015) Neural correlates of 'distracting' from emotion during autobiographical recollection. Soc Cogn Affect Neurosci 10(2):219–230
Depue BE, Banich MT, Curran T (2006) Suppression of emotional and nonemotional content in memory: effects of repetition on cognitive control. Psychol Sci 17(5):441–447
Depue BE, Curran T, Banich MT (2007) Prefrontal regions orchestrate suppression of emotional memories via a two-phase process. Science 317(5835):215–219
Desbordes G, Negi LT, Pace TW, Wallace BA, Raison CL, Schwartz EL (2012) Effects of mindful-attention and compassion meditation training on amygdala response to emotional stimuli in an ordinary, non-meditative state. Front Hum Neurosci 6:292
Diana RA, Van den Boom W, Yonelinas AP, Ranganath C (2011) ERP correlates of source memory: unitized source information increases familiarity-based retrieval. Brain Res 1367:278–286
Dillon DG, Ritchey M, Johnson BD, LaBar KS (2007) Dissociable effects of conscious emotion regulation strategies on explicit and implicit memory. Emotion 7(2):354–365
Doerksen S, Shimamura AP (2001) Source memory enhancement for emotional words. Emotion 1(1):5–11
Dolan RJ (2007) The human amygdala and orbital prefrontal cortex in behavioural regulation. Philos Trans R Soc Lond Ser B Biol Sci 362(1481):787–799
Dolan RJ, Lane R, Chua P, Fletcher P (2000) Dissociable temporal lobe activations during emotional episodic memory retrieval. NeuroImage 11(3):203–209
Dolcos F (2014) The fast and the slow sides of cortisol's effects on emotional interference and sustained attention. Front Neurosci 8:268
Dolcos F, Denkova E (2008) Neural correlates of encoding emotional memories: a review of functional neuroimaging evidence. Cell Sci Rev 5:78–122
Dolcos F, LaBar KS, Cabeza R (2004a) Dissociable effects of arousal and valence on prefrontal activity indexing emotional evaluation and subsequent memory: an event-related fMRI study. NeuroImage 23(1):64–74
Dolcos F, LaBar KS, Cabeza R (2004b) Interaction between the amygdala and the medial temporal lobe memory system predicts better memory for emotional events. Neuron 42(5):855–863
Dolcos F, LaBar KS, Cabeza R (2005) Remembering one year later: role of the amygdala and the medial temporal lobe memory system in retrieving emotional memories. Proc Natl Acad Sci 102(7):2626–2631
Dolcos F, Kragel P, Wang L, McCarthy G (2006a) Role of the inferior frontal cortex in coping with distracting emotions. Neuroreport 17(15):1591–1594
Dolcos F, LaBar KS, Cabeza R (2006b) The memory-enhancing effect of emotion: functional neuroimaging evidence. In: Uttl B, Ohta N, Siegenthaler AL (eds) Memory and emotion: interdisciplinary perspectives. Blackwell Publishing, Malden, pp 105–134
Dolcos F, Iordan AD, Dolcos S (2011) Neural correlates of emotion-cognition interactions: a review of evidence from brain imaging investigations. J Cogn Psychol 23(6):669–694
Dolcos F, Denkova E, Dolcos S (2012) Neural correlates of emotional memories: a review of evidence from brain imaging studies. Psychologia 55(2):80–111
Dolcos F, Katsumi Y, Weymar M, Moore M, Tsukiura T, Dolcos S (2017) Emerging directions in emotional episodic memory. Front Psychol. 8:1867
Donaldson DI, Buckner RL (2001) Effective paradigm design. In: Jezzard P, Matthews PM, Smith SM (eds) Functional MRI: an introduction to methods. Oxford University Press, New York, pp 177–198

Dorfel D, Lamke JP, Hummel F, Wagner U, Erk S, Walter H (2014) Common and differential neural networks of emotion regulation by detachment, reinterpretation, distraction, and expressive suppression: a comparative fMRI investigation. NeuroImage 101:298–309
Dougal S, Phelps EA, Davachi L (2007) The role of medial temporal lobe in item recognition and source recollection of emotional stimuli. Cogn Affect Behav Neurosci 7(3):233–242
Easterbrook JA (1959) The effect of emotion on cue utilization and the organization of behavior. Psychol Rev 66(3):183–201
Edelson M, Sharot T, Dolan RJ, Dudai Y (2011) Following the crowd: brain substrates of long-term memory conformity. Science 333(6038):108–111
Eippert F, Veit R, Weiskopf N, Erb M, Birbaumer N, Anders S (2007) Regulation of emotional responses elicited by threat-related stimuli. Hum Brain Mapp 28(5):409–423
Erk S, Martin S, Walter H (2005) Emotional context during encoding of neutral items modulates brain activation not only during encoding but also during recognition. NeuroImage 26(3):829–838
Erk S, von Kalckreuth A, Walter H (2010) Neural long-term effects of emotion regulation on episodic memory processes. Neuropsychologia 48(4):989–996
Fabiansson EC, Denson TF, Moulds ML, Grisham JR, Schira MM (2012) Don't look back in anger: neural correlates of reappraisal, analytical rumination, and angry rumination during recall of an anger-inducing autobiographical memory. NeuroImage 59(3):2974–2981
Fanselow MS, LeDoux JE (1999) Why we think plasticity underlying Pavlovian fear conditioning occurs in the basolateral amygdala. Neuron 23(2):229–232
Flor H, Nees F (2014) Learning, memory and brain plasticity in posttraumatic stress disorder: context matters. Restor Neurol Neurosci 32(1):95–102
Flores LE, Berenbaum H (2017) The effect of the social regulation of emotion on emotional long-term memory. Emotion 17(3):547–556
Fossati P, Hevenor SJ, Lepage M, Graham SJ, Grady C, Keightley ML, Craik F, Mayberg H (2004) Distributed self in episodic memory: neural correlates of successful retrieval of self-encoded positive and negative personality traits. NeuroImage 22(4):1596–1604
Gagnepain P, Henson RN, Anderson MC (2014) Suppressing unwanted memories reduces their unconscious influence via targeted cortical inhibition. Proc Natl Acad Sci 111(13):E1310–E1319
Gilron R, Gutchess AH (2012) Remembering first impressions: effects of intentionality and diagnosticity on subsequent memory. Cogn Affect Behav Neurosci 12(1):85–98
Glascher J, Adolphs R (2003) Processing of the arousal of subliminal and supraliminal emotional stimuli by the human amygdala. J Neurosci 23(32):10274–10282
Goldin PR, McRae K, Ramel W, Gross JJ (2008) The neural bases of emotion regulation: reappraisal and suppression of negative emotion. Biol Psychiatry 63(6):577–586
Gotlib IH, Joormann J (2010) Cognition and depression: current status and future directions. Annu Rev Clin Psychol 6:285–312
Graf P, Schacter DL (1989) Unitization and grouping mediate dissociations in memory for new associations. J Exp Psychol Learn Mem Cogn 15(5):930–940
Grecucci A, Theuninck A, Frederickson J, Job R (2015) Mechanisms of social emotion regulation: from neuroscience to psychotherapy. In: Bryant ML (ed) Emotion regulation: processes, cognitive effects and social consequences. Nova Science Publishers, New York, pp 57–84
Greenberg DL, Rice HJ, Cooper JJ, Cabeza R, Rubin DC, LaBar KS (2005) Co-activation of the amygdala, hippocampus and inferior frontal gyrus during autobiographical memory retrieval. Neuropsychologia 43(5):659–674
Greenwald AG, Banaji MR (1989) The self as a memory system: powerful, but ordinary. J Pers Soc Psychol 57(1):41
Gross JJ (2002) Emotion regulation: affective, cognitive, and social consequences. Psychophysiology 39(3):281–291
Gross JJ (2008) Emotion regulation. In: Lewis M, Haviland-Jones JM, Barrett LF (eds) Handbook of emotions. Guilford, New York, pp 497–512
Gross JJ (2015) Emotion regulation: current status and future prospects. Psychol Inq 26(1):1–26

Guillet R, Arndt J (2009) Taboolds: the effect of emotion on memory for peripheral information. Mem Cogn 37(6):866–879

Gutchess AH, Sokal R, Coleman JA, Gotthilf G, Grewal L, Rosa N (2015) Age differences in self-referencing: evidence for common and distinct encoding strategies. Brain Res 1612:118–127

Haas BW, Canli T (2008) Emotional memory function, personality structure and psychopathology: a neural system approach to the identification of vulnerability markers. Brain Res Rev 58(1):71–84

Hamann S (2001) Cognitive and neural mechanisms of emotional memory. Trends Cogn Sci 5(9):394–400

Hamann SB, Ely TD, Grafton ST, Kilts CD (1999) Amygdala activity related to enhanced memory for pleasant and aversive stimuli. Nat Neurosci 2(3):289–293

Harvey PO, Fossati P, Lepage M (2007) Modulation of memory formation by stimulus content: specific role of the medial prefrontal cortex in the successful encoding of social pictures. J Cogn Neurosci 19(2):351–362

Haskins AL, Yonelinas AP, Quamme JR, Ranganath C (2008) Perirhinal cortex supports encoding and familiarity-based recognition of novel associations. Neuron 59(4):554–560

Hayes JP, Morey RA, Petty CM, Seth S, Smoski MJ, McCarthy G, LaBar KS (2010) Staying cool when things get hot: emotion regulation modulates neural mechanisms of memory encoding. Front Hum Neurosci 4:230

Heeren A, Van Broeck N, Philippot P (2009) The effects of mindfulness on executive processes and autobiographical memory specificity. Behav Res Ther 47(5):403–409

Heinzel A, Northoff G (2009) Emotional feeling and the orbitomedial prefrontal cortex: theoretical and empirical considerations. Philos Psychol 22(4):443–464

Hermann A, Bieber A, Keck T, Vaitl D, Stark R (2014) Brain structural basis of cognitive reappraisal and expressive suppression. Soc Cogn Affect Neurosci 9(9):1435–1442

Hermans EJ, Henckens MJ, Joels M, Fernandez G (2014) Dynamic adaptation of large-scale brain networks in response to acute stressors. Trend Neurosci 37(6):304–314

Holland AC, Kensinger EA (2013) The neural correlates of cognitive reappraisal during emotional autobiographical memory recall. J Cogn Neurosci 25(1):87–108

Iordan AD (2016) The impact of emotional distraction on cognition: from basic brain responses to large-scale network interactions. Doctoral dissertation, University of Illinois at Urbana-Champaign

Jaeger A, Rugg MD (2012) Implicit effects of emotional contexts: an ERP study. Cogn Affect Behav Neurosci 12(4):748–760

Jaeger A, Johnson JD, Corona M, Rugg MD (2009) ERP correlates of the incidental retrieval of emotional information: effects of study-test delay. Brain Res 1269:105–113

Jing HG, Madore KP, Schacter DL (2016) Worrying about the future: an episodic specificity induction impacts problem solving, reappraisal, and well-being. J Exp Psychol Gen 145(4):402–418

Joels M, Fernandez G, Roozendaal B (2011) Stress and emotional memory: a matter of timing. Trends Cogn Sci 15(6):280–288

Kalokerinos EK, Greenaway KH, Denson TF (2015) Reappraisal but not suppression downregulates the experience of positive and negative emotion. Emotion 15(3):271–275

Kanske P, Heissler J, Schonfelder S, Bongers A, Wessa M (2011) How to regulate emotion? Neural networks for reappraisal and distraction. Cereb Cortex 21(6):1379–1388

Kapur S, Tulving E, Cabeza R, McIntosh AR, Houle S, Craik FI (1996) The neural correlates of intentional learning of verbal materials: a PET study in humans. Brain Res Cogn Brain Res 4(4):243–249

Katsumi Y, Denkova E, Dolcos S (2017) Personality and memory. In: Zeigler-Hill V, Shackelford TK (eds) Encyclopedia of personality and individual differences. Springer International Publishing, New York

Kensinger EA (2004) Remembering emotional eliences: the contribution of valence and arousal. Rev Neurosci 15(4):241–251

Kensinger EA (2009) Remembering the details: effects of emotion. Emot Rev 1(2):99–113

Kensinger EA, Corkin S (2004) Two routes to emotional memory: distinct neural processes for valence and arousal. Proc Natl Acad Sci 101(9):3310–3315
Kensinger EA, Schacter DL (2005) Retrieving accurate and distorted memories: neuroimaging evidence for effects of emotion. NeuroImage 27(1):167–177
Kensinger EA, Schacter DL (2006a) Amygdala activity is associated with the successful encoding of item, but not source, information for positive and negative stimuli. J Neurosci 26(9):2564–2570
Kensinger EA, Schacter DL (2006b) Processing emotional pictures and words: effects of valence and arousal. Cogn Affect Behav Neurosci 6(2):110–126
Kensinger EA, Schacter DL (2008) Neural processes supporting young and older adults' emotional memories. J Cogn Neurosci 20(7):1161–1173
Kensinger EA, Garoff-Eaton RJ, Schacter DL (2007) Effects of emotion on memory specificity: memory trade-offs elicited by negative visually arousing stimuli. J Mem Lang 56(4):575–591
Kensinger EA, Addis DR, Atapattu RK (2011) Amygdala activity at encoding corresponds with memory vividness and with memory for select episodic details. Neuropsychologia 49(4):663–673
Kilpatrick L, Cahill L (2003) Amygdala modulation of parahippocampal and frontal regions during emotionally influenced memory storage. NeuroImage 20(4):2091–2099
Kim SH, Hamann S (2012) The effect of cognitive reappraisal on physiological reactivity and emotional memory. Int J Psychophysiol 83(3):348–356
Kleinhans NM, Johnson LC, Mahurin R, Richards T, Stegbauer KC, Greenson J, Dawson G, Aylward E (2007) Increased amygdala activation to neutral faces is associated with better face memory performance. Neuroreport 18(10):987–991
Konkel A, Cohen NJ (2009) Relational memory and the hippocampus: representations and methods. Front Neurosci 3(2):166–174
Kosslyn SM, Shin LM, Thompson WL, McNally RJ, Rauch SL, Pitman RK, Alpert NM (1996) Neural effects of visualizing and perceiving aversive stimuli: a PET investigation. Neuroreport 7(10):1569–1576
Kross E, Davidson ML, Weber M, Ochsner K (2009) Coping withltions past: the neural bases of regulating affect associated with negative autobiographical memories. Biol Psychiatry 65(5):361–366
Kupper CS, Benoit RG, Dalgleish T, Anderson MC (2014) Direct suppression as a mechanism for controlling unpleasant memories in daily life. J Exp Psychol Gen 143(4):1443–1449
LaBar KS, Cabeza R (2006) Cognitive neuroscience of emotional memory. Nat Rev Neurosci 7(1):54–64
LaBar KS, Gatenby JC, Gore JC, LeDoux JE, Phelps EA (1998) Human amygdala activation during conditioned fear acquisition and extinction: a mixed-trial fMRI study. Neuron 20(5):937–945
Lang PJ, Bradley MM (2010) Emotion and the motivational brain. Biol Psychol 84(3):437–450
Lang PJ, Greenwald MK, Bradley MM, Hamm AO (1993) Looking at pictures: affective, facial, visceral, and behavioral reactions. Psychophysiology 30:261–273
Laurita AC, Spreng RN (2017) The hippocampus and social cognition. In: Hannula DE, Duff MC (eds) The hippocampus from cells to systems: Structure, connectivity, and functional contributions to memory and flexible cognition. Springer International Publishing, Cham, pp 537–558
LeDoux J (2000) Emotion circuits in the brain. Annu Rev Neurosci 23:155–184
LeDoux J (2007) The amygdala. Curr Biol 17(20):R868–R874
Lee TH, Sakaki M, Cheng R, Velasco R, Mather M (2014) Emotional arousal amplifies the effects of biased competition in the brain. Soc Cogn Affect Neurosci 9(12):2067–2077
Levine LJ, Schmidt S, Kang HS, Tinti C (2012) Remembering the silver lining: reappraisal and positive bias in memory for emotion. Cognit Emot 26(5):871–884
Li W, Mai X, Liu C (2014) The default mode network and social understanding of others: what do brain connectivity studies tell us. Front Hum Neurosci 8:74
Lieberman MD, Eisenberger NI, Crockett MJ, Tom SM, Pfeifer JH, Way BM (2007) Putting feelings into words: affect labeling disrupts amygdala activity in response to affective stimuli. Psychol Sci 18(5):421–428

Lieberman MD, Inagaki TK, Tabibnia G, Crockett MJ (2011) Subjective responses to emotional stimuli during labeling, reappraisal, and distraction. Emotion 11(3):468–480
Liu F, Cui L, Zhang Q (2015) The influences of reappraisal and suppression instructions on memory for neutral words in negative background. Neuroreport 26(17):1023–1031
Liu ZX, Grady C, Moscovitch M (2016) Effects of prior-knowledge on brain activation and connectivity during associative memory encoding. Cereb Cortex 27(3):1991–2009
Luck D, Leclerc M-E, Lepage M (2014) The potentiation of associative memory by emotions: an event-related fMRI study. Adv Neurosci 2014:9
Macrae CN, Moran JM, Heatherton TF, Banfield JF, Kelley WM (2004) Medial prefrontal activity predicts memory for self. Cereb Cortex 14(6):647–654
Madore KP, Schacter DL (2016) Remembering the past and imagining the future: selective effects of an episodic specificity induction on detail generation. Q J Exp Psychol 69(2):285–298
Madore KP, Addis DR, Schacter DL (2015) Creativity and memory: effects of an episodic-specificity induction on divergent thinking. Psychol Sci 26(9):1461–1468
Maguire EA, Frith CD (2003) Lateral asymmetry in the hippocampal response to the remoteness of autobiographical memories. J Neurosci 23(12):5302–5307
Mano Y, Sugiura M, Tsukiura T, Chiao JY, Yomogida Y, Jeong H, Sekiguchi A, Kawashima R (2011) The representation of social interaction in episodic memory: a functional MRI study. NeuroImage 57(3):1234–1242
Maratos EJ, Dolan RJ, Morris JS, Henson RN, Rugg MD (2001) Neural activity associated with episodic memory for emotional context. Neuropsychologia 39(9):910–920
Maren S, Phan KL, Liberzon I (2013) The contextual brain: implications for fear conditioning, extinction and psychopathology. Nat Rev Neurosci 14(6):417–428
Markowitsch HJ, Thiel A, Reinkemeier M, Kessler J, Koyuncu A, Heiss WD (2000) Right amygdalar and temporofrontal activation during autobiographic, but not during fictitious memory retrieval. Behav Neurol 12(4):181–190
Markowitsch HJ, Vandekerckhove MM, Lanfermann H, Russ MO (2003) Engagement of lateral and medial prefrontal areas in the ecphory of sad and happy autobiographical memories. Cortex 39(4–5):643–665
Martin SJ, Grimwood PD, Morris RG (2000) Synaptic plasticity and memory: an evaluation of the hypothesis. Annu Rev Neurosci 23:649–711
Mather M (2007) Emotional arousal and memory binding: an object-based framework. Perspect Psychol Sci 2(1):33–52
Mather M, Knight M (2008) The emotional harbinger effect: poor context memory for cues that previously predicted something arousing. Emotion 8(6):850–860
Mather M, Nesmith K (2008) Arousal-enhanced location memory for pictures. J Mem Lang 58(2):449–464
Mather M, Sutherland MR (2011) Arousal-biased competition in perception and memory. Perspect Psychol Sci 6(2):114–133
Mather M, Clewett D, Sakaki M, Harley CW (2015) Norepinephrine ignites local hot spots of neuronal excitation: how arousal amplifies selectivity in perception and memory. Behav Brain Sci 39:e200
Mattarozzi K, Todorov A, Codispoti M (2015) Memory for faces: the effect of facial appearance and the context in which the face is encountered. Psychol Res 79(2):308–317
Mayberg HS (1997) Limbic-cortical dysregulation: a proposed model of depression. J Neuropsychiatry Clin Neurosci 9(3):471–481
McFarland CP, Primosch M, Maxson CM, Stewart BT (2017) Enhancing memory and imagination improves problem solving among individuals with depression. Mem Cogn 45(6):932–939
McGaugh JL (2000) Memory—a century of consolidation. Science 287(5451):248–251
McGaugh JL (2004) The amygdala modulates the consolidation of memories of emotionally arousing experiences. Annu Rev Neurosci 27:1–28
McRae K, Hughes B, Chopra S, Gabrieli JD, Gross JJ, Ochsner KN (2010) The neural bases of distraction and reappraisal. J Cogn Neurosci 22(2):248–262

Mende-Siedlecki P, Said CP, Todorov A (2013) The social evaluation of faces: a meta-analysis of functional neuroimaging studies. Soc Cogn Affect Neurosci 8(3):285–299

Meyer ML, Lieberman MD (2012) Social working memory: neurocognitive networks and directions for future research. Front Psychol 3:571

Meyer ML, Taylor SE, Lieberman MD (2015) Social working memory and its distinctive link to social cognitive ability: an fMRI study. Soc Cogn Affect Neurosci 10(10):1338–1347

Mickley KR, Kensinger EA (2008) Emotional valence influences the neural correlates associated with remembering and knowing. Cogn Affect Behav Neurosci 8(2):143–152

Mickley Steinmetz KR, Kensinger EA (2009) The effects of valence and arousal on the neural activity leading to subsequent memory. Psychophysiology 46(6):1190–1199

Mickley Steinmetz KR, Addis DR, Kensinger EA (2010) The effect of arousal on the emotional memory network depends on valence. NeuroImage 53(1):318–324

Miskovic V, Keil A (2012) Acquired fears reflected in cortical sensory processing: a review of electrophysiological studies of human classical conditioning. Psychophysiology 49(9):1230–1241

Mitchell JP, Macrae CN, Banaji MR (2004) Encoding-specific effects of social cognition on the neural correlates of subsequent memory. J Neurosci 24(21):4912–4917

Morris JS, Ohman A, Dolan RJ (1999) A subcortical pathway to the right amygdala mediating "unseen" fear. Proc Natl Acad Sci 96(4):1680–1685

Murray BD, Kensinger EA (2013) A review of the neural and behavioral consequences for unitizing emotional and neutral information. Front Behav Neurosci 7:42

Murray BD, Kensinger EA (2014) The route to an integrative associative memory is influenced by emotion. PLoS One 9(1):e82372

Muscatell KA, Addis DR, Kensinger EA (2010) Self-involvement modulates the effective connectivity of the autobiographical memory network. Soc Cogn Affect Neurosci 5(1):68–76

Nashiro K, Mather M (2011) How arousal affects younger and older adults' memory binding. Exp Aging Res 37(1):108–128

Northoff G, Heinzel A, de Greck M, Bermpohl F, Dobrowolny H, Panksepp J (2006) Self-referential processing in our brain – a meta-analysis of imaging studies on the self. NeuroImage 31(1):440–457

Northoff G, Schneider F, Rotte M, Matthiae C, Tempelmann C, Wiebking C, Bermpohl F, Heinzel A, Danos P, Heinze HJ, Bogerts B, Walter M, Panksepp J (2009) Differential parametric modulation of self-relatedness and emotions in different brain regions. Hum Brain Mapp 30(2):369–382

O'Doherty JP (2004) Reward representations and reward-related learning in the human brain: insights from neuroimaging. Curr Opin Neurobiol 14(6):769–776

Ochsner KN, Knierim K, Ludlow D, Hanelin J, Ramachandran T, Mackey S (2004) Reflecting upon feelings: an fMRI study of neural systems supporting the attribution of emotion to self and other. J Cogn Neurosci 16(10):1748–1772

Oddo S, Lux S, Weiss PH, Schwab A, Welzer H, Markowitsch HJ, Fink GR (2008) Specific role of medial prefrontal cortex in retrieving recent autobiographical memories: an fMRI study of young female subjects. Cortex 46(1):29–39

Okada G, Okamoto Y, Kunisato Y, Aoyama S, Nishiyama Y, Yoshimura S, Onoda K, Toki S, Yamashita H, Yamawaki S (2011) The effect of negative and positive emotionality on associative memory: an fMRI study. PLoS One 6(9):e24862

Olatunji BO, Berg HE, Zhao Z (2015) Emotion regulation of fear and disgust: differential effects of reappraisal and suppression. Cognit Emot 31(2):403–410

Olson IR, McCoy D, Klobusicky E, Ross LA (2013) Social cognition and the anterior temporal lobes: a review and theoretical framework. Soc Cogn Affect Neurosci 8(2):123–133

Owen AM, Herrod NJ, Menon DK, Clark JC, Downey SP, Carpenter TA, Minhas PS, Turkheimer FE, Williams EJ, Robbins TW, Sahakian BJ, Petrides M, Pickard JD (1999) Redefining the functional organization of working memory processes within human lateral prefrontal cortex. Eur J Neurosci 11(2):567–574

Paller KA, Wagner AD (2002) Observing the transformation of experience into memory. Trends Cogn Sci 6(2):93–102

Paller KA, Kutas M, Shimamura AP, Squire LR (1987) Brain responses to concrete and abstract words reflect processes that correlate with later performance on a test of stem-completion priming. Electroencephalogr Clin Neurophysiol Suppl 40:360–365
Pasupathi M (2003) Emotion regulation during social remembering: differences between emotions elicited during an event and emotions elicited when talking about it. Memory 11(2):151–163
Petrides M (1995) Functional organization of the human frontal cortex for mnemonic processing. Evidence from neuroimaging studies. Ann N Y Acad Sci 769:85–96
Phan KL, Wager T, Taylor SF, Liberzon I (2002) Functional neuroanatomy of emotion: a meta-analysis of emotion activation studies in PET and fMRI. NeuroImage 16(2):331–348
Phelps EA (2004) Human emotion and memory: interactions of the amygdala and hippocampal complex. Curr Opin Neurobiol 14(2):198–202
Phelps EA, LeDoux J (2005) Contributions of the amygdala to emotion processing: from animal models to human behavior. Neuron 48(2):175–187
Phelps EA, LaBar KS, Anderson AK, O'Connor KJ, Fulbright RK, Spencer DD (1998) Specifying the contributions of the human amygdala to emotional memory: a case study. Neurocase 4(6):527–540
Phelps EA, O'Connor KJ, Gatenby JC, Gore JC, Grillon C, Davis M (2001) Activation of the left amygdala to a cognitive representation of fear. Nat Neurosci 4(4):437–441
Piefke M, Weiss PH, Zilles K, Markowitsch HJ, Fink GR (2003) Differential remoteness and emotional tone modulate the neural correlates of autobiographical memory. Brain 126(3):650–668
Pierce BH, Kensinger EA (2011) Effects of emotion on associative recognition: valence and retention interval matter. Emotion 11(1):139–144
Poldrack RA, Wagner AD, Prull MW, Desmond JE, Glover GH, Gabrieli JD (1999) Functional specialization for semantic and phonological processing in the left inferior prefrontal cortex. NeuroImage 10(1):15–35
Prince SE, Daselaar SM, Cabeza R (2005) Neural correlates of relational memory: successful encoding and retrieval of semantic and perceptual associations. J Neurosci 25(5):1203–1210
Prince SE, Dennis NA, Cabeza R (2009) Encoding and retrieving faces and places: distinguishing process- and stimulus-specific differences in brain activity. Neuropsychologia 47(11):2282–2289
Raes F, Williams JM, Hermans D (2009) Reducing cognitive vulnerability to depression: a preliminary investigation of MEmory Specificity Training (MEST) in inpatients with depressive symptomatology. J Behav Ther Exp Psychiatry 40(1):24–38
Ranganath C (2010) Binding items and contexts the cognitive neuroscience of episodic memory. Curr Dir Psychol Sci 19(3):131–137
Ray RD, McRae K, Ochsner KN, Gross JJ (2010) Cognitive reappraisal of negative affect: converging evidence from EMG and self-report. Emotion 10(4):587–592
Richards JM, Gross JJ (2000) EmoI regulation and memory: the cognitive costs of keeping one's cool. J Pers Soc Psychol 79(3):410–424
Richards JM, Butler EA, Gross JJ (2003) Emotion regulatiIn romantic relationships: the cognitive consequences of concealing feelings. J Soc Pers Relat 20(5):599–620
Richardson MP, Strange BA, Dolan RJ (2004) Encoding of emotional memories depends on amygdala and hippocampus and their interactions. Nat Neurosci 7(3):278–285
Rimmele U, Davachi L, Petrov R, Dougal S, Phelps EA (2011) Emotion enhances the subjective feeling of remembering, despite lower accuracy for contextual details. Emotion 11(3):553–562
Ritchey M, Dolcos F, Cabeza R (2008) Role of amygdala connectivity in the persistence of emotional memories over time: an event-related fMRI investigation. Cereb Cortex 18(11):2494–2504
Ritchey M, LaBar KS, Cabeza R (2011) Level of processing modulates the neural correlates of emotional memory formation. J Cogn Neurosci 23(4):757–771
Rosen BR, Buckner RL, Dale AM (1998) Event-related functional MRI: past, present, and future. Proc Natl Acad Sci 95(3):773–780
Rubin DC, Dennis MF, Beckham JC (2011) AutobiograIal memory for stressful events: the role of autobiographical memory in posttraumatic stress disorder. Conscious Cogn 20(3):840–856

Rugg MD, Curran T (2007) Event-related potentials and recognition memory. Trends Cogn Sci 11(6):251–257

Russell J (1980) A circumplex model of affect. J Pers Soc Psychol 39(6):1161–1178

Said CP, Baron SG, Todorov A (2009) Nonlinear amygdala response to face trustworthiness: contributions of high and low spatial frequency information. J Cogn Neurosci 21(3):519–528

Sakaki M, Niki K, Mather M (2012) Beyond arousal and valence: the importance of the biological versus social relevance of emotional stimuli. Cogn Affect Behav Neurosci 12(1):115–139

Sakaki M, Fryer K, Mather M (2014) Emotion strengthens high-priority memory traces but weakens low-priority memory traces. Psychol Sci 25(2):387–395

Schilbach L, Eickhoff SB, Rotarska-Jagiela A, Fink GR, Vogeley K (2008) Minds at rest? Social cognition as the default mode of cognizing and its putative relationship to the "default system" of the brain. Conscious Cogn 17(2):457–467

Schilbach L, Bzdok D, Timmermans B, Fox PT, Laird AR, Vogeley K, Eickhoff SB (2012) Introspective minds: using ALE meta-analyses to study commonalities in the neural correlates of emotional processing, social & unconstrained cognition. PLoS One 7(2):e30920

Schiller D, Freeman JB, Mitchell JP, Uleman JS, Phelps EA (2009) A neural mechanism of first impressions. Nat Neurosci 12(4):508–514

Schmeichel BJ, Volokhov RN, Demaree HA (2008) Working memory capacity and the self-regulation of emotional expression and experience. J Pers Soc Psychol 95(6):1526–1540

Schweizer S, Grahn J, Hampshire A, Mobbs D, Dalgleish T (2013) Training the emotional brain: improving affective control through emotional working memory training. J Neurosci 33(12):5301–5311

Sergerie K, Lepage M, Armony JL (2006) A process-specific functional dissociation of the amygdala in emotional memory. J Cogn Neurosci 18(8):1359–1367

Shafer A, Iordan A, Cabeza R, Dolcos F (2011) Brain imaging investigation of the memory-enhancing effect of emotion. J Vis Exp 51:2433

Shafer AT, Matveychuk D, Penney T, O'Hare AJ, Stokes J, Dolcos F (2012) Processing of emotional distraction is both automatic and modulated by attention: evidence from an event-related fMRI investigation. J Cogn Neurosci 24(5):1233–1252

Shallice T, Fletcher P, Frith CD, Grasby P, Frackowiak RS, Dolan RJ (1994) Brain regions associated with acquisition and retrieval of verbal episodic memory. Nature 368(6472):633–635

Sharot T (2011) The optimism bias. Curr Biol 21(23):R941–R945

Sharot T, Delgado MR, Phelps EA (2004) How emotion enhances the feeling of remembering. Nat Neurosci 7(12):1376–1380

Sharot T, Martorella EA, Delgado MR, Phelps EA (2007a) How personal experience modulates the neural circuitry of memories of September 11. Proc Natl Acad Sci 104(1):389–394

Sharot T, Riccardi AM, Raio CM, Phelps EA (2007b) Neural mechanisms mediating optimism bias. Nature 450(7166):102–105

Sheppes G, Gross JJ (2012) Emotion regulation effectiveness: what works when. In: Tennen HA, Suls JM (eds) Handbook of psychology: personality and social psychology, vol 5. Wiley, New York

Sheppes G, Scheibe S, Suri G, Radu P, Blechert J, Gross JJ (2014) Emotion regulation choice: a conceptual framework and supporting evidence. J Exp Psychol Gen 143(1):163–181

Sheppes G, Suri G, Gross JJ (2015) Emotion regulation and psychopathology. Annu Rev Clin Psychol 11:379–405

Singer T, Kiebel SJ, Winston JS, Dolan RJ, Frith CD (2004) Brain responses to the acquired moral status of faces. Neuron 41(4):653–662

Smith AP, Dolan RJ, Rugg MD (2004a) Event-related potential correlates of the retrieval of emotional and nonemotional context. J Cogn Neurosci 16(5):760–775

Smith AP, Henson RN, Dolan RJ, Rugg MD (2004b) fMRI correlates of the episodic retrieval of emotional contexts. NeuroImage 22(2):868–878

Smith AP, Henson RN, Rugg MD, Dolan RJ (2005) Modulation of retrieval processing reflects accuracy of emotional source memory. Learn Mem 12(5):472–479

Smith AP, Stephan KE, Rugg MD, Dolan RJ (2006) Task and content modulate amygdala-hippocampal connectivity in emotional retrieval. Neuron 49(4):631–638
Somerville LH, Wig GS, Whalen PJ, Kelley WM (2006) Dissociable medial temporal lobe contributions to social memory. J Cogn Neurosci 18(8):1253–1265
Spreng RN, Andrews-Hanna JR (2015) The default network and social cognition. In: Brain mapping: an encyclopedic reference, vol 3. Elsevier/Academic Press, Amsterdam, pp 165–169
Spreng RN, Mar RA (2012) I remember you: a role for memory in social cognition and the functional neuroanatomy of their interaction. Brain Res 1428:43–50
Staresina BP, Davachi L (2010) Object unitization and associative memory formation are supported by distinct brain regions. J Neurosci 30(29):9890–9897
Strange BA, Dolan RJ (2004) Beta-adrenergic modulation of emotional memory-evoked human amygdala and hippocampal responses. Proc Natl Acad Sci 101(31):11454–11458
Strange BA, Hurlemann R, Dolan RJ (2003) An emotion-induced retrograde amnesia in humans is amygdala- and beta-adrenergic-dependent. Proc Natl Acad Sci 100(23):13626–13631
Takashima A, van der Ven F, Kroes MCW, Fernández G (2016) Retrieved emotional context influences hippocampal involvement during recognition of neutral memories. NeuroImage 143:280–292
Takeuchi H, Taki Y, Nouchi R, Sekiguchi A, Hashizume H, Sassa Y, Kotozaki Y, Miyauchi CM, Yokoyama R, Iizuka K, Nakagawa S, Nagase T, Kunitoki K, Kawashima R (2013) Resting state functional connectivity associated with trait emotional intelligence. NeuroImage 83:318–328
Talarico JM, LaBar KS, Rubin DC (2004) Emotional intensity predicts autobiographical memory experience. Mem Cogn 32(7):1118–1132
Tambini A, Rimmele U, Phelps EA, Davachi L (2017) Emotional brain states carry over and enhance future memory formation. Nat Neurosci 20(2):271–278
Taylor SF, Liberzon I, Fig LM, Decker LR, Minoshima S, Koeppe RA (1998) The effect of emotional content on visual recognition memory: a PET activation study. NeuroImage 8(2):188–197
Thiruchselvam R, Blechert J, Sheppes G, Rydstrom A, Gross JJ (2011) The temporal dynamics of emotion regulation: an EEG study of distraction and reappraisal. Biol Psychol 87(1):84–92
Thoresen C, Jensen J, Sigvartsen NP, Bolstad I, Server A, Nakstad PH, Andreassen OA, Endestad T (2012) Arousal modulates activity in the medial temporal lobe during a short-term relational memory task. Front Hum Neurosci 5:177
Touryan SR, Marian DE, Shimamura AP (2007b) Effect of negative emotional pictures on associative memory for peripheral information. Memory 15(2):154–166
Tsukiura T (2012) Neural mechanisms underlying the effects of face-based affective signals on memory for faces: a tentative model. Front Integr Neurosci 6:50
Tsukiura T, Cabeza R (2008) Orbitofrontal and hippocampal contributionI memory for face-name associations: the rewarding power of a smile. Neuropsychologia 46(9):2310–2319
Tsukiura T, Cabeza R (2011a) Remembering beauty: roles of orbitofrontal and hippocampal regions in successful memory encoding of attractive faces. NeuroImage 54(1):653–660
Tsukiura T, Cabeza R (2011b) Shared brain activity for aesthetic and moral judgments: implications for the beauty-is-good stereotype. Soc Cogn Affect Neurosci 6(1):138–148
Tsukiura T, Suzuki C, Shigemune Y, Mochizuki-Kawai H (2008) Differential contributions of the anterior temporal and medial temporal lobe to the retrieval of memory for person identity information. Hum Brain Mapp 29(12):1343–1354
Tsukiura T, Mano Y, Sekiguchi A, Yomogida Y, Hoshi K, Kambara T, Takeuchi H, Sugiura M, Kawashima R (2010) Dissociable roles of the anterior temporal regions in successful encoding of memory for person identity information. J Cogn Neurosci 22(10):2226–2237
Tsukiura T, Sekiguchi A, Yomogida Y, Nakagawa S, Shigemune Y, Kambara T, Akitsuki Y, Taki Y, Kawashima R (2011) Effects of aging on hippocampal and anterior temporal activations during successful retrieval of memory for face-name associations. J Cogn Neurosci 23(1):200–213
Tsukiura T, Shigemune Y, Nouchi R, Kambara T, Kawashima R (2013) Insular and hippocampal contributions to remembering people with an impression of bad personality. Soc Cogn Affect Neurosci 8(5):515–522
Tulving E (1985) Memory and consciousness. Can Psychol 26(1):1–12

Van Ast VA, Cornelisse S, Marin MF, Ackermann S, Garfinkel SN, Abercrombie HC (2013) Modulatory mechanisms of cortisol effects on emotional learning and memory: novel perspectives. Psychoneuroendocrinology 38(9):1874–1882

Van den Stock J, de Gelder B (2012) Emotional information in body and background hampers recognition memory for faces. Neurobiol Learn Mem 97(3):321–325

Vandekerckhove MM, Markowitsch HJ, Mertens M, Woermann FG (2005) Bi-hemispheric engagement in the retrieval of autobiographical episodes. Behav Neurol 16(4):203–210

Vazdarjanova A, McGaugh JL (1998) Basolateral amygdala is not critical for cognitive memory of contextual fear conditioning. Proc Natl Acad Sci 95(25):15003–15007

Ventura-Bort C, Löw A, Wendt J, Dolcos F, Hamm AO, Weymar M (2016a) When neutral turns significant: brain dynamics of rapidly formed associations between neutral stimuli and emotional contexts. Eur J Neurosci 44(5):2176–2183

Ventura-Bort C, Löw A, Wendt J, Molto J, Poy R, Dolcos F, Hamm AO, Weymar M (2016b) Binding neutral information to emotional contexts: brain dynamics of long-term recognition memory. Cogn Affect Behav Neurosci 16(2):234–247

Vilberg KL, Rugg MD (2009) Functional significance of retrieval-related activity in lateral parietal cortex: Evidence from fMRI and ERPs. Hum Brain Mapp 30(5): 1490-1501

Vogel S, Fernandez G, Joels M, Schwabe L (2016) Cognitive adaptation under stress: a case for the mineralocorticoid receptor. Trends Cogn Sci 20(3):192–203

Vrtička P, Sander D, Vuilleumier P (2012) Lateralized interactive social content and valence processing within the human amygdala. Front Hum Neurosci 6:358

Vuilleumier P, Richardson MP, Armony JL, Driver J, Dolan RJ (2004) Distant influences of amygdala lesion on visual cortical activation during emotional face processing. Nat Neurosci 7(11):1271–1278

Wagner DD, Haxby JV, Heatherton TF (2012) The representation of self and person knowledge in the medial prefrontal cortex. Wiley Interdiscip Rev Cogn Sci 3(4):451–470

Wagner U, Handke L, Walter H (2015) The relationship between trait empathy and memory formation for social vs. non-social information. BMC Psychol 3(1):2

Waring JD, Kensinger EA (2011) How emotion leads to selective memory: neuroimaging evidence. Neuropsychologia 49(7):1831–1842

Watkins ER, Baeyens CB, Read R (2009) Concreteness training reduces dysphoria: proof-of-principle for repeated cognitive bias modification in depression. J Abnorm Psychol 118(1):55–64

Webb TL, Miles E, Sheeran P (2012) Dealing with feeling: a meta-analysis of the effectiveness of strategies derived from the process model of emotion regulation. Psychol Bull 138(4):775–808

Weis S, Klaver P, Reul J, Elger CE, Fernandez G (2004) Temporal and cerebellar brain regions that support both declarative memory formation and retrieval. Cereb Cortex 14(3):256–267

Weymar M, Löw A, Melzig CA, Hamm AO (2009) Enhanced long-term recollection for emotional pictures: evidence from high-density ERPs. Psychophysiology 46(6):1200–1207

Weymar M, Hamm AO (2013) Electrophysiological signature of emotional memories. In: Linden M, Rutkowski K (eds) Hurting memories and beneficial forgetting: posttraumatic stress disorders, biographical developments and social conflicts. Elsevier, Saint Louis, pp 21–35

Weymar M, Löw A, Hamm AO (2011) Emotional memories are resilient to time: evidence from the parietal ERP old/new effect. Hum Brain Mapp 32(4):632–640

Weymar M, Bradley MM, Hamm AO, Lang PJ (2013) When fear forms memories: threat of shock and brain potentials during encoding and recognition. Cortex 49(3):819–826

Weymar M, Bradley MM, Hamm AO, Lang PJ (2014) Encoding and reinstatement of threat: recognition potentials. Neurobiol Learn Mem 107:87–92

Wieser MJ, Keil A (2014) Fearful faces heighten the cortical representation of contextual threat. NeuroImage 86:317–325

Wilker S, Elbert T, Kolassa IT (2014) The downside of strong emotional memories: how human memory-related genes influence the risk for posttraumatic stress disorder—a selective review. Neurobiol Learn Mem 112:75–86

Williams AD, Moulds ML (2010) The impact of ruminative processing on the experience of self-referent intrusive memories in dysphoria. Behav Ther 41(1):38–45

Williams JM, Teasdale JD, Segal ZV, Soulsby J (2000) Mindfulness-based cognitive therapy reduces overgeneral autobiographical memory in formerly depressed patients. J Abnorm Psychol 109(1):150–155

Yang XF, Bossmann J, Schiffhauer B, Jordan M, Immordino-Yang MH (2013) Intrinsic default mode network connectivity predicts spontaneous verbal descriptions of autobiographical memories during social processing. Front Psychol 3:592

Yaoi K, Osaka M, Osaka N (2015) Neural correlates of the self-reference effect: evidence from evaluation and recognition processes. Front Hum Neurosci 9:383

Yonelinas AP, Ritchey M (2015) The slow forgetting of emotional episodic memories: an emotional binding account. Trends Cogn Sci 19(5):259–267

Chapter 6
Modulatory Mechanism of Three-Layered Hierarchical Structure in Memories Associated with Rewards and Punishments

Yayoi Shigemune and Takashi Tsukiura

Abstract Memory for reward- and punishment-related information is important to effectively obtain rewards and avoid punishments. Functional neuroimaging findings of studies investigating memory enhancement for reward- and punishment-related information are methodologically categorized into two main types: conditioning and stimuli associated with a reward or punishment. In conditioning, memory enhancement involves interactions between regions related to reward and punishment, including the substantia nigra, ventral tegmental area, striatum and insula, and regions related to memory, including the hippocampus and parahippocampal cortex. Memory enhancement for stimuli associated with a reward or punishment involves interactions between the orbitofrontal cortex, which is related to reward outcomes, or the insula, which is related to punishment outcomes, and the hippocampus, which is important for memory. In addition, findings from functional neuroimaging studies have revealed that the memory enhancement in conditioning is arranged in a three-layered hierarchical structure of psychological components, involving interactions between reward-/punishment-related regions and memory-related regions, modulations of cognitive factors that are associated with the prefrontal cortex, and modulations of social factors that are associated with the ventromedial prefrontal cortex, the temporo-parietal junction, and the posterior superior temporal sulcus. This three-layered hierarchical structure (i.e., modulatory effects of cognitive

Y. Shigemune (✉)
Department of Psychology, Graduate School of Letters, Chuo University,
Tokyo 192-0393, Japan
e-mail: gemune@gmail.com

T. Tsukiura
Department of Cognitive and Behavioral Sciences, Graduate School of Human and Environmental Studies, Kyoto University, Kyoto 606-8501, Japan

T. Tsukiura, S. Umeda (eds.), *Memory in a Social Context*,
https://doi.org/10.1007/978-4-431-56591-8_6

and social factors on the interaction between reward/punishment and memory processes) could be critical for human beings, who are social animals, in obtaining rewards and avoiding punishments.

Keywords Reward • Punishment • Episodic memory • Functional neuroimaging • Cognitive factors • Social factors • Anticipation • Outcome

6.1 Introduction

Experience or knowledge of rewards and punishments is important for obtaining rewards and avoiding punishments. Therefore, reward- or punishment-related information is remembered better, and this information modulates human or animal behavior. Functional neuroimaging studies have demonstrated the neural mechanisms underlying the interactions between reward/punishment and episodic memory (Adcock et al. 2006; Murty et al. 2012; Shigemune et al. 2014; Wittmann et al. 2005, 2013). These studies are categorized into two main types: conditioning and processing of stimuli associated with a reward or punishment (Fig. 6.1). In addition, evidence from conditioning studies suggests a three-layered hierarchical structure of psychological components that includes interactions between the processes of reward/punishment and memory; cognitive factors that modulate the interactions between the processes of reward/punishment and memory; and social factors that modulate the interactions between processes of reward/punishment and memory (Fig. 6.2). The aim of this chapter is to review the neural mechanisms involved in the interactions between processes of reward/punishment and memory by categorizing the methods employed in previous studies. In addition, we review functional neuroimaging findings according to the three layers of the possible hierarchical structure.

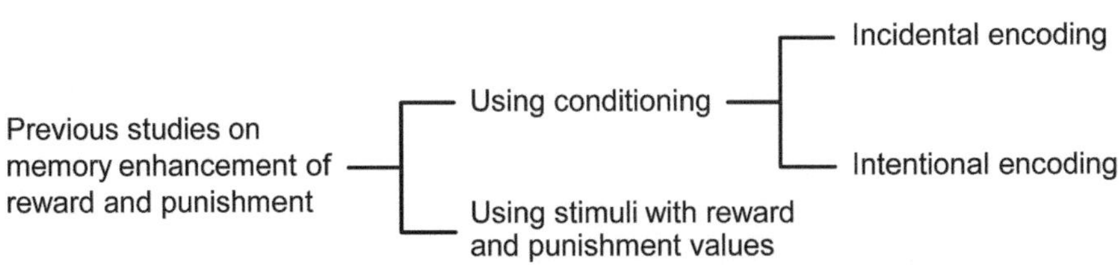

Fig. 6.1 Categorization of previous studies on reward- or punishment-related memory enhancement

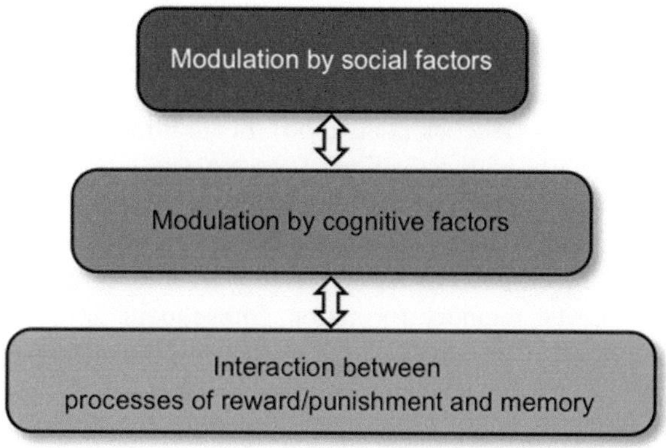

Fig. 6.2 Three-layered hierarchical structure of memory enhancement by rewards or punishments in conditioning

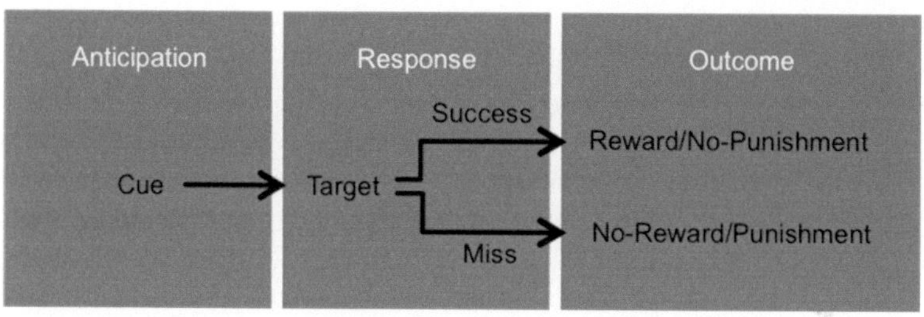

Fig. 6.3 General procedures of the monetary incentive delay (*MID*) task

6.2 Methodological Categorization of Studies on the Interaction Between Reward/Punishment and Memory

The first functional neuroimaging study investigating the reward-dependent enhancement of memory using the conditioning method was reported by Wittmann and colleagues (Wittmann et al. 2005). In their study, they employed the monetary incentive delay (MID) task (Fig. 6.3), which included different phases of reward or punishment anticipation, responses, and outcomes, and investigated how the neural activation of brain regions associated with reward anticipation during incidental encoding predicted subsequent memory retrieval. The results of this study showed that reward-associated cues were remembered better than non-rewarded cues, and that successful memory encoding was associated with activation of the hippocampus and substantia nigra/ventral tegmental area (SN/VTA). In addition, activation of the SN/VTA was significantly greater during the anticipation of rewards compared with non-rewards. Another study using the MID task demonstrated that target pictures intentionally encoded with high reward cues were remembered better than target pictures encoded with low reward cues, and successful memory encoding of pictures with high reward cues was associated with significantly greater activation of the SN/VTA, nucleus accumbens (NA), and hippocampus than that of pictures with low reward cues (Adcock et al. 2006). In addition, activation of the SN/VTA correlated significantly with activation of the NA and hippocampus, and with memory accuracy. These studies suggest that the mechanism of reward-associated memory enhancement in conditioning could involve interactions between the activation of the SN/VTA and NA related to reward anticipation and the hippocampus related to episodic memory formation. Subsequent to the publication of these studies, functional neuroimaging studies have reported the neural mechanisms underlying the effect of both punishments and rewards on episodic memory formation (Bauch et al. 2014; Murty et al. 2012; Shigemune et al. 2014). In addition, other studies have shown that the interaction between reward/punishment and memory is associated with emotion, which enhances memory (Shigemune et al. 2010; Wittmann et al. 2008), or salience, which is important for reward-seeking (Murty and Adcock 2014). These studies indicate that memory enhancements by rewards or punishments in conditioning are subdivided into two types according to different experimental procedures. One type investigated the neural mechanisms during the incidental encoding of stimuli in the MID task, and the other type investigated the neural mechanisms related to the processing of rewards during the intentional encoding.

One potential problem with studies investigating memory enhancement for stimuli associated with a reward or punishment is that reward-related enhancement of memory might be mixed with emotion-related enhancement of memory. However, neurophysiological studies in experimental animals have reported that cell activity of the orbitofrontal cortex (OFC) and amygdala is increased by visual, olfactory, and gustatory stimuli associated with the primary reward cue of food (Nishijo et al. 1988; Rolls et al. 1990; Sanghera et al. 1979), and that the activity in these brain regions is modulated by hunger or satiety (Critchley and Rolls 1996; Rolls et al. 1989; Scott et al. 1995). These findings suggest that activation of the OFC and amygdala could reflect the motivational significance of rewards, but not simple perceptual properties. Based on this concept, Morris and colleagues compared memories of food and non-food pictures in participants with different levels of motivation for food by controlling the fasting states (Morris and Dolan 2001). In the behavioral data, food pictures were remembered more accurately than non-food pictures during fasting, but no difference in memory performance was observed between the food and non-food pictures during satiating. In the positron emission tomography data, activation of the OFC was associated with memories for both food and nonfood pictures, whereas activation of the amygdala was only observed for memories of food pictures. In addition, activation of the OFC correlated positively with activation of the amygdala during the encoding of food pictures, whereas OFC activation correlated negatively with amygdala activation during the encoding of non-food pictures. The contribution of interacting mechanisms between the OFC and hippocampus to reward-dependent memory enhancement has been identified during the processing of social rewards, including face-based social signals such as smiling facial expressions or facial attractiveness (Tsukiura and Cabeza 2008, 2011). The effect of face based social signals on memory for faces was also reported for social punishments, and the enhancement of facial memories by social punishment was modulated by interactions between the insula and the hippocampus (Tsukiura et al. 2013). These findings suggest that interactions between the OFC and hippocampus could contribute to memory enhancement of stimuli associated with rewards, whereas interactions between the insula and hippocampus could be critical for memory enhancement of stimuli associated with punishments.

The neural mechanisms associated with memory enhancement of reward or punishment stimuli, which involve interactions between the OFC/insula and the hippocampus, are different from the neural mechanisms identified in conditioning, which involve a reward-dependent enhancement of memory through interactions of the reward-related regions of the SN/VTA and striatum with the memory-related hippocampus. The reward- and punishment-dependent enhancement of memories for stimuli associated with a reward/punishment could be generated by the outcomes of rewards and punishments, whereas the reward- and punishment-dependent enhancement of memories in conditioning could be mediated by the anticipation of rewards and punishments. This theory is well supported by previous functional neuroimaging studies, in which the SN/VTA and striatum are significantly activated during reward anticipation, whereas activation of the SN/VTA and striatum is decreased and activation of the OFC is increased during receiving reward outcomes (Knutson et al. 2001, 2003; O'Doherty et al. 2002).

6.3 First Layer: Interaction Between Reward/Punishment and Memory

The first layer of the hierarchical structure, which was proposed by functional neuroimaging studies investigating reward- or punishment-dependent enhancement of memories in conditioning, is the interaction between reward-/punishment-related regions, such as the SN/VTA, striatum, amygdala, and insula, and memory-related regions, such as the hippocampus and parahippocampal cortex (PHC). Electrophysiological studies in non-human primates have reported that neural activation of midbrain dopaminergic neurons of the SN/VTA reflects the reward prediction error in learning stimulus–reward associations (Schultz 1998; Schultz et al. 1997). Functional neuroimaging studies in humans have demonstrated that activation of the SN/VTA and striatum is associated with the reward prediction error (D'Ardenne et al. 2008; McClure et al. 2003; O'Doherty et al. 2003). In addition, the SN/VTA region is anatomically connected to the striatum region (Haber and Knutson 2010), and task performance and striatum activation in reward-dependent association learning were enhanced by a dopamine agonist, but were inhibited by a dopamine antagonist (Pessiglione et al. 2006). These findings suggest that dopaminergic neuronal projections to the striatum could reinforce reward-dependent behavioral learning and that activation of the SN/VTA and striatum, which was identified in previous studies investigating the reward-related enhancement of memory, could reflect dopaminergic neuronal activity related to reward-dependent behavioral learning (Dillon et al. 2014; Loh et al. 2015; Shigemune et al. 2014; Wittmann et al. 2013; Wolosin et al. 2012, 2013).

Some electrophysiological studies in experimental animals have reported that there is significant activation of midbrain dopaminergic neurons during the processing of punishments (Guarraci and Kapp 1999; Joshua et al. 2008; Matsumoto and Hikosaka 2009; Mirenowicz and Schultz 1996), whereas other studies have shown a decrease in this activation (Coizet et al. 2006; Ungless et al. 2004). Functional neuroimaging studies in humans have revealed increased activation of the SN/VTA and striatum during the anticipation of both rewards and punishments (Carter et al. 2009; Knutson et al. 2000; Samanez-Larkin et al. 2007; Wrase et al. 2007), which is similar to the activation pattern observed in studies investigating punishment-induced memory enhancement in conditioning (Bauch et al. 2014; Shigemune et al. 2014; Wittmann et al. 2013). Given that the NA is part of the striatum, which was significantly activated in the processing of avoidable punishments and was significantly deactivated in the processing of unavoidable punishment (Cooper and Knutson 2008), and that neural responses of the anticipation of punishment avoidance are similar to those of the anticipation of reward acceptance (Kim et al. 2006), activation of the SN/VTA and striatum during the anticipation of punishments could reflect the anticipation of rewards induced by avoiding punishments (Shigemune et al. 2014). Moreover, functional neuroimaging studies have shown that the amygdala and insula could contribute to punishment-induced memory enhancement in conditioning (Bauch et al. 2014; Murty et al. 2012; Shigemune et al. 2014). Functional neuroimaging studies have consistently shown that the amygdala is involved in emotional memory enhancement (Cahill et al. 1996; Canli et al. 1999, 2000; Hamann et al. 1999), and the insula has been shown to be important for social punishment-induced memory enhancement (Tsukiura et al. 2013). Other functional magnetic resonance imaging (fMRI) studies have demonstrated that the amygdala is associated with the processing of emotions (Breiter et al. 1996; Morris et al. 1996), and that the insula is important for the processing of pain (Coghill et al. 1994; Ploghaus et al. 1999) and aversion (Nitschke et al. 2006; Phillips et al. 1997; Wicker et al. 2003). These findings suggest that punishment prediction could generate emotions and aversion, and could be associated with punishment-induced memory enhancement. However, it is possible that punishment-induced memory enhancement through activation of the amygdala might be caused by primary punishment such as electrical shock (Bauch et al. 2014; Murty et al. 2012), but not by secondary punishment such as monetary losses. This possibility should be carefully considered in future studies.

Memory enhancement by rewards/punishments in conditioning has been shown to involve activation of the hippocampus and PHC. Some fMRI studies have consistently reported activation of the hippocampus in interactions between reward/punishment processes and memory processes (Adcock et al. 2006; Bauch et al. 2014; Murty et al. 2012; Shigemune et al. 2014; Wittmann et al. 2005, 2013; Wolosin et al. 2012, 2013). Other fMRI studies, which investigated neural activation related to association processes between target memory and source memory (Shigemune et al. 2014), between paired objects (Wolosin et al. 2012, 2013), and between target memory and information regarding whether or not targets are accompanied by rewards (Dillon et al. 2014), have shown that reward/punishment-related memory enhancement is associated with increased activation of the PHC. Memory for items and contexts is processed separately in the perirhinal cortex and PHC, and is integrated into the hippocampus through the entorhinal cortex (Davachi 2006; Diana et al. 2007; Eichenbaum et al. 2007). In addition, memory-related brain regions, such as the hippocampus and PHC, are anatomically and functionally connected with reward-/punishment-related brain regions including the SN/VTA, NA, amygdala, and insula (Haber and Knutson 2010; Lisman and Grace 2005; Pikkarainen et al. 1999; Pribram and Maclean 1953; Rutecki et al. 1989; Suzuki and Amaral 1994), and significant interactions between activation in these regions have been found in functional neuroimaging studies showing memory enhancement by the processing of rewards/punishments in conditioning (Adcock et al. 2006; Gruber et al. 2016; Kuhl et al. 2010; Murty et al. 2012; Shigemune et al. 2014; Wolosin et al. 2012) and that for stimuli associated with rewards or punishments (Tsukiura and Cabeza 2008, 2011; Tsukiura et al. 2013). Thus, the first layer of the hierarchical structure could be involved in interactions between memory-related regions, such as the hippocampus and PHC, and the SN/VTA and striatum, which process both rewards and punishments, or the amygdala and insula, which only process punishments.

6.4 Second Layer: Modulation of Reward–Memory Interactions by Cognitive Factors

The second layer of the hierarchical structure of memory enhancement by the processing of rewards or punishments in conditioning involves modulation of the first layer of the hierarchical structure by cognitive factors. Previous studies have reported that the NA is anatomically connected to the amygdala, OFC, anterior cingulate cortex (ACC), prefrontal cortex (PFC), SN/VTA, and the hippocampus (Haber and Knutson 2010), and that activation of the SN/VTA and NA is dynamically affected by activation of the dorsolateral PFC during the anticipation of monetary gains and losses (Ballard et al. 2011). Thus, interactions between reward/punishment processes and memory processes could be modulated by psychological processes represented in these regions. In the MID task, functional neuroimaging studies have shown functional modulation by cognitive factors in each component of cue, target, response, and outcome (Fig. 6.3).

Previous fMRI studies have consistently shown that cognitive factors in the cue component of the MID task modulate reward-induced memory enhancement. For example, cues associated with predictable reward were remembered more accurately than cues associated with unpredictable reward, and the memory enhancement by reward predictability was involved in the ventromedial PFC (vmPFC), the NA, and the hippocampus (Bialleck et al. 2011). In addition, the optimization of memory strategy to obtain greater rewards was observed only when the reward values were high, and the behavioral optimization by higher reward values was associated with activation of the inferior frontal gyrus and posterior lateral temporal cortex (Cohen et al. 2014). In another study, memory accuracy was decreased by reward-induced anxiety only when the reward value was relatively high, and was associated with an activation of the ACC and middle frontal gyrus (Callan and Schweighofer 2008). Functional neuroimaging evidence has also shown that activation of the reward-/punishment-related regions and memory-related regions is directly modulated by several cognitive factors. For example, reward-related memory enhancement was higher when the rewarding context was perceptually similar to the non-rewarding context than when the context was not perceptually similar, and these effects involved higher functional connectivity between the SN/VTA and the hippocampus (Loh et al. 2015). Another study demonstrated that in a learning task, which required participants to choose a target picture that was related to the higher probability of a reward, memory accuracy for the pictures was decreased with increases in learning-related activation in the striatum (Wimmer et al. 2014).

Functional neuroimaging studies have shown that the neural mechanisms underlying the modulatory effects of cognitive factors in the target of the MID task are different according to features of the stimuli. For example, one fMRI study reported that reward-related memory enhancement was significantly increased only for salient stimuli and that interactions between the VTA and the medial PFC, ventrolateral PFC, and occipital cortex predicted activation of the hippocampus that was associated with memory enhancement (Murty and Adcock 2014). In addition, other fMRI studies have found that reward-related memory enhancement for emotionally positive stimuli was associated with activation of the striatum and amygdala (Wittmann et al. 2008), and that significant activation of the amygdala reflected emotion-related memory enhancement in both the high and low reward conditions (Shigemune et al. 2010).

One fMRI study, which investigated how memories for target pictures were modulated by factors of "action" (Go or No-go) and "valence" (obtaining rewards or avoiding punishments) (Koster et al. 2015), showed that the responses in the MID task modulate memory. Memory accuracy was significantly enhanced when the participants were required to choose the "Go" response for a reward, and activation of the SN/VTA, basal ganglia, and hippocampus was important for memory enhancement.

Few fMRI studies have investigated the modulatory effects of the reward outcomes in the MID task, whereas previous fMRI studies have revealed that activation of the SN/VTA and striatum associated with reward anticipation is modulated by the cue presentation, target presentation, and responses in the MID task. Activation of the OFC, but not the SN/VTA and striatum, has been shown to be involved in the processing of reward outcomes (Knutson et al. 2001, 2003; O'Doherty et al. 2002), and interactions between the OFC and the hippocampus have been identified in the memory enhancement of stimuli with rewarding values (Morris and Dolan 2001; Tsukiura and Cabeza 2008, 2011). Thus, the OFC may contribute to the modulation of memory by reward outcomes. However, given that the striatum and lateral PFC were significantly activated in the outcome phase when the information values were high in an associative learning task (Tricomi and Fiez 2012), further studies would be required to investigate whether the reward-related regions such as the SN/VTA and striatum contribute to the modulation of memory by reward outcomes.

The reverse effect, in which reward-enhanced memories affect some cognitive processes, has also been identified in previous fMRI studies. For example, one fMRI study revealed that functional connectivity between the vmPFC and the hippocampus was important for the modulation of reward-associated memories in learning-guided behaviors or value-based decisions (Gerraty et al. 2014; Wimmer and Shohamy 2012). In another study, future events were more vividly imagined by cued words associated with rewards, and this enhanced vividness of future thinking involved functional connectivity among the hippocampus, striatum, and SN/VTA (Bulganin and Wittmann 2015). These findings suggest that the modulation of reward-related memory enhancement by cognitive factors could involve the prefrontal regions related to the processing and comparison of reward values (Basten et al. 2010; Hare et al. 2009; Kable and Glimcher 2007; Lim et al. 2013; Wang et al. 2014) or to cognitive controls (D'Esposito et al. 1998; Owen 1997; Smith and Jonides 1999). These prefrontal regions could interact with the first layer of the hierarchical structure, which is related to interactions between reward/punishment processes and memory processes.

6.5 Third Layer: Modulation of Reward–Memory Interactions by Social Factors

The third layer of the hierarchical structure involves modulation of the first and second layers of the hierarchal structure by social factors. Psychological studies have shown that the retrieval accuracy of memory for words remembered by another person is significantly better than the memory for words chosen by a non-human entity, such as a computer (Reysen 2003; Reysen and Adair 2008). Functional neuroimaging studies investigating the effects of social interactions during the processing of monetary rewards or punishments have demonstrated that task performance is modulated by several social factors, which involves activation of the striatum, vmPFC, temporo-parietal junction (TPJ), and posterior superior temporal sulcus (pSTS) (Bault et al. 2015; Delgado et al. 2005, 2008; Fareri et al. 2012, 2015; Fareri and Delgado 2014; Halko et al. 2009; Hampton et al. 2008, 2015; Le Bouc and Pessiglione 2013; Li et al. 2013; Mobbs et al. 2009; Phan et al. 2010; Rilling et al. 2002, 2004a, b; Seid-Fatemi and Tobler 2015; van den Bos et al. 2013; Votinov et al. 2015). For example, one fMRI study reported that the trend of selecting cooperation behaviors was higher when the participants performed the prisoner's dilemma game with other people rather than computers, and greater activation was observed in the striatum and vmPFC when the choice of cooperation behaviors was successful rather than unsuccessful (Rilling et al. 2004b). Given that the activation of the vmPFC is related to the processing of both self-gained monetary rewards and those gained by other persons, and that the activation of the vmPFC was greater in the generous judgments than in the selfish judgments, the vmPFC could contribute to the comparison between and integration of gaining values of rewards for oneself and for other persons (Hutcherson et al. 2015). Thus, reward-/punishment-related memory enhancement in conditioning could be modulated by social factors. Furthermore, the modulation of memory enhancement by social factors could also be mediated by the vmPFC, which is associated with the comparison and integration between reward values by oneself and by other persons (Bault et al. 2015; Behrens et al. 2008; Hutcherson et al. 2015), and by the TPJ and pSTS, which are associated with the inference of others' thoughts (Mitchell 2009; Schurz et al. 2014; Van Overwalle 2009; Van Overwalle and Baetens 2009). However, there is limited evidence from functional neuroimaging studies showing the modulation of reward-/punishment–memory interactions by social factors. Further investigations are required in the future.

6.6 Conclusions

Neural mechanisms underlying reward- or punishment-induced memory enhancement have been investigated in functional neuroimaging studies, and the findings are divided into two main categories. One category involves the enhancement of memory by rewards or punishments in conditioning. The memory enhancement in

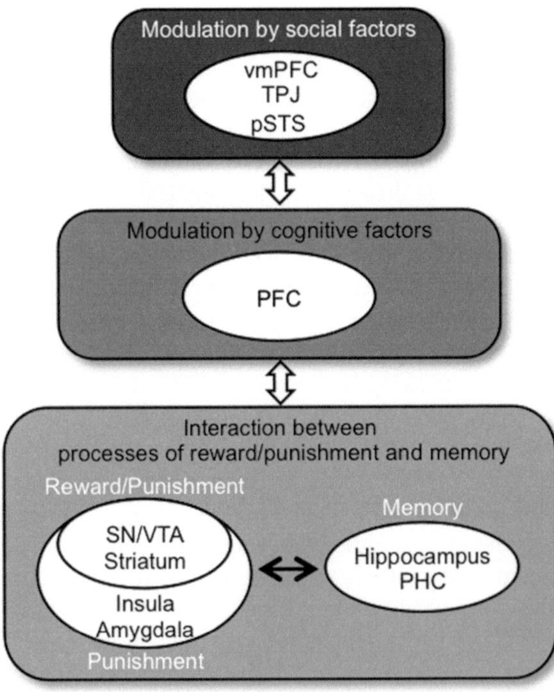

Fig. 6.4 Hypothetical model of the neural mechanisms underlying the three-layered hierarchical structure of memory enhancement by rewards or punishments in conditioning

this category is supported by interactions between regions associated with the processing of both rewards and punishments, such as the SN/VTA and striatum, or those associated only with the processing of punishments, such as the insula and amygdala, and regions associated with memory, such as the hippocampus and PHC. The second category involves memory enhancement for stimuli associated with a reward or punishment, and the memory enhancement is supported by interactions between the reward-related OFC, punishment-related insula, and memory-related hippocampus. In addition, reward-/punishment-related memory enhancement in conditioning could be explained by the three-layered hierarchical structure, in which the first layer involves the interaction between reward-/punishment-related regions and memory-related regions, the second layer involves modulation by cognitive factors, which is mediated by the PFC, and the third layer involves modulation by social factors, which is mediated by the vmPFC, TPJ, and pSTS (Fig. 6.4). In this hierarchy, the processes in each layer could interact with the processes in the other layers. In human society, rewards or punishments are exchanged through money, objects, time, effort, etc., in interpersonal relationships. The modulation of interactions between reward/punishment processes and memory processes by social and cognitive factors could be critical for humans, who are social animals, in obtaining rewards and avoiding punishments.

References

Adcock RA, Thangavel A, Whitfield-Gabrieli S, Knutson B, Gabrieli JD (2006) Reward-motivated learning: mesolimbic activation precedes memory formation. Neuron 50(3):507–517

Ballard IC, Murty VP, Carter RM, MacInnes JJ, Huettel SA, Adcock RA (2011) Dorsolateral prefrontal cortex drives mesolimbic dopaminergic regions to initiate motivated behavior. J Neurosci 31(28):10340–10346

Basten U, Biele G, Heekeren HR, Fiebach CJ (2010) How the brain integrates costs and benefits during decision making. Proc Natl Acad Sci U S A 107(50):21767–21772

Bauch EM, Rausch VH, Bunzeck N (2014) Pain anticipation recruits the mesolimbic system and differentially modulates subsequent recognition memory. Hum Brain Mapp 35(9):4594–4606

Bault N, Pelloux B, Fahrenfort JJ, Ridderinkhof KR, van Winden F (2015) Neural dynamics of social tie formation in economic decision-making. Soc Cogn Affect Neurosci 10(6):877–884

Behrens TE, Hunt LT, Woolrich MW, Rushworth MF (2008) Associative learning of social value. Nature 456(7219):245–249

Bialleck KA, Schaal HP, Kranz TA, Fell J, Elger CE, Axmacher N (2011) Ventromedial prefrontal cortex activation is associated with memory formation for predictable rewards. PLoS One 6(2):e16695

Breiter HC, Etcoff NL, Whalen PJ, Kennedy WA, Rauch SL, Buckner RL, Strauss MM, Hyman SE, Rosen BR (1996) Response and habituation of the human amygdala during visual processing of facial expression. Neuron 17(5):875–887

Bulganin L, Wittmann BC (2015) Reward and novelty enhance imagination of future events in a motivational-episodic network. PLoS One 10(11):e0143477

Cahill L, Haier RJ, Fallon J, Alkire MT, Tang C, Keator D, Wu J, McGaugh JL (1996) Amygdala activity at encoding correlated with long-term, free recall of emotional information. Proc Natl Acad Sci U S A 93(15):8016–8021

Callan DE, Schweighofer N (2008) Positive and negative modulation of word learning by reward anticipation. Hum Brain Mapp 29(2):237–249
Canli T, Zhao Z, Desmond JE, Glover G, Gabrieli JDE (1999) fMRI identifies a network of structures correlated with retention of positive and negative emotional memory. Psychobiology 27(4):441–452
Canli T, Zhao Z, Brewer J, Gabrieli JD, Cahill L (2000) Event-related activation in the human amygdala associates with later memory for individual emotional experience. J Neurosci 20(19):RC99
Carter RM, Macinnes JJ, Huettel SA, Adcock RA (2009) Activation in the VTA and nucleus accumbens increases in anticipation of both gains and losses. Front Behav Neurosci 3:21
Coghill RC, Talbot JD, Evans AC, Meyer E, Gjedde A, Bushnell MC, Duncan GH (1994) Distributed processing of pain and vibration by the human brain. J Neurosci 14(7):4095–4108
Cohen MS, Rissman J, Suthana NA, Castel AD, Knowlton BJ (2014) Value-based modulation of memory encoding involves strategic engagement of fronto-temporal semantic processing regions. Cogn Affect Behav Neurosci 14(2):578–592
Coizet V, Dommett EJ, Redgrave P, Overton PG (2006) Nociceptive responses of midbrain dopaminergic neurones are modulated by the superior colliculus in the rat. Neuroscience 139(4):1479–1493
Cooper JC, Knutson B (2008) Valence and salience contribute to nucleus accumbens activation. NeuroImage 39(1):538–547
Critchley HD, Rolls ET (1996) Hunger and satiety modify the responses of olfactory and visual neurons in the primate orbitofrontal cortex. J Neurophysiol 75(4):1673–1686
D'Ardenne K, McClure SM, Nystrom LE, Cohen JD (2008) BOLD responses reflecting dopaminergic signals in the human ventral tegmental area. Science 319(5867):1264–1267
D'Esposito M, Aguirre GK, Zarahn E, Ballard D, Shin RK, Lease J (1998) Functional MRI studies of spatial and nonspatial working memory. Brain Res Cogn Brain Res 7(1):1–13
Davachi L (2006) Item, context and relational episodic encoding in humans. Curr Opin Neurobiol 16(6):693–700
Delgado MR, Frank RH, Phelps EA (2005) Perceptions of moral character modulate the neural systems of reward during the trust game. Nat Neurosci 8(11):1611–1618
Delgado MR, Schotter A, Ozbay EY, Phelps EA (2008) Understanding overbidding: using the neural circuitry of reward to design economic auctions. Science 321(5897):1849–1852
Diana RA, Yonelinas AP, Ranganath C (2007) Imaging recollection and familiarity in the medial temporal lobe: a three-component model. Trends Cogn Sci 11(9):379–386
Dillon DG, Dobbins IG, Pizzagalli DA (2014) Weak reward source memory in depression reflects blunted activation of VTA/SN and parahippocampus. Soc Cogn Affect Neurosci 9(10):1576–1583
Eichenbaum H, Yonelinas AP, Ranganath C (2007) The medial temporal lobe and recognition memory. Annu Rev Neurosci 30:123–152
Fareri DS, Delgado MR (2014) Differential reward responses during competition against in- and out-of-network others. Soc Cogn Affect Neurosci 9(4):412–420
Fareri DS, Niznikiewicz MA, Lee VK, Delgado MR (2012) Social network modulation of reward-related signals. J Neurosci 32(26):9045–9052
Fareri DS, Chang LJ, Delgado MR (2015) Computational substrates of social value in interpersonal collaboration. J Neurosci 35(21):8170–8180
Gerraty RT, Davidow JY, Wimmer GE, Kahn I, Shohamy D (2014) Transfer of learning relates to intrinsic connectivity between hippocampus, ventromedial prefrontal cortex, and large-scale networks. J Neurosci 34(34):11297–11303
Gruber MJ, Ritchey M, Wang SF, Doss MK, Ranganath C (2016) Post-learning hippocampal dynamics promote preferential retention of rewarding events. Neuron 89(5):1110–1120
Guarraci FA, Kapp BS (1999) An electrophysiological characterization of ventral tegmental area dopaminergic neurons during differential pavlovian fear conditioning in the awake rabbit. Behav Brain Res 99(2):169–179

Haber SN, Knutson B (2010) The reward circuit: linking primate anatomy and human imaging. Neuropsychopharmacology 35(1):4–26

Halko ML, Hlushchuk Y, Hari R, Schurmann M (2009) Competing with peers: mentalizing-related brain activity reflects what is at stake. NeuroImage 46(2):542–548

Hamann SB, Ely TD, Grafton ST, Kilts CD (1999) Amygdala activity related to enhanced memory for pleasant and aversive stimuli. Nat Neurosci 2(3):289–293

Hampton AN, Bossaerts P, O'Doherty JP (2008) Neural correlates of mentalizing-related computations during strategic interactions in humans. Proc Natl Acad Sci U S A 105(18):6741–6746

Hare TA, Camerer CF, Rangel A (2009) Self-control in decision-making involves modulation of the vmPFC valuation system. Science 324(5927):646–648

Hutcherson CA, Bushong B, Rangel A (2015) A neurocomputational model of altruistic choice and its implications. Neuron 87(2):451–462

Joshua M, Adler A, Mitelman R, Vaadia E, Bergman H (2008) Midbrain dopaminergic neurons and striatal cholinergic interneurons encode the difference between reward and aversive events at different epochs of probabilistic classical conditioning trials. J Neurosci 28(45):11673–11684

Kable JW, Glimcher PW (2007) The neural correlates of subjective value during intertemporal choice. Nat Neurosci 10(12):1625–1633

Kim H, Shimojo S, O'Doherty JP (2006) Is avoiding an aversive outcome rewarding? Neural substrates of avoidance learning in the human brain. PLoS Biol 4(8):e233

Knutson B, Westdorp A, Kaiser E, Hommer D (2000) FMRI visualization of brain activity during a monetary incentive delay task. NeuroImage 12(1):20–27

Knutson B, Fong GW, Adams CM, Varner JL, Hommer D (2001) Dissociation of reward anticipation and outcome with event-related fMRI. Neuroreport 12(17):3683–3687

Knutson B, Fong GW, Bennett SM, Adams CM, Hommer D (2003) A region of mesial prefrontal cortex tracks monetarily rewarding outcomes: characterization with rapid event-related fMRI. NeuroImage 18(2):263–272

Koster R, Guitart-Masip M, Dolan RJ, Duzel E (2015) Basal ganglia activity mirrors a benefit of action and reward on long-lasting event memory. Cereb Cortex 25(12):4908–4917

Kuhl BA, Shah AT, DuBrow S, Wagner AD (2010) Resistance to forgetting associated with hippocampus-mediated reactivation during new learning. Nat Neurosci 13(4):501–506

Le Bouc R, Pessiglione M (2013) Imaging social motivation: distinct brain mechanisms drive effort production during collaboration versus competition. J Neurosci 33(40):15894–15902

Li P, Shen Y, Sui X, Chen C, Feng T, Li H, Holroyd C (2013) The neural basis of responsibility attribution in decision-making. PLoS One 8(11):e80389

Lim SL, O'Doherty JP, Rangel A (2013) Stimulus value signals in ventromedial PFC reflect the integration of attribute value signals computed in fusiform gyrus and posterior superior temporal gyrus. J Neurosci 33(20):8729–8741

Lisman JE, Grace AA (2005) The hippocampal-VTA loop: controlling the entry of information into long-term memory. Neuron 46(5):703–713

Loh E, Kumaran D, Koster R, Berron D, Dolan R, Duzel E (2015) Context-specific activation of hippocampus and SN/VTA by reward is related to enhanced long-term memory for embedded objects. Neurobiol Learn Mem 134(Pt A):65–77

Matsumoto M, Hikosaka O (2009) Two types of dopamine neuron distinctly convey positive and negative motivational signals. Nature 459(7248):837–841

McClure SM, Berns GS, Montague PR (2003) Temporal prediction errors in a passive learning task activate human striatum. Neuron 38(2):339–346

Mirenowicz J, Schultz W (1996) Preferential activation of midbrain dopamine neurons by appetitive rather than aversive stimuli. Nature 379(6564):449–451

Mitchell JP (2009) Inferences about mental states. Philos Trans R Soc Lond Ser B Biol Sci 364(1521):1309–1316

Mobbs D, Yu R, Meyer M, Passamonti L, Seymour B, Calder AJ, Schweizer S, Frith CD, Dalgleish T (2009) A key role for similarity in vicarious reward. Science 324(5929):900

Morris JS, Dolan RJ (2001) Involvement of human amygdala and orbitofrontal cortex in hunger-enhanced memory for food stimuli. J Neurosci 21(14):5304–5310
Morris JS, Frith CD, Perrett DI, Rowland D, Young AW, Calder AJ, Dolan RJ (1996) A differential neural response in the human amygdala to fearful and happy facial expressions. Nature 383(6603):812–815
Murty VP, Adcock RA (2014) Enriched encoding: reward motivation organizes cortical networks for hippocampal detection of unexpected events. Cereb Cortex 24(8):2160–2168
Murty VP, Labar KS, Adcock RA (2012) Threat of punishment motivates memory encoding via amygdala, not midbrain, interactions with the medial temporal lobe. J Neurosci 32(26):8969–8976
Nishijo H, Ono T, Nishino H (1988) Single neuron responses in amygdala of alert monkey during complex sensory stimulation with affective significance. J Neurosci 8(10):3570–3583
Nitschke JB, Sarinopoulos I, Mackiewicz KL, Schaefer HS, Davidson RJ (2006) Functional neuroanatomy of aversion and its anticipation. NeuroImage 29(1):106–116
O'Doherty JP, Deichmann R, Critchley HD, Dolan RJ (2002) Neural responses during anticipation of a primary taste reward. Neuron 33(5):815–826
O'Doherty JP, Dayan P, Friston K, Critchley H, Dolan RJ (2003) Temporal difference models and reward-related learning in the human brain. Neuron 38(2):329–337
Owen AM (1997) The functional organization of working memory processes within human lateral frontal cortex: the contribution of functional neuroimaging. Eur J Neurosci 9(7):1329–1339
Pessiglione M, Seymour B, Flandin G, Dolan RJ, Frith CD (2006) Dopamine-dependent prediction errors underpin reward-seeking behaviour in humans. Nature 442(7106):1042–1045
Phan KL, Sripada CS, Angstadt M, McCabe K (2010) Reputation for reciprocity engages the brain reward center. Proc Natl Acad Sci U S A 107(29):13099–13104
Phillips ML, Young AW, Senior C, Brammer M, Andrew C, Calder AJ, Bullmore ET, Perrett DI, Rowland D, Williams SC, Gray JA, David AS (1997) A specific neural substrate for perceiving facial expressions of disgust. Nature 389(6650):495–498
Pikkarainen M, Ronkko S, Savander V, Insausti R, Pitkanen A (1999) Projections from the lateral, basal, and accessory basal nuclei of the amygdala to the hippocampal formation in rat. J Comp Neurol 403(2):229–260
Ploghaus A, Tracey I, Gati JS, Clare S, Menon RS, Matthews PM, Rawlins JN (1999) Dissociating pain from its anticipation in the human brain. Science 284(5422):1979–1981
Pribram KH, Maclean PD (1953) Neuronographic analysis of medial and basal cerebral cortex. II. Monkey. J Neurophysiol 16(3):324–340
Reysen MB (2003) The effects of social pressure on group recall. Mem Cogn 31(8):1163–1168
Reysen MB, Adair SA (2008) Social processing improves recall performance. Psychon Bull Rev 15(1):197–201
Rilling J, Gutman D, Zeh T, Pagnoni G, Berns G, Kilts C (2002) A neural basis for social cooperation. Neuron 35(2):395–405
Rilling JK, Sanfey AG, Aronson JA, Nystrom LE, Cohen JD (2004a) The neural correlates of theory of mind within interpersonal interactions. NeuroImage 22(4):1694–1703
Rilling JK, Sanfey AG, Aronson JA, Nystrom LE, Cohen JD (2004b) Opposing BOLD responses to reciprocated and unreciprocated altruism in putative reward pathways. Neuroreport 15(16):2539–2543
Rolls ET, Sienkiewicz ZJ, Yaxley S (1989) Hunger modulates the responses to gustatory stimuli of single neurons in the caudolateral orbitofrontal cortex of the macaque monkey. Eur J Neurosci 1(1):53–60
Rolls ET, Yaxley S, Sienkiewicz ZJ (1990) Gustatory responses of single neurons in the caudolateral orbitofrontal cortex of the macaque monkey. J Neurophysiol 64(4):1055–1066
Rutecki PA, Grossman RG, Armstrong D, Irish-Loewen S (1989) Electrophysiological connections between the hippocampus and entorhinal cortex in patients with complex partial seizures. J Neurosurg 70(5):667–675

Samanez-Larkin GR, Gibbs SE, Khanna K, Nielsen L, Carstensen LL, Knutson B (2007) Anticipation of monetary gain but not loss in healthy older adults. Nat Neurosci 10(6):787–791
Sanghera MK, Rolls ET, Roper-Hall A (1979) Visual responses of neurons in the dorsolateral amygdala of the alert monkey. Exp Neurol 63(3):610–626
Schultz W (1998) Predictive reward signal of dopamine neurons. J Neurophysiol 80(1):1–27
Schultz W, Dayan P, Montague PR (1997) A neural substrate of prediction and reward. Science 275(5306):1593–1599
Schurz M, Radua J, Aichhorn M, Richlan F, Perner J (2014) Fractionating theory of mind: a meta-analysis of functional brain imaging studies. Neurosci Biobehav Rev 42:9–34
Scott TR, Yan J, Rolls ET (1995) Brain mechanisms of satiety and taste in macaques. Neurobiology 3(3-4):281–292
Seid-Fatemi A, Tobler PN (2015) Efficient learning mechanisms hold in the social domain and are implemented in the medial prefrontal cortex. Soc Cogn Affect Neurosci 10(5):735–743
Shigemune Y, Abe N, Suzuki M, Ueno A, Mori E, Tashiro M, Itoh M, Fujii T (2010) Effects of emotion and reward motivation on neural correlates of episodic memory encoding: a PET study. Neurosci Res 67(1):72–79
Shigemune Y, Tsukiura T, Kambara T, Kawashima R (2014) Remembering with gains and losses: effects of monetary reward and punishment on successful encoding activation of source memories. Cereb Cortex 24(5):1319–1331
Smith EE, Jonides J (1999) Storage and executive processes in the frontal lobes. Science 283(5408):1657–1661
Suzuki WA, Amaral DG (1994) Perirhinal and parahippocampal cortices of the macaque monkey: cortical afferents. J Comp Neurol 350(4):497–533
Tricomi E, Fiez JA (2012) Information content and reward processing in the human striatum during performance of a declarative memory task. Cogn Affect Behav Neurosci 12(2):361–372
Tsukiura T, Cabeza R (2008) Orbitofrontal and hippocampal contributions to memory for face-name associations: the rewarding power of a smile. Neuropsychologia 46(9):2310–2319
Tsukiura T, Cabeza R (2011) Remembering beauty: roles of orbitofrontal and hippocampal regions in successful memory encoding of attractive faces. NeuroImage 54(1):653–660
Tsukiura T, Shigemune Y, Nouchi R, Kambara T, Kawashima R (2013) Insular and hippocampal contributions to remembering people with an impression of bad personality. Soc Cogn Affect Neurosci 8(5):515–522
Ungless MA, Magill PJ, Bolam JP (2004) Uniform inhibition of dopamine neurons in the ventral tegmental area by aversive stimuli. Science 303(5666):2040–2042
Van den Bos W, Talwar A, McClure SM (2013) Neural correlates of reinforcement learning and social preferences in competitive bidding. J Neurosci 33(5):2137–2146
Van Overwalle F (2009) Social cognition and the brain: a meta-analysis. Hum Brain Mapp 30(3):829–858
Van Overwalle F, Baetens K (2009) Understanding others' actions and goals by mirror and mentalizing systems: a meta-analysis. NeuroImage 48(3):564–584
Votinov M, Pripfl J, Windischberger C, Sailer U, Lamm C (2015) Better you lose than I do: neural networks involved in winning and losing in a real time strictly competitive game. Sci Rep 5:11017
Wang Q, Luo S, Monterosso J, Zhang J, Fang X, Dong Q, Xue G (2014) Distributed value representation in the medial prefrontal cortex during intertemporal choices. J Neurosci 34(22):7522–7530
Wicker B, Keysers C, Plailly J, Royet JP, Gallese V, Rizzolatti G (2003) Both of us disgusted in my insula: the common neural basis of seeing and feeling disgust. Neuron 40(3):655–664
Wimmer GE, Shohamy D (2012) Preference by association: how memory mechanisms in the hippocampus bias decisions. Science 338(6104):270–273
Wimmer GE, Braun EK, Daw ND, Shohamy D (2014) Episodic memory encoding interferes with reward learning and decreases striatal prediction errors. J Neurosci 34(45):14901–14912

Wittmann BC, Schott BH, Guderian S, Frey JU, Heinze HJ, Duzel E (2005) Reward-related FMRI activation of dopaminergic midbrain is associated with enhanced hippocampus-dependent long-term memory formation. Neuron 45(3):459–467

Wittmann BC, Schiltz K, Boehler CN, Duzel E (2008) Mesolimbic interaction of emotional valence and reward improves memory formation. Neuropsychologia 46(4):1000–1008

Wittmann BC, Tan GC, Lisman JE, Dolan RJ, Duzel E (2013) DAT genotype modulates striatal processing and long-term memory for items associated with reward and punishment. Neuropsychologia 51(11):2184–2193

Wolosin SM, Zeithamova D, Preston AR (2012) Reward modulation of hippocampal subfield activation during successful associative encoding and retrieval. J Cogn Neurosci 24(7):1532–1547

Wolosin SM, Zeithamova D, Preston AR (2013) Distributed hippocampal patterns that discriminate reward context are associated with enhanced associative binding. J Exp Psychol Gen 142(4):1264–1276

Wrase J, Kahnt T, Schlagenhauf F, Beck A, Cohen MX, Knutson B, Heinz A (2007) Different neural systems adjust motor behavior in response to reward and punishment. NeuroImage 36(4):1253–1262

Chapter 7
The Self-Trait Evaluation Task: Exodus from the Cortical Midline Structure Dogma

Motoaki Sugiura

Abstract The self-trait evaluation (STE) task, typically a self-descriptiveness judgment on a personality trait adjective, was initially introduced into cognitive neuroscience to address the memory enhancement effect of the self-referential (SR) process. It has now become a standard paradigm in a variety of basic psychological and clinical fields relevant to the SR process. A decade of neuroimaging research using the STE task was, however, influenced by the cortical midline structure (CMS) dogma, which oversimplified the known involvement of the CMS in the SR process to an exclusive structure–function relationship. The meta-analysis included in this chapter demonstrates that activation outside the CMS also seems to be sensitive to various task, stimulus, and participant factors of the STE task as follows: the lateral prefrontal cortex is sensitive to mental disorders, the lateral social- and body-related regions are sensitive to self-relatedness, and the body-related and visual areas are sensitive to psychological adaptation and aging. Future exploration of such variant components of the SR process over the entire cortex is warranted to enrich our cognitive neuroscientific knowledge of this process, in addition to its related psychological and clinical phenomena.

Keywords Self • Self-reference • Self-evaluation • Social evaluation • Social value • fMRI • Cortical midline structures • Culture • Schizophrenia • Depression

---

M. Sugiura (✉)
Department of Human Brain Science, Institute of Development, Aging and Cancer, Tohoku University, 4-1 Seiryo-cho, Aoba-ku, Sendai 980-8575, Japan

International Research Institute of Disaster Science, Tohoku University, 6-6-4 Aoba, Aoba-ku, Sendai 980-8579, Japan
e-mail: sugiura@tohoku.ac.jp

T. Tsukiura, S. Umeda (eds.), *Memory in a Social Context*,
https://doi.org/10.1007/978-4-431-56591-8_7

7.1 Introduction

The self-trait evaluation (STE) task represents the intimate relationship between the cognitive neuroscientific study of memory and social cognition. This task was initially brought into functional neuroimaging studies to address an issue specific to memory research, and subsequently became a standard paradigm of social cognitive neuroscience in both basic and application studies. As of mid-2016, the task has been used in more than 100 studies. However, an integrated overview of the accumulated findings appears difficult to obtain. This chapter attempts to determine the apparent divergence in research concerning the STE task and proposes future directions for research.

7.1.1 The STE Task

The STE task is very simple. A subject is presented with a single personality trait adjective and judges whether or not the adjective is descriptive of him/herself (Fig. 7.1a). It is often coupled with one or a few control tasks that also require a judgment on a trait adjective, such as descriptive judgment on another person (Fig. 7.1b), judgment on whether or not the trait is socially preferred (Fig. 7.1c), or judgment on whether the adjective is printed with uppercase or lowercase letters (Fig. 7.1d).

The STE task was first introduced in behavioral experiments to demonstrate the self-referential (SR) effect on memory. It was shown that the adjectival words presented in the STE task were better remembered in the later retrieval test compared with the words presented in other control tasks, such as semantic, phonetic, and

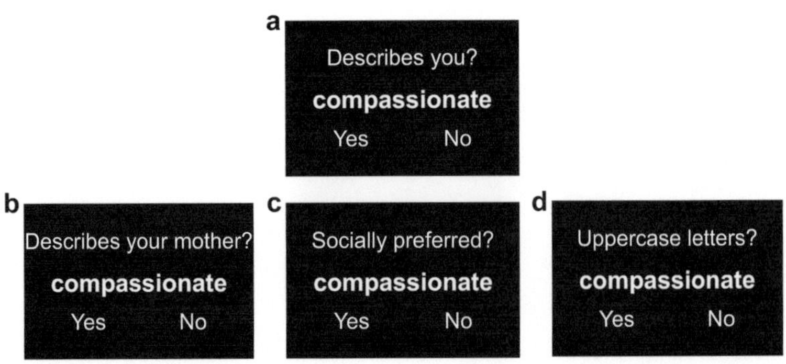

Fig. 7.1 Self-trait evaluation (STE) task. (**a**) Subjects are asked to judge whether or not the presented trait adjective is descriptive of themselves, or to rate the degree of descriptiveness. With only minimal modifications of the task instructions, various control tasks may be used, including (**b**) social–referential tasks on any familiar other, and different types of baseline tasks, such as (**c**) the semantic (or valence) task and (**d**) the structural (or perceptual) task

structural tasks (Rogers et al. 1977). That is, the SR process during descriptiveness judgment enhanced the encoding of the adjective word. Subsequent studies replicated this finding and added to the evidence highlighting the advantage of the SR process in memory enhancement over the reference process to other individuals (Symons and Johnson 1997). The self was considered "a well-developed and often-used construct that promotes elaboration and organization of encoded information." Therefore, it was natural for neuropsychologists and memory cognitive neuroscientists to adopt the STE task in functional neuroimaging.

The neural correlate of the SR process was first explored using positron emission tomography (PET) by comparing activation during the execution of the STE task and that during three control tasks (Craik et al. 1999). A subsequent study used functional magnetic resonance imaging (fMRI), which allowed the analysis of event-related neural responses, taking advantage of its temporal resolution (Kelley et al. 2002). Needless to say, the retrieval test was conducted after brain scanning and better retrieval of the words encoded in the STE task than in control tasks was demonstrated. The neural correlate of the SR process identified was then compared with the existing neuropsychological knowledge of memory encoding, which was at that time centered on the lateralized role of the prefrontal cortex. However, the STE task was too attractive to be monopolized by memory researchers. The self and the SR process were some of the core subjects in other domains of cognitive science, psychology, and psychiatry. Researchers started using the STE task to explore brain activation related to the SR process outside its effects on memory enhancement. They explored activation that overlapped with activation during different SR tasks (Kircher et al. 2002), activation that was affected by valence of the trait adjective (Fossati et al. 2003), and activation that depended on the types of source knowledge (Lieberman et al. 2004). They paid particular attention to the self-relevance of the other individuals referenced in the control task (Schmitz et al. 2004) and explored the SR process from the viewpoint of others (Ochsner et al. 2005). These studies pioneered the use of the STE task as a standard paradigm in basic social-cognitive neuroscience and its application in a variety of fields, including cultural, developmental, and patient studies (Fig. 7.2a).

The benefits of the STE task are abundant. In addition to its memory enhancement effects, it has elegance in its simplicity and in stimulus control and provides feasibility for context manipulation and performance monitoring. These advantages became obvious when the STE task was compared with other types of SR tasks, in which stimuli were qualitatively different between conditions (Johnson et al. 2002; Zysset et al. 2002, 2003) or in which tasks were open questions (Kjaer et al. 2002).

7.1.2 Background

Despite an increase in the use of the STE task, an overview of the studies that used this task may cause confusions, as the research frameworks of these studies were inconsistent. Some authors confidently attributed the neural basis of the SR process

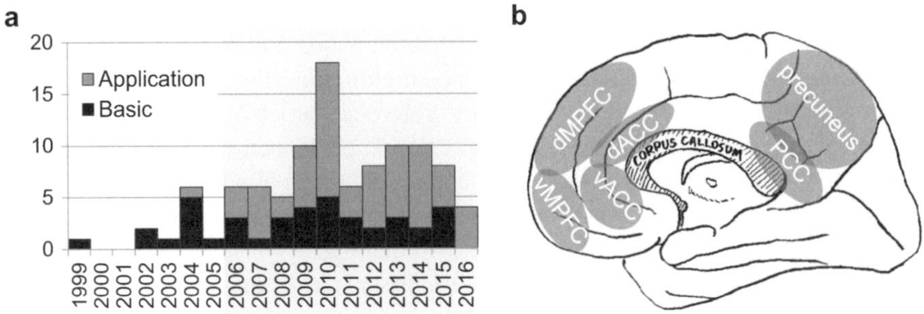

Fig. 7.2 Studies using the STE task and the cortical midline structures (CMS). (**a**) The number of published papers that reported CMS activation during the STE task up to mid-2016. In total, 102 studies (40 basic and 62 application studies) were identified (see the text on the method used for the meta-analysis for details). An increase in the number of studies coincided with the period of time during which the CMS dogma became influential (shaded background; see text for the details). (**b**) A typical anatomical delineation of the CMS. *vMPFC* ventromedial prefrontal cortex, *dMPFC* dorsomedial prefrontal cortex, *vACC* ventral part of the anterior cingulate cortex, *dACC* dorsal part of the anterior cingulate cortex, *PCC* posterior cingulate cortex

to a specific region or set of regions and used activation in these regions as a neural index. However, the ROIs were not consistent across studies. Others considered the neural basis of the SR process a matter of conceptual and neuroscientific scrutiny and searched for the effects of experimental manipulation on activation relevant to SR processing over the entire brain.

This divergence stems from two important factors. First, researchers have different conceptions regarding the nature of the SR process and its relationship with other cognitive processes. Some consider the SR process fundamental, forming a basis of our conscious awareness and high-level cognitive functioning (Northoff and Bermpohl 2004; Northoff et al. 2006; Qin and Northoff 2011). Others consider the SR process metacognitive, substantialized by the integration of multiple basic processes, including mnemonic, affective, and judgmental processes (Lieberman et al. 2004; Schmitz and Johnson 2006, 2007; Sugiura 2013; van der Meer et al. 2010). The former, *self-fundamentalists*, are primarily interested in identifying the unitary region or network dedicated to the SR process by focusing on common activation across a wide range of SR tasks, including the STE task. The latter, *self-deconstructors*, endeavor to attribute the regions activated during the STE task to different basic processes, referring to their activation profile across different experimental conditions. As the two positions on the SR process stem from different preverbal, irrebuttable, personal intuitions, their reconciliation may not be possible. In addition, many authors avoid clarifying their position, obscuring this critical distinction of the conceptual basis of the SR process in the literature.

Second, nearly all studies reported since 2006 are under the influence of a specific conceptual framework, which is here referred to as the "cortical midline structure (CMS) dogma." This dogma refers to the notion that the SR process is supported by the CMS, including the medial prefrontal cortex (MPFC), the anterior cingulate cortex (ACC), the posterior cingulate cortex (PCC), and the precuneus (Fig. 7.2b). This notion stems from an early self-fundamentalist view generated from the review of findings on the wide range of SR tasks at that time (Northoff and Bermpohl 2004; Northoff et al. 2006). Although self-fundamentalists have adopted an updated view (Qin and Northoff 2011), the CMS dogma remains influential in the field. In fact, the idea that the CMS plays a critical role in the SR process or the execution of the STE task is no doubt correct. The problematic aspect of the dogma is the notion that the role of the CMS in the SR process is "exclusive"; that is, brain structures other than the CMS have little involvement. This notion has rarely been explicitly addressed, but implicitly dominates the literature, as many studies reported after 2006 have focused their analyses on the CMS. This seems to have been self-propagated via two dynamics. The first dynamic is a symbolic–esthetic one, simply that the idea of "the core self at the center of the brain" is elegant and impressive. The second dynamic is a social–academic one, in that many authors had a practical reason for relying on the dogma. One reason is that the dogma justifies the use of ROIs for analysis. It allowed researchers to test activation data at a liberal statistical threshold, at which data would otherwise not reach statistical significance and be published. Strong statements for such justifications in the introduction of several papers may have contributed to the propagation of the CMS dogma.

The two factors described above are of course interrelated. The conception of the SR process strongly affects the stance that researchers adopt regarding the CMS dogma. The self-fundamentalists take the CMS dogma for granted, as the CMS is the very candidate area that they search for neural substrates that specifically support the SR process. The meaning of the CMS dogma may vary across self-deconstructors; the conservatives prefer to update the dogma by differentiating the roles of the regions within the CMS, whereas the radicals are eager to abolish the dogma by extending the neural substrate of the SR process outside the CMS. This chapter is written by a radical self-deconstructor.

In this chapter, the findings of basic and application functional neuroimaging studies in which the STE task was used are reviewed, paying particular attention to the influence of the CMS dogma. A meta-analysis is then performed to preview the world after the dominance of the CMS dogma.

7.2 Basic Research

Several lines of study have used the STE task to elucidate the neural basis of the SR process. Studies that particularly contributed to updating the neural model of the SR process were those that focused on the self-specificity of the neural response during the STE task. However, studies that examined the commonality of such neural responses across the SR task and that investigated the relationship between the SR process and resting state also contributed to the update. In addition, many studies that examined the modulation of the response by a stimulus factor and the perspective of the STE task enriched this update. Research into the memory enhancement effect of the STE task has been conducted independently of such updates.

7.2.1 Self-Specificity, Commonality, and the Update

To examine the self-specificity of activation during the STE task, it was contrasted with the judgment on close others, such as the subject's mother or close friends (Heatherton et al. 2006; Jenkins et al. 2008; Krienen et al. 2010; Moran et al. 2011; Schmitz et al. 2004; Vanderwal et al. 2008; Wang et al. 2012; Yaoi et al. 2009) rather than famous people, such as politicians, as used in early studies (Craik et al. 1999; Kelley et al. 2002). To the discouragement of those who supported the CMS dogma, differential activation was not stably observed in the CMS across studies. Although some authors identified differential activation in the anterior CMS (Heatherton et al. 2006), others presented its absence (Schmitz et al. 2004; Wang et al. 2012), its common involvement between the self and close-other conditions (Jenkins et al. 2008; Moran et al. 2011; Vanderwal et al. 2008; Yaoi et al. 2009), and its correlation with personal distance between the self and close-others (Krienen et al. 2010; Wang et al. 2012). Furthermore, differential activation was frequently observed outside of the CMS (Heatherton et al. 2006; Schmitz et al. 2004). A meta-analysis focusing on such a contrast revealed that only a small portion of the anterior CMS showed differential activation; most of the CMS was commonly activated for the self and others, and some posterior regions showed a larger activation during the other-referential condition (Araujo et al. 2013).

Conversely, the exclusive involvement of the CMS in the SR process has become doubted for the self-nonspecific contrast, that is, for comparisons between the self and baseline conditions, such as for social preferability, semantic, and structural decisions. Activation outside the CMS, including the anterior insula, lateral frontal, temporal, and parietal cortices, has become commonly recognized (Araujo et al. 2013; Modinos et al. 2009; Morin and Hamper 2012).

The findings described above stimulated the update of the SR process neural model in various ways. Together with the accumulated neuroimaging findings of other SR tasks, self-fundamentalists limited the neural basis of the SR process to the pregenual ACC (pACC), stating the involvement of other CMS for common substrates for the SR process and referencing others (Qin and Northoff 2011). Overlapping with this view, some conservative self-deconstructors proposed a view that the anterior and posterior components of the CMS are involved in the SR and the other-referential process respectively (Araujo et al. 2013; Murray et al. 2015). These two views, which inherited the simplicity and symbolic strength of the CMS dogma, have been supported by a commonality study; SR activation in the pACC was observed during the STE task and two other types of SR tasks (Jenkins and Mitchell 2011). Furthermore, several studies demonstrated the relationship between the SR process and resting state in terms of activation and connectivity of the CMS (Saverino et al. 2015; van Buuren et al. 2010), which is compatible with the fundamental roles of the SR process; the relationship differed between the anterior and posterior nodes (Murray et al. 2015; Qin and Northoff 2011; Whitfield-Gabrieli et al. 2011).

On the other hand, many radical self-deconstructors interpreted the findings of self-specificity in a different way. They share a view that the neural networks for the SR and other-referential processes are largely common and are composed not only of the CMS, but also of additional lateral cortices. They often divide the network into ventral and dorsal components; the former typically includes the ventral MPFC/ACC (vMPFC/vACC) as a key node and subserves affective or evaluative aspects of the SR process, whereas the latter includes the dorsal MPFC (dMPFC) and plays a role in cognitive aspects (Denny et al. 2012; Lieberman et al. 2004; Murray et al. 2012; Schmitz and Johnson 2006, 2007; M. Sugiura 2013; van der Meer et al. 2010). They attributed this network to the social evaluation process relevant to both the self- and other-referential processes rather than to the self-specific process across a variety of referential tasks. Supporting this position, a commonality study demonstrated that the CMS is involved in the STE task, but not in the agentic self (Powell et al. 2010).

Some authors assign particular emphasis to the self-specificity of the vMPFC and discuss its functional segregation (Flagan and Beer 2013) and its specific roles in self-related processing (D'Argembeau 2013). Although such views seem to agree with the self-fundamentalist view, they may actually harmonize better with the self-deconstructors' view, as they consider the SR process decomposable to more basic processes.

7.2.2 *Modulation by Stimulus Factors and Perspectives*

The enhanced activation by self-relatedness, including self-schematicity (e.g., athletes judging on the athlete-relevant trait rather than on the actor-relevant trait) and the personal importance of the trait adjectives, was investigated as an alternative approach to the SR process neural substrate. The effect was frequently identified in the vMPFC/ACC, lateral temporal cortex (LTC), and striatum, whereas other CMS, lateral cortical, and subcortical regions were also reported by various studies (D'Argembeau et al. 2012; Kircher et al. 2002; Lieberman et al. 2004; Rameson et al. 2010).

Many researchers were interested in the effect of the emotional valence of trait adjectives based on the intimate relationship between self-evaluation and affect. The accumulated results show a tendency toward more pronounced activation of positive stimuli, with limited consistency of the implicated regions or support for self-specificity of the positive-valence effect. A larger SR response for positive relative to negative trait words has typically been reported in studies using the self-baseline contrast in the CMS, insular, and lateral regions, although regions varied across studies (Fossati et al. 2003; Hoefler et al. 2015; Moran et al. 2006). The studies using the self–other contrast (Gutchess et al. 2007; Meffert et al. 2013; Yoshimura et al. 2009) failed to replicate this positive-valence effect, except for a recent study investigating 40 subjects, in which the effect was observed in the PCC and right supramarginal gyrus (SMG) (Kim et al. 2016). A study focusing on the amygdala using ROI analysis identified opposite valence effects in the left and right hemispheres (Yoshimura et al. 2009).

Several studies have examined the effect of the evaluation perspective by comparing the STE from an individual's own view (direct) and from the view of another person (reflected), based on an interest in the development of self-concept and self–other dissociation during the appraisal process. During the self-baseline contrast, greater activation for the reflected than for the direct STE task was reported in the LTC and precuneus predominantly in the left hemisphere (D'Argembeau et al. 2007; McAdams and Krawczyk 2014; Ochsner et al. 2005; Veroude et al. 2014). The results of the opposite contrast were inconsistent across studies. During the self–other contrast, greater activation for the reflected STE was reported only in one study at a liberal threshold in the dACC (D'Argembeau et al. 2007). One study also examined the effect of the time perspective by comparing the evaluation of the present and past self, and that of the present and past friend (D'Argembeau et al. 2008). A larger activation for the past self and the past friend overlapped with the findings on the reflected self, whereas greater activation for the present was self-specifically observed in the CMS and the left inferior frontal gyrus (IFG).

7.2.3 SR Effect on Memory

Memory researchers extended their investigation of the neural correlates of the SR effect on memory enhancement in two directions: the examination of retrieval activation and that of the effect of a subsequent memory on encoding activation. For the former, three studies consistently reported greater differential self-other activation in the temporoparietal junction (TPJ) during the retrieval of words that were encoded during the STE task (Lou et al. 2004; Pauly et al. 2013; Yaoi et al. 2015), whereas two studies reported the involvement of additional areas (Pauly et al. 2013; Yaoi et al. 2015). Two studies that examined the effect of trait valence on the SR effect during retrieval were in accordance with each other, in that negative traits had dominant effects over positive ones, and the effect appeared in the right dMPFC and occipital cortices (Fossati et al. 2004; Pauly et al. 2013). For the latter, the subsequent memory paradigm, a surprise recognition test was applied after scanning during the STE task and activation of the items that were remembered and those forgotten was compared. The paradigm consistently reported an association of the vMPFC/vACC and the medial temporal lobe (MTL; i.e., hippocampus and parahippocampal gyrus), whereas the location of the activation foci varied and other regions were also involved (Macrae et al. 2004; L. Zhu et al. 2012). Using a variation of the paradigm in which remembering of the encoding context (i.e., source memory) was asked about at the later surprise test, an association of differential activation in the vMPFC with the dissimilarity between the self and other was demonstrated (Benoit et al. 2010).

A recent study showed greater functional connectivity among the vMPFC, PCC, and hippocampus during the STE task than in the baseline task for encoding, but not during the retrieval phase (Morel et al. 2014).

7.2.4 *Summary*

The accumulated findings on the self-specificity of activation during the STE task have dissolved the classical CMS dogma and resulted in an update of the neural SR processing model in three directions. The self-fundamentalists focused on the pACC as the neural basis of the SR process. The conservative self-deconstructors assigned the anterior and posterior CMS to self- and other-referential processing. The radical self-deconstructors expanded the mechanism outside the CMS and initiated a discussion of its functional segregation. The findings on the modulation of SR activation by stimulus factors and the perspectives of the STE task supported all three views; the effect sometimes appeared in the pACC, often separately in the anterior and posterior CMS, and in the lateral cortices. The overall findings are not very consistent with each other and convergence of the findings in these research domains is limited. Research investigating the memory enhancement effect of the STE has yet to take into account these updates of the neural SR processing model.

7.3 Application Research

It is understandable that the CMS dogma was enthusiastically welcomed in many psychological and clinical fields in which the involvement of the SR process is considered critical in the phenomenon or disease. Because measuring the SR process was difficult because of its subjective nature, the possibility of measuring it objectively using functional neuroimaging and the SR task seemed revolutionary. Expecting that activation in the CMS could be used as an objective index of the SR process, the STE task was used to explore issues of psychology on culture, religion, development, mental health, and aging. In a similar vein, the task was used to demonstrate abnormalities in the SR process in the pathophysiology of various psychiatric diseases and to evaluate the effects of cognitive interventions. It is interesting to note the differences in the ways in which oversimplification of the CMS dogma has been accepted across fields.

7.3.1 *Psychological Applications*

Psychologists examined multiple aspects of the effects of culture on the neural SR process. This line of study was triggered by the demonstration of a lack of self-mother activation in the CMS in Chinese individuals (Zhang et al. 2006; Zhu et al. 2007), which was assumed to be associated with the less clear self-other distinction in those with a collectivist cultural orientation. Most research was subsequently developed using this individualism–collectivism framework. Although demonstrating differences between cultural groups required a large sample size (Ma et al.

2014a), several within-group studies succeeded in identifying the relationship between self-other activation in the CMS and the self-construal score, which is the index of the individualism–collectivism orientation (Chiao et al. 2009; Ray et al. 2010; Sul et al. 2012). The effects of the interaction of cultural orientation and the context of the STE task (Chiao et al. 2009, 2010), cultural motivation (Chen et al. 2013), and genetic polymorphism (Ma et al. 2014b) on self–other activation were also demonstrated. Moreover, an association between short- (Ng et al. 2010) and long-term (Chen et al. 2015) changes in cultural orientation and CMS activation was demonstrated. Several studies showed the effects of religious belief on CMS activation during the STE task (Han et al. 2008, 2010; Wu et al. 2010). However, it is worth noting that these apparently CMS-dominant findings were likely due to the ROI approach adopted in these studies, exemplifying self-propagation of the CMS dogma. In the results of a voxel-wise exploration, a CMS-dominant finding was rare (Ray et al. 2010), and most of the studies identified a significant culture-related effect in widespread cortical areas (Chen et al. 2015; Chiao et al. 2009, 2010; Ma et al. 2014a, b).

On the other hand, researchers studying developmental psychology were exceptionally open to the update of the SR process neural models. The application of the STE task to developing populations started from using CMS activation as some arbitrary index of the SR process. Self-other activation in the CMS was shown to be greater for children than for adults and attributed to "some" developmental processes in one study (Pfeifer et al. 2007), whereas it was associated with an index of developmental individuation in another study (Ray et al. 2009). Recent studies have expanded their focus outside the CMS, as interest shifted from self-specificity to the relationship between the self and the other, specifically considering the critical impact of views of others (particularly, those of peers) in the development of the self-concept. Comparing the direct and reflected STE tasks, adolescents were shown to use the dMPFC and TPJ during the direct STE task, which the adults used during the reflected STE task (Pfeifer et al. 2009). The ventral striatum was also activated in adolescents during the reflected STE task (Jankowski et al. 2014). In contrast, the "canonical" adult-typical SR process seems to be attributable to the vMPFC; a longitudinal (from 10 to 13 years) increase in self–other activation was associated with biological pubertal development (Pfeifer et al. 2013a), whereas a longitudinal decrease was observed in the hippocampus. No age difference was observed between late adolescents (18–19 years) and adults (23–25 years), but gender differences existed in the social–referential process regardless of the target (i.e., self or other) or perspective (Veroude et al. 2014).

Healthy adults typically show positive bias in the STE task, that is, they endorse more positive than negative traits. This phenomenon is a long-standing issue of the social psychology of the self and is discussed in association with psychological adaptation. Several studies have attempted to reveal the neural underpinnings of this phenomenon using the STE task, expecting the involvement of the CMS, particularly its anterior portion. One study compared responses to traits with broad meaning (i.e., easy to self-flatter) against those with narrow meaning (Beer et al. 2010). Another line of study examined activation related to STE task bias in response to social–evaluative threat using manipulated social feedback (Hoefler et al. 2015; Hughes and Beer 2013) or age–stereotype trait adjectives in elderly subjects (Colton et al. 2013). The results were not consistent; the neural correlates of positive bias decreased (Beer et al. 2010) or increased (Hughes and Beer 2013) vMPFC activation, increased dACC activation (Hoefler et al. 2015), increased posterior CMS activation (Colton et al. 2013), and increased (Hughes and Beer 2013) or decreased (Hoefler et al. 2015) activation in the regions outside the CMS. Other studies examined the correlation of activation during the STE task with individual degree of self-esteem (Yang et al. 2012, 2016) or life satisfaction (Kim et al. 2016). This line of research also resulted in divergent findings across studies.

Changes in neural SR processing during normal aging have been examined in various contexts, but the findings have hardly converged. Activation during the STE task has been shown to be age-invariant (Gutchess et al. 2007), or to decrease in the CMS (Ruby et al. 2009) or decrease in widespread cortical regions (Grady et al. 2012) with aging. The subsequent memory paradigm demonstrated an enigmatic age-related reverse of the SR memory effect in widespread cortical areas (Gutchess et al. 2010), but this finding was not shown in an experiment in which the other-referential task was eliminated (Gutchess et al. 2015). Studies addressing emotional regulation demonstrated age-related enhancement of the frontal response for positive traits relative to negative traits (Gutchess et al. 2007) and of various posterior cortices for age-stereotype traits (Colton et al. 2013). Moreover, a study reported decreased functional connectivity within the CMS (i.e., MPFC-PCC) after the STE task in elderly individuals, which was associated with a larger positivity bias (Saverino et al. 2015).

7.3.2 Clinical Applications

In schizophrenia, abnormal SR processing (activation in the CMS) is expected based on the assumption that the disturbed self-relevance attribution may underlie its symptoms, including various forms of hallucinations and delusions. Neuroimaging findings using the STE task are relatively convergent in showing decreased activation of the anterior CMS in patients (Bedford et al. 2012; Blackwood et al. 2004; Holt et al. 2011; Pankow et al. 2016; Tan et al. 2015) and adolescents with a genetic risk factor for schizophrenia (Schneider et al. 2012), whereas negative (Brent et al. 2014; Murphy et al. 2010; Pauly et al. 2014; van der Meer et al. 2013) and contradictory (Modinos et al. 2011) findings have also been reported. Studies investigating the posterior areas, including both the CMS and lateral areas, are highly inconsistent. Some studies have reported increased activation (Bedford et al. 2012; Blackwood et al. 2004; Holt et al. 2011), whereas others have reported decreased activation (Bedford et al. 2012; Brent et al. 2014; Murphy et al. 2010; Pauly et al. 2014; Schneider et al. 2012; van der Meer et al. 2013) in association with the disease.

The cognitive theory of depression postulates enhanced SR processing ("self-focus"), particularly on negative stimuli; it is characterized by an evaluative, analytical style (i.e., a "thinking about" experience), rather than a direct, intuitive, experiential awareness of an experience in the moment. Insights into the SR process of this disease may have driven the early adaptation of the self-deconstructors' view, which takes into account the detailed functional segregation of the CMS (Lemogne et al. 2012). Consistent with the assumed characteristics of depressive self-focus, increased response of the dMPFC in patients during the STE task, particularly to negative traits, has commonly been reported (Lemogne et al. 2009, 2010; Sarsam et al. 2013; Yoshimura et al. 2010, 2014). Among these studies, some have also reported an increased response of the dorsolateral prefrontal cortex (Lemogne et al. 2009, 2010) and vACC (Yoshimura et al. 2010, 2014). However, a recent study on nonmedicated adolescent patients did not replicate the involvement of the dMPFC, which raises the possibility that it is due to the effect of medications. The study instead reported an increased response of the PCC to positive traits (Bradley et al. 2016). A study that examined the effects of cognitive behavioral therapy reported a decrease in the patients' responses to negative traits in the dMPFC, vACC, and the left TPJ (Yoshimura et al. 2014).

In people with autism, self–mother activation did not significantly differ from the controls (Kennedy and Courchesne 2008), but self–Queen (Lombardo et al. 2010) and self–Harry Potter (Pfeifer et al. 2013b) activation was lower in the vMPFC and middle cingulate cortex (MCC), suggesting the importance of the choice of the "other" as a control in successfully delineating neural abnormalities in this disease. In contrast to the second study, which used an ROI approach, the last study explored the entire cortex and identified group differences in several additional regions, indicating the necessity of developing this line of research outside the CMS dogma.

There are many other diseases and clinical states that have been addressed using the STE task, but an integrated picture of these has yet to be provided. Abnormal activation associated with age-related pathological conditions, such as mild cognitive impairment (Ries et al. 2006, 2007) and Alzheimer's disease (Ruby et al. 2009), and its risk factors (Johnson et al. 2007), has been reported inconsistently across widespread cortical areas. The effects of meditation-based stress reduction on social anxiety disorder have been examined (Goldin et al. 2009) and compared with those of aerobic exercise (Goldin et al. 2012). Although an increased response of the precuneus/PCC and the right SMG for negative adjectives was consistent, changes in a variety of regions were inconsistent in the two studies. A recent study addressed anorexia nervosa and showed abnormalities in the precuneus, dACC, and the left lateral prefrontal cortex (LPFC) (McAdams and Krawczyk 2014). A common interest that is addressed across many pathological states using the STE task is decreased clinical insight, or anosognosia. However, findings on traumatic brain injury (Schmitz et al. 2006), mild cognitive impairment (Ries et al. 2007), and schizophrenia (Bedford et al. 2012; van der Meer et al. 2013) do not converge.

7.3.3 Summary

The CMS dogma has made a sizable contribution to the flourish of functional neuroimaging using the STE task in many psychological and clinical fields. The accumulated findings, most of which have yet to be integrated, show promise in contributing to the exploration of new neural and cognitive models of SR processing and its related psychological and clinical phenomena. On the other hand, some negative aspects of the CMS dogma, such as statistical bias due to the use of the ROI approach in many studies, resulting in skewed conclusions with an unbalanced focus on the CMS, remain influential in these research fields.

7.4 Exodus from the CMS Dogma

Although basic research shows that the CMS dogma has generally been phased out of cognitive neuroscience using the STE task, an alternative framework has yet to be described. In this section, the reasons for moving away from the CMS dogma and ways of doing this are outlined, preliminary support from a casual meta-analysis is obtained, and future steps are summarized.

7.4.1 Why and How to Move Away from the CMS Dogma

The "self" varies across individuals, cultures, and developmental stages. In understanding the SR neural process, its variant component is no less important than its invariant component. In personality, cultural, and developmental psychologies in particular, and in psychiatry, the variant rather than the invariant component of the self should be of interest. Selfish individuals and self-transcendent individuals may rely on different SR processes during the STE task. Individuals from collective and individualistic cultures may use different parts of the brain during the STE task. Children and adolescents, whose self-concept is under development, may resort to processes that adults no longer use. A range of self-related abnormalities in patients across diseases, and those within a single disease construct, may stem from a variety of neural SR processes. Thus, it is extremely important to remember that the CMS dogma has emerged from an attempt to identify the neural basis of the invariant component of the SR process by self-fundamentalists; it should not be expected that the variant component of the SR process should be solely restricted to the CMS.

Regarding the variant component of the SR process, it does not necessarily seem reasonable to consider the process self-specific or independent of other "general" cognitive processes. It is not strange to assume that the lateral prefrontal executive system plays a role in the STE task when there is a reason for scrutiny or effort in an evaluative decision. The STE task may require access to relevant experiences, which may result in activation of brain networks related to social cognition, bodily action, and sensation. The degree of such access to different domains of experience may vary among individuals depending on their life experience, sociocultural background, and behavioral prospects in the near and far future. However, the notion of the importance of non-CMS regions in the SR process thus far remains speculative and our knowledge concerning the roles of these regions in this process is very limited. Apart from some regions related to emotional (e.g., anterior insula) and mnemonic (i.e., MTL) processes, hypotheses regarding the roles of non-CMS regions are scarce or at best sporadic and have yet to be validated. Still, there are accumulated findings on non-CMS activation relevant to the STE task. Although it is possible that they are simply statistical noise due to a liberal statistical threshold and large number of studies, they may also represent the reflection of cognitive processes that would provide meaningful explanations for the variant component of the SR process. The next section contains a casual meta-analysis exploring activation relevant to the variant component of the SR process reported in studies using the STE task, and points to the latter interpretation being a possibility.

7.4.2 *Meta-analysis: Methods*

In this meta-analysis, as many contrasts as possible were included that examined some effect of task manipulation or individual difference on self–baseline or self–other activation. The contrast of self–baseline or self–other per se, which conceptually addresses the invariant component of the SR process and has already been addressed in a previous meta-analysis (Araujo et al. 2013), was not included. Only the contrasts that used an explicit judgment task on trait adjectives or verbal descriptions with some evaluative nuance were included; those that used implicit tasks or pictures were not. The results of voxel-wise analysis only were included.

First, a search in PubMed was conducted to identify studies that cited any of the following three "seed" papers: the first fMRI study using the STE task (Kelley et al. 2002) and two recent meta-analyses (Araujo et al. 2013; van der Meer et al. 2010). Then, the studies they cited were added. The contrasts were identified from these papers. As a result, 139 legitimate contrasts from 51 papers were identified. The remaining 51 of the 102 studies presented in Fig. 7.2a address only contrasts of self–baseline or self–other per se, or ROI analysis. This survey was not exhaustive; studies that did not cite any of the seed papers or are not registered in PubMed, and not cited by other relevant papers were not included. This may explain the apparent decrease in the number of papers in 2015 and 2016 (Fig. 7.2a).

Each contrast was labeled and coded based on the anatomical location of the activation peak. The contrast labels included the effects of stimulus factors (self-relatedness, emotional valence, context specificity, and familiarity in judgment) and perspective (person and time), the relevance to memory enhancement, development, aging, and psychological adaptation, and the effects of disease (schizophrenia, depression, autism, Alzheimer's disease, social anxiety disorder, and anorexia nervosa).

Regarding the results of cultural effects, culturally self-schematic judgment was labeled as self-related, and nonself-schematic judgment was labeled as less familiar judgment; cultural adaptation was attributed to psychological adaptation.

Anatomical coding targeted only cortical activation peaks; subcortical activation was not coded. The cortical areas were divided into 29 regions, not considering laterality; these regions were intended to be later classified into six groups. Two of these groups were the CMS and other regions related to emotion or memory. For the CMS, six classical members (the vMPFC, dMPFC, vACC, dACC, PCC, and precuneus) as well as the supplementary motor area and the MCC were included. Regions related to emotion (the anterior insula, orbitofrontal cortex, and amygdala) and mnemonic processes (i.e., MTL) are referred to here as "extended CMS" for brevity. The other four groups were composed of non-CMS structures. The first is the LPFC group, which may be associated with executive processes or cognitive control during the STE task and includes the IFG, middle frontal gyrus (MFG), superior frontal gyrus (SFG), and anterior prefrontal cortex (aPFC). The second is the lateral cortical areas relevant to social cognition, which include two major cortical components for social cognitive processes, namely, the TPJ and temporal pole (Frith and Frith 2003), as well as the LTC for ventral and dorsal sectors (vLTC and dLTC respectively) and the posteriorly adjacent angular gyrus. The third group is composed of the lateral areas related to body processing, which include the precentral gyrus, postcentral gyrus, SMG, posterior insula, and intraparietal sulcus extending to the adjacent superior parietal lobule. Finally, the last group is composed of the visual areas, including the medial occipital, lateral occipital, and posterior ventral temporal cortex.

Each contrast was classified by the anatomical group that dominated its activation in terms of occupancy of the regions with activation peaks (percentage of the region occupied). For example, if a contrast gave peaks in three of the four LPFC groups (e.g., the IFG, MFG, SFG, and aPFC), the occupancy of the LPFC group was 75%; if this occupancy was the highest among the six anatomical groups, this contrast was classified as LPFC-dominant. The contrasts with the highest occupancy in two or more groups were not assigned to any class. Then, for each class, the characteristics were examined in terms of the contrast labels that comprised the class. If two or more contrasts (from different studies) with a specific label were included and the contrast with the opposite label (i.e., positive versus negative effect) was negligible, the label was considered a characteristic of the class.

7.4.3 *Meta-analysis: Results and Discussion*

The results of the meta-analysis are summarized in Table 7.1. Although the CMS-dominant class included the largest number of contrasts (52 contrasts), a considerable number of contrasts were classified into a non-CMS group; the LPFC-, social cognition-, body-processing-, and visual-dominant classes were composed of 16, 13, 20, and 16 contrasts respectively, which exceeded the number of members in the

Table 7.1 Variation of activation during the self-trait evaluation (*STE*) task

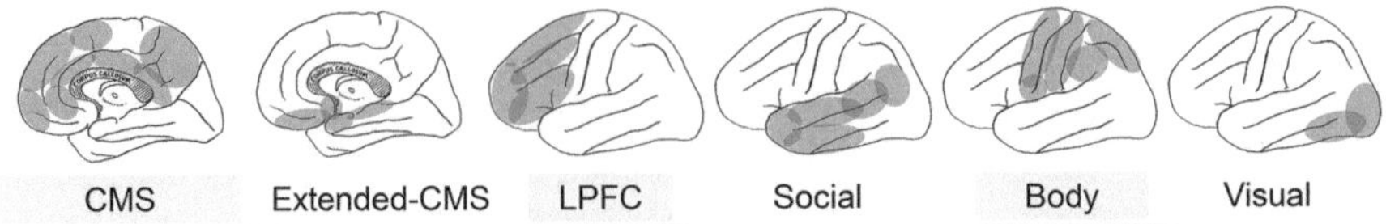

Paper	label	vMPFC	dMPFC	vACC	dACC	SMA	MCC	PCC	precuneus	aIns	OFC	amygdala	MTL	IFG	MFG	SFG	aPFC	TP	vLTC	dLTC	TPJ	AG	preCG	postCG	SMG	IPS/SPL	pIns	mOC	lOC	pVTC
Pfeifer07jocn19_1323	Yng+		●	●					●																					
Pfeifer09ChildDev80_1016	Yng+	●	●		●																									
Bedford12BMCPsci12_106	Scz-		●					●																						
Bedford12BMCPsci12_106	Scz+			●					●																					
Holt11BiolPsci69_415	Scz+						●	●																						
McAdams14scan9_12	Ref+							●	●																					
Ray09ChildDev80_1232	Mem+	●		●	●	●				●				●																
Ray10scan5_318	Slf+		●					●																						
Ruby09NrbiolAging30_1637	Age-		●						●																					
Schneider12DevCgNsci2_277	Scz-			●	●																									
Schneider12DevCgNsci2_277	Scz-	●		●																										
Schneider12DevCgNsci2_277	Scz-				●		●																							
Pfeifer09ChildDev80_1016	Yng+	●	●		●																●									
Bedford12BMCPsci12_106	Scz-		●																											
Bedford12BMCPsci12_106	Scz-		●																											
Blackwood04PscoMed34_591	Scz-				●																									
Blackwood04PscoMed34_591	Scz+							●																						
D'Argembeau07jocn19_935	Ref+				●																									
Holt11BiolPsci69_415	Scz-	●																												
Lieberman04JPrsScPsco87_421	Slf-	●																												
Lieberman04JPrsScPsco87_421	Slf+	●																												
Lieberman04JPrsScPsco87_421	Slf-	●																												
Ma14scan9_73	Slf-			●																										
Ma14scan9_73	Slf+			●																										
Ma14scan9_73	Slf-			●																										
Macrae04cc14_647	Slf+	●																												
Macrae04cc14_647	Slf-				●																									
McAdams14scan9_12	Ann-								●																					
McAdams14scan9_12	Ann+				●																									
McAdams14scan9_12	Ann-				●																									
McAdams14scan9_12	Ref+								●																					
McAdams14scan9_12	Ref+								●																					
McAdams14scan9_12	Ref-				●																									
Moran06jocn18_1	Val+			●																										
Moran06jocn18_1	Slf+			●	●	●					●			●																
Moran09SocNsci4_197	Slf+	●																												
Ochsner05ni28_797	Ref-							●																						
Pfeifer07jocn19_1323	Yng-								●																					
Pfeifer09ChildDev80_1016	Yng+	●	●		●												●													
Pfeifer09ChildDev80_1016	Fem+								●																					
Pfeifer13jns33_7415	Yng-	●																												
Pfeifer13JAutismDevDisord43_272	Asd-	●			●		●			●					●								●							
Ruby09NrbiolAging30_1637	Alz-		●																											
Sarsam13POne8_e78844	Dep+		●																											
Tan15PlosOne10_e0138737	Scz-				●																									
Veroude14scan9_513	Fem-							●																						
Yang12nsy50_1267	Adp+				●																									
Hoefler15POne10_e136027	Val+				●			●													●					●				
Schmitz06nsy44_762	Tbi+			●					●									●												
Yoshimura14scan9_487	Dep-		●	●																	●									
McAdams14scan9_12	Ref+					●		●	●											●								●		
D'Argembeau12cc22_659	Slf+	●	●		●		●	●			●		●			●						●			●	●	●			
Ma14scan9_1360	Slf-		●							●			●		●				●											
Pfeifer13jns33_7415	Yng+												●																	
Zhu12Hippo22_1540	Mem+												●																	
Hughes13jocn25_613	Adp+	●							●	●	●	●	●	●	●			●	●		●		●	●	●	●			●	
Hoefler15POne10_e136027	Adp-		●										●																	
Lieberman04JPrsScPsco87_421	Slf-		●										●																	
Ma14cc24_2421	Val-				●					●																				
Colton13FrnHumNsci7_537	Age+						●		●	●		●								●	●				●		●		●	
Ochsner05ni28_797	Ref+										●		●													●	●	●		
Lieberman04JPrsScPsco87_421	Slf+	●										●								●						●				
Pauly14scan9_1779	Scz-									●								●												
Lieberman04JPrsScPsco87_421	Slf+	●						●				●								●										

(continued)

Table 7.1 (continued)

Paper	label	vMPFC	dMPFC	vACC	dACC	SMA	MCC	PCC	precuneus	aIns	OFC	amygdala	MTL	IFG	MFG	SFG	aPFC	TP	vLTC	dLTC	TPJ	AG	preCG	postCG	SMG	IPS/SPL	pIns	mOC	lOC	pVTC
Goldin12FrnHumNsci6_295	Sad-	●							●					●	●					●					●					
Lieberman04JPrsScPsco87_421	Slf-														●															
McAdams14scan9_12	Ann+														●															
Pfeifer13JAutismDevDisord43_272	Asd+									●				●	●					●			●							
Schmitz06nsy44_762	Tbi-															●														
Yang16SRep6_20274	Adp+																●													
Bedford12BMCPsci12_106	Scz+		●												●															
Benoit10ni50_1340	Slf+	●													●															
Kim16PlosOne11_e0149554	Adp-		●													●														
Lemogne09scan4_305	Dep+		●												●															
Ma14scan9_73	Slf+								●						●															
Ma14scan9_1360	Slf-		●													●														
McAdams14scan9_12	Ref-				●										●															
Chiao09hbm30_2813	Ctx+		●		●			●		●				●	●				●	●								●		
Ma14scan9_1360	Slf+														●						●					●				
Ma14scan9_1360	Slf+								●						●						●				●					
D'Argembeau08scan3_244	Now+		●					●						●																
Goldin09JCgnPscoThr23_242	Sad+	●	●											●																
Goldin12FrnHumNsci6_295	Sad-	●	●								●	●		●		●		●												
Gutchess07SocNsci2_117	Age+		●			●									●			●												
Lieberman04JPrsScPsco87_421	Slf-												●		●															
Lieberman04JPrsScPsco87_421	Slf+				●								●		●															
Ma14cc24_2421	Val-		●		●					●				●																
Ma14cc24_2421	Val-		●		●					●				●																
Macrae04cc14_647	Mem+	●											●				●													
Rameson10ni50_701	Slf+	●	●	●			●	●	●			●	●	●	●	●			●	●									●	
Ochsner05ni28_797	Ref+																	●	●											
Fossati03AmJPsychi160_1938	Val+																			●	●	●					●		●	
Pauly14scan9_1779	Scz-																		●											
Yang16SRep6_20274	Adp+														●				●	●										
Zhu12Hippo22_1540	Mem+		●						●		●							●	●											
D'Argembeau12cc22_659	Fam+	●	●				●		●		●					●			●	●	●			●	●					
Lieberman04JPrsScPsco87_421	Slf+		●			●	●		●				●	●	●			●	●		●					●				
Ma14scan9_73	Slf+								●												●									
Pfeifer09ChildDev80_1016	Yng-							●													●									
Veroude14scan9_513	Ref+								●												●									
Veroude14scan9_513	Fem-								●												●									
Murphy10SczRes116_252	Scz-								●												●	●						●		
Zhu12Hippo22_1540	Mem+		●								●			●				●	●									●		
Chiao09hbm30_2813	Ctx-																						●	●						
Meffert13FrntHumNsci7_46	Val-																							●			●			
McAdams14scan9_12	Ref-		●																					●		●				
Meffert13FrntHumNsci7_46	Val+						●																	●			●			
Ochsner05ni28_797	Ref-								●																	●	●			
Bedford12BMCPsci12_106	Scz-																								●					
Chen15CultBrain3_39	Adp+	●					●		●										●		●		●	●	●				●	
Goldin12FrnHumNsci6_295	Sad-																										●			
Kim16PlosOne11_e0149554	Adp+																								●					
Ma14scan9_1360	Slf+																								●					
Pauly14scan9_1779	Scz+																						●							
Ruby09NrbiolAging30_1637	Alz+																									●				
Kircher02npla40_683	Slf+							●	●					●										●		●				
Moran06jocn18_1	Slf+	●						●																●		●				
Gutchess10nsy48_211	Age+		●		●				●					●	●					●				●	●	●		●		
Bedford12BMCPsci12_106	Scz+			●																					●					
Goldin09JCgnPscoThr23_242	Sad-								●																●					
Gutchess07SocNsci2_117	Age+					●																		●						
Kim16PlosOne11_e0149554	Val+							●																	●					
D'Argembeau12cc22_659	Slf+	●	●						●							●					●				●		●		●	
Chen15CultBrain3_39	Adp+																												●	●
Colton13FrnHumNsci7_537	Age+							●	●					●						●		●						●	●	●
Chiao09hbm30_2813	Fam+			●				●					●						●										●	●
Bedford12BMCPsci12_106	Scz-																												●	
Bradley16DevCogNsci19_87	Dep-																												●	
Veroude14scan9_513	Ref-																											●		
Zhu12Hippo22_1540	Mem+																													●
Colton13FrnHumNsci7_537	Age+							●	●				●							●	●								●	●
Bradley16DevCogNsci19_87	Dep+							●																				●		
Ma14cc24_2421	Val+								●																			●		
Chiao10jocn22_1	Fam-				●		●							●		●				●	●				●				●	●
Yang16SRep6_20274	Adp+								●				●	●	●					●	●							●	●	
Chiao09hbm30_2813	Fam+												●							●									●	
Chiao10jocn22_1	Fam+		●						●												●									●
Goldin12FrnHumNsci6_295	Sad+		●	●										●												●		●		
Zhu12Hippo22_1540	Slf+			●									●															●		

This meta-analysis summarizes the cortical regions where effects of task manipulation or individual difference on self-baseline or self-other activation were identified in a voxel-wise search of 139 contrasts. For each contrast, the source paper ([family name of the first author] [last two digits of the published year] [abbreviated journal name] [volume number] "_" [first page]), label (*Slf* sv

(continued)

Table 7.1 (continued)

elf-relatedness, *Val* emotional valence (positive), *Ctx* context specificity, *Fam* familiarity in judgment, *Ref* reflected (versus direct) judgment, *Now* present (versus past) self, *Mem* memory enhancement, *Yng* young subjects (versus adults), *Age* elderly subjects (versus young adults), *Adp* psychologically adaptive response or population, *Scz* schizophrenia, *Dep* depression, *Asd* autism spectrum disorder, *Alz* Alzheimer's disease, *Sad* social anxiety disorder, *Ann* anorexia nervosa; + associated increase, − associated decrease), and location of the activation peaks (*closed circle*) are shown. The locations were divided into six groups, each composed of 3–8 cortical regions (the two hemispheres were combined). The CMS group included the ventromedial and dorsomedial prefrontal cortices (*vMPFC* and *dMPFC* respectively), the ventral and dorsal parts of the anterior cingulate cortices (*vACC* and *dACC* respectively), the supplementary motor area (*SMA*), the middle and posterior cingulate cortices (*PCC* and *MCC* respectively), and the precuneus. The extended CMS included the anterior insula (*aIns*), orbitofrontal cortex (*OFC*), amygdala, and medial temporal lobe (*MTL*), including the hippocampus and parahippocampal gyrus. The lateral prefrontal cortex (*LPFC*) group included the inferior, middle, and superior frontal gyri (*IFG*, *MFG*, and *SFG* respectively) and the anterior prefrontal cortex (*aPFC*). The lateral cortical area group relevant to social cognition included the temporal pole (*TP*), ventral and dorsal sectors of the lateral temporal cortex (*vLTC* and *dLTC* respectively), temporoparietal junction (*TPJ*), and angular gyrus (*AG*). The group that was composed of the lateral areas related to body processing included the precentral gyrus (*preCG*), postcentral gyrus (*postCG*), supramarginal gyrus (*SMG*), intraparietal sulcus extending to the adjacent superior parietal lobule (*IPS/SPL*), and the posterior insula (*pIns*). The visual group was composed of the occipital and temporal areas including the medial occipital (*mOC*), lateral occipital (*lOC*), and posterior ventral temporal cortices (*pVTC*). The contrasts were classified using the anatomical group that dominated activation and ordered so that those in the same class were together (encompassed by a *thick gray frame*). Classification was based on occupancy of the regions with activation peaks (percentage of region occupied); those with the highest occupancy in two or more groups were not assigned to a class

extended-CMS-dominant class (11 contrasts). Thus, these results clearly show that activation of non-CMS regions during the STE task is by no means uncommon for the variant component of the SR process.

In the LPFC-dominant class, activation peaks were most commonly observed in the MFG. The contrasts distinctive to this class were associated with mental disorders, such as depression (Lemogne et al. 2009), schizophrenia (Bedford et al. 2012), autism (Pfeifer et al. 2013b), and anorexia nervosa (McAdams and Krawczyk 2014). It appears reasonable to assume that abnormal SR processes in these mental disorders are related to the increased demand of the prefrontal executive function for scrutiny or effort during the STE task. Although two contrasts can be labeled as oppositely associated with a disease, one represents increased activation after mindfulness-based stress reduction in patients with social anxiety disorder (Goldin et al. 2012) and the other represents a decrease in patients with traumatic brain injury (Schmitz et al. 2006).

Increased activation associated with self-relatedness was characteristic of both the social-cognition- (Lieberman et al. 2004; Ma et al. 2014a) and body-processing-dominant (D'Argembeau et al. 2012; Kircher et al. 2002; Ma et al. 2014b; Moran et al. 2006) classes. This may be explained by the association of an increased sense of self-relatedness with successful access to social and body-related experiences respectively. The distinction of characteristics between the two classes is made by the response to the reflected STE; the social cognition areas have increased activation during the reflected STE task (Ochsner et al. 2005; Veroude et al. 2014), whereas the body-processing areas have increased activation during the direct STE task (McAdams and Krawczyk 2014; Ochsner et al. 2005). This distinction is reasonable as the reflected STE task is highly dependent on social experiences and less dependent on direct body-related experiences.

The social cognition-dominant class was also characterized by a reduced activation in schizophrenia (Murphy et al. 2010; Pauly et al. 2014). Because both contrasts included are relative to a baseline condition (i.e., self–baseline), this finding seems to reflect impairment of the social–referential rather than the SR process in this disease (Murphy et al. 2010). The body-processing-dominant class was characterized by an increased activation associated with psychological adaptation (Chen et al. 2015; Kim et al. 2016) and aging (Gutchess et al. 2007, 2010). Although the association of the two labels is reasonable in terms of increased positive bias in elderly individuals (Saverino et al. 2015), their association with body-related processes remains unknown. An increase in activation associated with psychological adaptation was also seen in the visual-dominant class (Chen et al. 2015; Yang et al. 2016). One possible explanation for this finding may be the relationship between psychological adaptation and an attitude of being aware of physical experiences (i.e., bodily state and external environment) during the STE task, which may be the opposite of the depressive self-focus.

In summary, there is an abundance of evidence demonstrating that the variant component of the SR process involves regions outside of the CMS, including areas in the entire cerebral cortex. These findings seem to reflect the involvement of the general cognitive processes inherent in these regions; that is, they are not statistical noise, but provide meaningful explanations for different types of modulation or variation of the SR process.

7.4.4 Future Directions

The results of the meta-analysis presented herein support the exploration of the entire cortex in search of neural correlates underlying the SR process, rather than focus being placed on only the CMS. This is opposite to the approach that was taken in many past studies that used the STE task under the influence of the CMS dogma. It may also be interesting to address the preliminary hypothesis of non-CMS regions raised by the current meta-analysis, such as the relationship between mental disorders and the LPFC, the association of self-relatedness with regions involved in social cognition and body processing, and the roles of the body-related and visual areas in psychological adaptation and aging.

Such an expansion of the scope of the SR process from the CMS to the entire cortex should accompany conceptual clarification of the different scopes of research areas that employ the STE task. Focusing on the CMS may be reasonable when research is driven by a motivation to identify the core process of the STE task, which is a self-specific and invariant component of the SR process. It should be the very process that self-fundamentalists are seeking. The interest in the variation of the STE task across different materials, contexts, and individuals should include a search of the entire brain to identify components of the SR process that may be relatively general, self-nonspecific cognitive processes. This approach largely fits with the views of self-deconstructors. These different views of the SR process can also be explained in terms of the original memory research studies that introduced the STE task into the field. The core process addressed in the self-fundamentalist view may correspond to the process that promotes elaboration and organization of encoded information into the self-construct. On the other hand, the self-construct is substantialized by organized information in addition to its self-organizing dynamics formed through life experience, which varies among individuals and is addressed in the self-deconstructor view.

Future research development addressing both views may be more fruitful if new perspectives are included. First, more attention should be placed on the motivations and subjective experiences of researchers attempting to identify the core SR process. Insufficient clarification of the differences in the basic views of the SR process in this field may be rooted in a lack of awareness at this level of individual differences among researchers. Besides the affective nuance of the core SR process conferred by the pACC, as argued by mainstream self-fundamentalists, some researchers underscore interoceptive experience supported in the anterior insula (Modinos et al. 2009, 2011) or inner speech in the left IFG (Morin and Hamper 2012). It is assumed that different subjective experiences underlie the different views of the core SR process. More explicit communication of such individual differences in intuitive conception or subjective experience may enhance the mutual understanding of motivations and constructive discussion. Second, an associationist perspective may help the conceptual development of the SR process model in an evolutionary context, such as determining the self-organization dynamics of the self-construct and its adaptive function. An example of such conception is a model that states that the self represents the functioning of forward prediction; this ability is afforded by the internal schema that represents the association between behavioral output and consequential input learned through life experience (Sugiura 2013). The model assumes that the schema is separately represented for physical, inter-personal, and social–evaluative domains, in body-related regions, social–cognitive regions, and the CMS respectively, which is consistent with the results of the meta-analysis presented herein. Furthermore, a cognitive model of belief formation and maintenance may speak to the adaptive nature of the SR process, combining the self-organizing nature of the self-construct and its adaptive modulation primarily by the prefrontal cortices (Sugiura et al. 2015).

Although this chapter is intended to highlight the negative influence of the CMS dogma in cognitive neuroscience involving the SR process, it is by no means meant to criticize specific studies or researchers involved in investigating the roles of the CMS in the SR process. In fact, there is no doubt that many regions in the CMS play important roles in the SR process, particularly in its core invariant component. The CMS dogma specifically denotes the simplistic assumption that the SR process is "exclusively" supported by the CMS, and that non-CMS regions play little role. This chapter is designed to highlight the notion that the CMS dogma is a "ghost" that emerged in this research field through social–academic dynamics.

Several technical considerations are necessary for the use of STE tasks in future studies. First, it is important to underline that self–baseline and self–other activations are different (Araujo et al. 2013), as previous application studies have often omitted this distinction. Self–baseline activation may reflect self-nonspecific social–referential processes and may provide a better opportunity to obtain significant results, whereas self–other activation has advantages of self-specificity at a higher risk for producing no significant findings. Second, further attention should be given to the selection of stimuli (i.e., trait words). It has been shown that valence, self-relatedness, and context dependence strongly affect activation during the STE task. Third, there is evidence that an assortment of control tasks affect activation during the STE task (Gutchess et al. 2010, 2015). Finally, the temporal or sequential characteristics of the experimental design seem to have an impact on the results. For example, subjects' strategies in approaching the STE task may differ between the block design, in which different adjectives are sequentially presented at a high pitch for a fixed reference target, and the event-related design, in which events under different conditions are randomly presented with a jittered inter-trial interval. Different results may also be explained by different levels of resting-state activity (Lemogne et al. 2012), which may be affected by subject factors (e.g., personality, disease) and by task manipulation itself (Saverino et al. 2015).

7.5 Conclusion

Since demonstration of the involvement of the CMS in the SR process in multiple domains, its simplistic notion, the CMS dogma, has been adopted in neuroimaging studies that use the STE task. The implicit assumption that the SR process is exclusively supported by the CMS may have motivated the application of this line of neuroimaging research in many psychological and clinical fields, but the CMS dogma seemed to hamper the exploration of potentially meaningful observations outside the CMS. The meta-analysis included in this chapter demonstrated the effects of various tasks/stimulus factors and individual difference in multiple domains, including cultural, personality, developmental, and pathological domains, on activation during the STE task outside the CMS. For example, activation of the LPFC was associated with mental disorders, that of the lateral cortical regions responsible for social cognition and body processing with self-relatedness, and that of body-related and visual areas with psychological adaptation and aging. For the comprehensive incorporation of such unexplored relationships between the self and non-CMS regions within the existing CMS-centered framework, future studies are warranted to broaden the conceptual scope of the SR process from its invariant to variant component and to update the STE task accordingly.

Acknowledgment The preparation of this manuscript was supported by KAKENHI 16H01873 from JSPS.

References

Araujo HF, Kaplan J, Damasio A (2013) Cortical midline structures and autobiographical-self processes: an activation-likelihood estimation meta-analysis. Front Hum Neurosci 7:548

Bedford NJ, Surguladze S, Giampietro V, Brammer MJ, David AS (2012) Self-evaluation in schizophrenia: an fMRI study with implications for the understanding of insight. BMC Psychiatry 12:106

Beer JS, Lombardo MV, Bhanji JP (2010) Roles of medial prefrontal cortex and orbitofrontal cortex in self-evaluation. J Cogn Neurosci 22(9):2108–2119

Benoit RG, Gilbert SJ, Volle E, Burgess PW (2010) When I think about me and simulate you: medial rostral prefrontal cortex and self-referential processes. NeuroImage 50(3):1340–1349

Blackwood NJ, Bentall RP, Ffytche DH, Simmons A, Murray RM, Howard RJ (2004) Persecutory delusions and the determination of self-relevance: an fMRI investigation. Psychol Med 34(4):591–596

Bradley KA, Colcombe S, Henderson SE, Alonso CM, Milham MP, Gabbay V (2016) Neural correlates of self-perceptions in adolescents with major depressive disorder. Dev Cogn Neurosci 19:87–97

Brent BK, Coombs G, Keshavan MS, Seidman LJ, Moran JM, Holt DJ (2014) Subclinical delusional thinking predicts lateral temporal cortex responses during social reflection. Soc Cogn Affect Neurosci 9(3):273–282

Chen P-HA, Wagner DD, Kelley WM, Powers KE, Heatherton TF (2013) Medial prefrontal cortex differentiates self from mother in Chinese: evidence from self-motivated immigrants. Cult Brain 1(1):3–15

Chen PA, Wagner DD, Kelley WM, Heatherton TF (2015) Activity in cortical midline structures is modulated by self-construal changes during acculturation. Cult Brain 3(1):39–52

Chiao JY, Harada T, Komeda H, Li Z, Mano Y, Saito D, Parrish TB, Sadato N, Iidaka T (2009) Neural basis of individualistic and collectivistic views of self. Hum Brain Mapp 30(9):2813–2820

Chiao JY, Harada T, Komeda H, Li Z, Mano Y, Saito D, Parrish TB, Sadato N, Iidaka T (2010) Dynamic cultural influences on neural representations of the self. J Cogn Neurosci 22(1):1–11

Colton G, Leshikar ED, Gutchess AH (2013) Age differences in neural response to stereotype threat and resiliency for self-referenced information. Front Hum Neurosci 7:537

Craik FIM, Moroz TM, Moscovitch M, Stuss DT, Winocur G, Tulving E, Kapur S (1999) In search of the self: a positron emission tomography study. Psychol Sci 10(1):26–34

D'Argembeau A (2013) On the role of the ventromedial prefrontal cortex in self-processing: the valuation hypothesis. Front Hum Neurosci 7:372

D'Argembeau A, Ruby P, Collette F, Degueldre C, Balteau E, Luxen A, Maquet P, Salmon E (2007) Distinct regions of the medial prefrontal cortex are associated with self-referential processing and perspective taking. J Cogn Neurosci 19(6):935–944

D'Argembeau A, Feyers D, Majerus S, Collette F, Van der Linden M, Maquet P, Salmon E (2008) Self-reflection across time: cortical midline structures differentiate between present and past selves. Soc Cogn Affect Neurosci 3(3):244–252

D'Argembeau A, Jedidi H, Balteau E, Bahri M, Phillips C, Salmon E (2012) Valuing one's self: medial prefrontal involvement in epistemic and emotive investments in self-views. Cereb Cortex 22(3):659–667

Denny BT, Kober H, Wager TD, Ochsner KN (2012) A meta-analysis of functional neuroimaging studies of self- and other judgments reveals a spatial gradient for mentalizing in medial prefrontal cortex. J Cogn Neurosci 24(8):1742–1752

Flagan T, Beer JS (2013) Three ways in which midline regions contribute to self-evaluation. Front Hum Neurosci 7:450

Fossati P, Hevenor SJ, Graham SJ, Grady C, Keightley ML, Craik F, Mayberg H (2003) In search of the emotional self: an fMRI study using positive and negative emotional words. Am J Psychiatry 160(11):1938–1945

Fossati P, Hevenor SJ, Lepage M, Graham SJ, Grady C, Keightley ML, Craik F, Mayberg H (2004) Distributed self in episodic memory: neural correlates of successful retrieval of self-encoded positive and negative personality traits. NeuroImage 22(4):1596–1604

Frith U, Frith CD (2003) Development and neurophysiology of mentalizing. Phil Trans R Soc Lond B 358: 459–473

Goldin P, Ramel W, Gross J (2009) Mindfulness meditation training and self-referential processing in social anxiety disorder: behavioral and neural effects. J Cogn Psychother 23(3):242–257

Goldin P, Ziv M, Jazaieri H, Gross JJ (2012) Randomized controlled trial of mindfulness-based stress reduction versus aerobic exercise: effects on the self-referential brain network in social anxiety disorder. Front Hum Neurosci 6:295

Grady CL, Grigg O, Ng C (2012) Age differences in default and reward networks during processing of personally relevant information. Neuropsychologia 50(7):1682–1697

Gutchess AH, Kensinger EA, Schacter DL (2007) Aging, self-referencing, and medial prefrontal cortex. Soc Neurosci 2(2):117–133

Gutchess AH, Kensinger EA, Schacter DL (2010) Functional neuroimaging of self-referential encoding with age. Neuropsychologia 48(1):211–219

Gutchess AH, Sokal R, Coleman JA, Gotthilf G, Grewal L, Rosa N (2015) Age differences in self-referencing: evidence for common and distinct encoding strategies. Brain Res 1612:118–127

Han S, Mao L, Gu X, Zhu Y, Ge J, Ma Y (2008) Neural consequences of religious belief on self-referential processing. Soc Neurosci 3(1):1–15

Han S, Gu X, Mao L, Ge J, Wang G, Ma Y (2010) Neural substrates of self-referential processing in Chinese Buddhists. Soc Cogn Affect Neurosci 5(2-3):332–339

Heatherton TF, Wyland CL, Macrae CN, Demos KE, Denny BT, Kelley WM (2006) Medial prefrontal activity differentiates self from close others. Soc Cogn Affect Neurosci 1(1):18–25

Hoefler A, Athenstaedt U, Corcoran K, Ebner F, Ischebeck A (2015) Coping with self-threat and the evaluation of self-related traits: an fMRI study. PLoS One 10(9):e0136027

Holt DJ, Cassidy BS, Andrews-Hanna JR, Lee SM, Coombs G, Goff DC, Gabrieli JD, Moran JM (2011) An anterior-to-posterior shift in midline cortical activity in schizophrenia during self-reflection. Biol Psychiatry 69(5):415–423

Hughes BL, Beer JS (2013) Protecting the self: the effect of social-evaluative threat on neural representations of self. J Cogn Neurosci 25(4):613–622

Jankowski KF, Moore WE, Merchant JS, Kahn LE, Pfeifer JH (2014) But do you think I'm cool? Developmental differences in striatal recruitment during direct and reflected social self-evaluations. Dev Cogn Neurosci 8:40–54

Jenkins AC, Mitchell JP (2011) Medial prefrontal cortex subserves diverse forms of self-reflection. Soc Neurosci 6(3):211–218

Jenkins AC, Macrae CN, Mitchell JP (2008) Repetition suppression of ventromedial prefrontal activity during judgments of self and others. Proc Natl Acad Sci U S A 105(11):4507–4512

Johnson SC, Baxter LC, Wilder LS, Pipe JG, Heiserman JE, Prigatano GP (2002) Neural correlates of self-reflection. Brain 125(Pt 8):1808–1814

Johnson SC, Ries ML, Hess TM, Carlsson CM, Gleason CE, Alexander AL, Rowley HA, Asthana S, Sager MA (2007) Effect of Alzheimer disease risk on brain function during self-appraisal in healthy middle-aged adults. Arch Gen Psychiatry 64(10):1163–1171

Kelley WM, Macrae CN, Wyland CL, Caglar S, Inati S, Heatherton TF (2002) Finding the self? An event-related fMRI study. J Cogn Neurosci 14(5):785–794

Kennedy DP, Courchesne E (2008) Functional abnormalities of the default network during self- and other-reflection in autism. Soc Cogn Affect Neurosci 3(2):177–190

Kim EJ, Kyeong S, Cho SW, Chun JW, Park HJ, Kim J, Kim J, Dolan RJ, Kim JJ (2016) Happier people show greater neural connectivity during negative self-referential processing. PLoS One 11(2):e0149554

Kircher TT, Brammer M, Bullmore E, Simmons A, Bartels M, David AS (2002) The neural correlates of intentional and incidental self processing. Neuropsychologia 40(6):683–692

Kjaer TW, Nowak M, Lou HC (2002) Reflective self-awareness and conscious states: PET evidence for a common midline parietofrontal core. NeuroImage 17(2):1080–1086

Krienen FM, Tu PC, Buckner RL (2010) Clan mentality: evidence that the medial prefrontal cortex responds to close others. J Neurosci 30(41):13906–13915

Lemogne C, le Bastard G, Mayberg H, Volle E, Bergouignan L, Lehericy S, Allilaire JF, Fossati P (2009) In search of the depressive self: extended medial prefrontal network during self-referential processing in major depression. Soc Cogn Affect Neurosci 4(3):305–312

Lemogne C, Mayberg H, Bergouignan L, Volle E, Delaveau P, Lehericy S, Allilaire JF, Fossati P (2010) Self-referential processing and the prefrontal cortex over the course of depression: a pilot study. J Affect Disord 124(1–2):196–201

Lemogne C, Delaveau P, Freton M, Guionnet S, Fossati P (2012) Medial prefrontal cortex and the self in major depression. J Affect Disord 136(1-2):e1–e11

Lieberman MD, Jarcho JM, Satpute AB (2004) Evidence-based and intuition-based self-knowledge: an FMRI study. J Pers Soc Psychol 87(4):421–435

Lombardo MV, Chakrabarti B, Bullmore ET, Sadek SA, Pasco G, Wheelwright SJ, Suckling J, Baron-Cohen S (2010) Atypical neural self-representation in autism. Brain 133(Pt 2):611–624

Lou HC, Luber B, Crupain M, Keenan JP, Nowak M, Kjaer TW, Sackeim HA, Lisanby SH (2004) Parietal cortex and representation of the mental self. Proc Natl Acad Sci U S A 101(17):6827–6832

Ma Y, Bang D, Wang C, Allen M, Frith C, Roepstorff A, Han S (2014a) Sociocultural patterning of neural activity during self-reflection. Soc Cogn Affect Neurosci 9(1):73–80

Ma Y, Wang C, Li B, Zhang W, Rao Y, Han S (2014b) Does self-construal predict activity in the social brain network? A genetic moderation effect. Soc Cogn Affect Neurosci 9(9):1360–1367

Macrae CN, Moran JM, Heatherton TF, Banfield JF, Kelley WM (2004) Medial prefrontal activity predicts memory for self. Cereb Cortex 14(6):647–654

McAdams CJ, Krawczyk DC (2014) Who am I? How do I look? Neural differences in self-identity in anorexia nervosa. Soc Cogn Affect Neurosci 9(1):12–21

Meffert H, Blanken L, Blair KS, White SF, Blair JR (2013) The influence of valence and decision difficulty on self-referential processing. Front Hum Neurosci 7:46

Modinos G, Ormel J, Aleman A (2009) Activation of anterior insula during self-reflection. PLoS One 4(2):e4618

Modinos G, Renken R, Ormel J, Aleman A (2011) Self-reflection and the psychosis-prone brain: an fMRI study. Neuropsychology 25(3):295–305

Moran JM, Macrae CN, Heatherton TF, Wyland CL, Kelley WM (2006) Neuroanatomical evidence for distinct cognitive and affective components of self. J Cogn Neurosci 18(9):1586–1594

Moran JM, Lee SM, Gabrieli JD (2011) Dissociable neural systems supporting knowledge about human character and appearance in ourselves and others. J Cogn Neurosci 23(9):2222–2230

Morel N, Villain N, Rauchs G, Gaubert M, Piolino P, Landeau B, Mezenge F, Desgranges B, Eustache F, Chetelat G (2014) Brain activity and functional coupling changes associated with self-reference effect during both encoding and retrieval. PLoS One 9(3):e90488

Morin A, Hamper B (2012) Self-reflection and the inner voice: activation of the left inferior frontal gyrus during perceptual and conceptual self-referential thinking. Open Neuroimaging J 6:78–89

Murphy ER, Brent BK, Benton M, Pruitt P, Diwadkar V, Rajarethinam RP, Keshavan MS (2010) Differential processing of metacognitive evaluation and the neural circuitry of the self and others in schizophrenia: a pilot study. Schizophr Res 116(2–3):252–258

Murray RJ, Schaer M, Debbane M (2012) Degrees of separation: a quantitative neuroimaging meta-analysis investigating self-specificity and shared neural activation between self- and other-reflection. Neurosci Biobehav Rev 36(3):1043–1059
Murray RJ, Debbane M, Fox PT, Bzdok D, Eickhoff SB (2015) Functional connectivity mapping of regions associated with self- and other-processing. Hum Brain Mapp 36(4):1304–1324
Ng SH, Han S, Mao L, Lai JC (2010) Dynamic bicultural brains: fMRI study of their flexible neural representation of self and significant others in response to culture primes. Asian J Soc Psychol 13(2):83–91
Northoff G, Bermpohl F (2004) Cortical midline structures and the self. Trends Cogn Sci 8(3):102–107
Northoff G, Heinzel A, de Greck M, Bermpohl F, Dobrowolny H, Panksepp J (2006) Self-referential processing in our brain – a meta-analysis of imaging studies on the self. NeuroImage 31(1):440–457
Ochsner KN, Beer JS, Robertson ER, Cooper JC, Gabrieli JD, Kihsltrom JF, D'Esposito M (2005) The neural correlates of direct and reflected self-knowledge. NeuroImage 28(4):797–814
Pankow A, Katthagen T, Diner S, Deserno L, Boehme R, Kathmann N, Gleich T, Gaebler M, Walter H, Heinz A, Schlagenhauf F (2016) Aberrant salience is related to dysfunctional self-referential processing in psychosis. Schizophr Bull 42(1):67–76
Pauly K, Finkelmeyer A, Schneider F, Habel U (2013) The neural correlates of positive self-evaluation and self-related memory. Soc Cogn Affect Neurosci 8(8):878–886
Pauly KD, Kircher TT, Schneider F, Habel U (2014) Me, myself and I: temporal dysfunctions during self-evaluation in patients with schizophrenia. Soc Cogn Affect Neurosci 9(11):1779–1788
Pfeifer JH, Lieberman MD, Dapretto M (2007) "I know you are but what am I?!": neural bases of self- and social knowledge retrieval in children and adults. J Cogn Neurosci 19(8):1323–1337
Pfeifer JH, Masten CL, Borofsky LA, Dapretto M, Fuligni AJ, Lieberman MD (2009) Neural correlates of direct and reflected self-appraisals in adolescents and adults: when social perspective-taking informs self-perception. Child Dev 80(4):1016–1038
Pfeifer JH, Kahn LE, Merchant JS, Peake SJ, Veroude K, Masten CL, Lieberman MD, Mazziotta JC, Dapretto M (2013a) Longitudinal change in the neural bases of adolescent social self-evaluations: effects of age and pubertal development. J Neurosci 33(17):7415–7419
Pfeifer JH, Merchant JS, Colich NL, Hernandez LM, Rudie JD, Dapretto M (2013b) Neural and behavioral responses during self-evaluative processes differ in youth with and without autism. J Autism Dev Disord 43(2):272–285
Powell LJ, Macrae CN, Cloutier J, Metcalfe J, Mitchell JP (2010) Dissociable neural substrates for agentic versus conceptual representations of self. J Cogn Neurosci 22(10):2186–2197
Qin P, Northoff G (2011) How is our self related to midline regions and the default-mode network? NeuroImage 57(3):1221–1233
Rameson LT, Satpute AB, Lieberman MD (2010) The neural correlates of implicit and explicit self-relevant processing. NeuroImage 50(2):701–708
Ray RD, Shelton AL, Hollon NG, Michel BD, Frankel CB, Gross JJ, Gabrieli JD (2009) Cognitive and neural development of individuated self-representation in children. Child Dev 80(4):1232–1242
Ray RD, Shelton AL, Hollon NG, Matsumoto D, Frankel CB, Gross JJ, Gabrieli JD (2010) Interdependent self-construal and neural representations of self and mother. Soc Cogn Affect Neurosci 5(2-3):318–323
Ries ML, Schmitz TW, Kawahara TN, Torgerson BM, Trivedi MA, Johnson SC (2006) Task-dependent posterior cingulate activation in mild cognitive impairment. NeuroImage 29(2):485–492
Ries ML, Jabbar BM, Schmitz TW, Trivedi MA, Gleason CE, Carlsson CM, Rowley HA, Asthana S, Johnson SC (2007) Anosognosia in mild cognitive impairment: relationship to activation of cortical midline structures involved in self-appraisal. J Int Neuropsychol Soc 13(3):450–461
Rogers TB, Kuiper NA, Kirker WS (1977) Self-reference and the encoding of personal information. J Pers Soc Psychol 35(9):677–688

Ruby P, Collette F, D'Argembeau A, Peters F, Degueldre C, Balteau E, Luxen A, Maquet P, Salmon E (2009) Perspective taking to assess self-personality: what's modified in Alzheimer's disease? Neurobiol Aging 30(10):1637–1651
Sarsam M, Parkes LM, Roberts N, Reid GS, Kinderman P (2013) The Queen and I: neural correlates of altered self-related cognitions in major depressive episode. PLoS One 8(10):e78844
Saverino C, Grigg O, Churchill NW, Grady CL (2015) Age differences in the default network at rest and the relation to self-referential processing. Soc Cogn Affect Neurosci 10(2):231–239
Schmitz TW, Johnson SC (2006) Self-appraisal decisions evoke dissociated dorsal-ventral aMPFC networks. NeuroImage 30(3):1050–1058
Schmitz TW, Johnson SC (2007) Relevance to self: a brief review and framework of neural systems underlying appraisal. Neurosci Biobehav Rev 31(4):585–596
Schmitz TW, Kawahara-Baccus TN, Johnson SC (2004) Metacognitive evaluation, self-relevance, and the right prefrontal cortex. NeuroImage 22(2):941–947
Schmitz TW, Rowley HA, Kawahara TN, Johnson SC (2006) Neural correlates of self-evaluative accuracy after traumatic brain injury. Neuropsychologia 44(5):762–773
Schneider M, Debbane M, Lagioia A, Salomon R, d'Argembeau A, Eliez S (2012) Comparing the neural bases of self-referential processing in typically developing and 22q11.2 adolescents. Dev Cogn Neurosci 2(2):277–289
Sugiura M (2013) Associative account of self-cognition: extended forward model and multi-layer structure. Front Hum Neurosci 7:535
Sugiura M, Seitz RJ, Angel H-F (2015) Models and neural bases of the believing process. J Behav Brain Sci 5(01):12
Sul S, Choi I, Kang P (2012) Cultural modulation of self-referential brain activity for personality traits and social identities. Soc Neurosci 7(3):280–291
Symons CS, Johnson BT (1997) The self-reference effect in memory: a meta-analysis. Psychol Bull 121(3):371–394
Tan S, Zhao Y, Fan F, Zou Y, Jin Z, Zen Y, Zhu X, Yang F, Tan Y, Zhou D (2015) Brain correlates of self-evaluation deficits in schizophrenia: a combined functional and structural MRI study. PLoS One 10(9):e0138737
van Buuren M, Gladwin TE, Zandbelt BB, Kahn RS, Vink M (2010) Reduced functional coupling in the default-mode network during self-referential processing. Hum Brain Mapp 31(8):1117–1127
van der Meer L, Costafreda S, Aleman A, David AS (2010) Self-reflection and the brain: a theoretical review and meta-analysis of neuroimaging studies with implications for schizophrenia. Neurosci Biobehav Rev 34(6):935–946
van der Meer L, de Vos AE, Stiekema AP, Pijnenborg GH, van Tol MJ, Nolen WA, David AS, Aleman A (2013) Insight in schizophrenia: involvement of self-reflection networks? Schizophr Bull 39(6):1288–1295
Vanderwal T, Hunyadi E, Grupe DW, Connors CM, Schultz RT (2008) Self, mother and abstract other: an fMRI study of reflective social processing. NeuroImage 41(4):1437–1446
Veroude K, Jolles J, Croiset G, Krabbendam L (2014) Sex differences in the neural bases of social appraisals. Soc Cogn Affect Neurosci 9(4):513–519
Wang G, Mao L, Ma Y, Yang X, Cao J, Liu X, Wang J, Wang X, Han S (2012) Neural representations of close others in collectivistic brains. Soc Cogn Affect Neurosci 7(2):222–229
Whitfield-Gabrieli S, Moran JM, Nieto-Castanon A, Triantafyllou C, Saxe R, Gabrieli JD (2011) Associations and dissociations between default and self-reference networks in the human brain. NeuroImage 55(1):225–232
Wu Y, Wang C, He X, Mao L, Zhang L (2010) Religious beliefs influence neural substrates of self-reflection in Tibetans. Soc Cogn Affect Neurosci 5(2-3):324–331
Yang J, Dedovic K, Chen W, Zhang Q (2012) Self-esteem modulates dorsal anterior cingulate cortical response in self-referential processing. Neuropsychologia 50(7):1267–1270
Yang J, Xu X, Chen Y, Shi Z, Han S (2016) Trait self-esteem and neural activities related to self-evaluation and social feedback. Sci Rep 6:20274

Yaoi K, Osaka N, Osaka M (2009) Is the self special in the dorsomedial prefrontal cortex? An fMRI study. Soc Neurosci 4(5):455–463
Yaoi K, Osaka M, Osaka N (2015) Neural correlates of the self-reference effect: evidence from evaluation and recognition processes. Front Hum Neurosci 9:383
Yoshimura S, Ueda K, Suzuki S, Onoda K, Okamoto Y, Yamawaki S (2009) Self-referential processing of negative stimuli within the ventral anterior cingulate gyrus and right amygdala. Brain Cogn 69(1):218–225
Yoshimura S, Okamoto Y, Onoda K, Matsunaga M, Ueda K, Suzuki S, Yamawaki S (2010) Rostral anterior cingulate cortex activity mediates the relationship between the depressive symptoms and the medial prefrontal cortex activity. J Affect Disord 122(1-2):76–85
Yoshimura S, Okamoto Y, Onoda K, Matsunaga M, Okada G, Kunisato Y, Yoshino A, Ueda K, Suzuki S, Yamawaki S (2014) Cognitive behavioral therapy for depression changes medial prefrontal and ventral anterior cingulate cortex activity associated with self-referential processing. Soc Cogn Affect Neurosci 9(4):487–493
Zhang L, Zhou T, Zhang J, Liu Z, Fan J, Zhu Y (2006) In search of the Chinese self: an fMRI study. Sci China C Life Sci 49(1):89–96
Zhu Y, Zhang L, Fan J, Han S (2007) Neural basis of cultural influence on self-representation. NeuroImage 34(3):1310–1316
Zhu L, Guo X, Li J, Zheng L, Wang Q, Yang Z (2012) Hippocampal activity is associated with self-descriptiveness effect in memory, whereas self-reference effect in memory depends on medial prefrontal activity. Hippocampus 22(7):1540–1552
Zysset S, Huber O, Ferstl E, von Cramon DY (2002) The anterior frontomedian cortex and evaluative judgment: an fMRI study. NeuroImage 15(4):983–991
Zysset S, Huber O, Samson A, Ferstl EC, von Cramon DY (2003) Functional specialization within the anterior medial prefrontal cortex: a functional magnetic resonance imaging study with human subjects. Neurosci Lett 335(3):183–186

Chapter 8
Voluntary Suppression and Involuntary Repression: Brain Mechanisms for Forgetting Unpleasant Memories

Hirokazu Kikuchi and Nobuhito Abe

Abstract The present chapter reviews the current accomplishments regarding investigations of the brain mechanisms that underlie active inhibitory control processes in mnemonic processing. We focus on both voluntary and involuntary inhibitory processes, namely, the suppression and repression of memory. The results from studies on voluntary memory suppression in healthy participants indicate that individuals may deteriorate or disrupt the encoding and retrieval of an unwanted memory by voluntary suppression. The neural basis of memory suppression comprises the activated dorsolateral prefrontal cortex, which operates as an inhibitor of the medial temporal lobe (MTL), and the deactivated MTL memory system, which enables motivated forgetting. The results from studies of involuntary memory repression in patients with dissociative amnesia indicate two potential brain mechanisms: "hyper-suppression" and "hypo-retrieval." The former appears to share a similar neural basis with voluntary suppression, in which interactions between prefrontal cortex activation and MTL deactivation are closely linked to memory impairment. In the latter, baseline activity in the regions responsible for the initiation of memory retrieval is decreased. A recently proposed model attempts to reconcile these two ideas. These lines of investigation have significant implications for understanding how our memory system may be controlled by active forgetting processes through interactions between specific brain areas.

Keywords Suppression • Repression • Dissociative amnesia • Motivated forgetting • Directed forgetting • Functional neuroimaging • Prefrontal cortex • Medial temporal lobe

---

H. Kikuchi (✉)
Department of Neurology, Tohoku Medical and Pharmaceutical University,
1-12-1 Fukumuro, Miyagino-ku, Sendai 983-8512, Japan
e-mail: kiku0215@hosp.tohoku-mpu.ac.jp

N. Abe
Kokoro Research Center, Kyoto University, Kyoto, Japan

T. Tsukiura, S. Umeda (eds.), *Memory in a Social Context*,
https://doi.org/10.1007/978-4-431-56591-8_8

8.1 Introduction

In human society, individuals have various experiences that may be pleasant or unpleasant. Through the processes of encoding, storage and retrieval, the experiences are formed into memories. However, not all memories are equally welcomed because they are, in some cases, accompanied by unpleasant emotions associated with the original experiences. To live calmly and smoothly in society, we need to regulate our emotional state and protect our sense of self by forgetting unpleasant memories. For example, to sustain positive emotions, concentration, confidence or optimism, it is necessary to reduce the accessibility of experiences that undermine these states. Therefore, in the context of living in human society, both memorizing and forgetting experiences are equally important. However, memory studies in the twentieth century largely or completely neglected the adaptive aspects of forgetting. Historically, forgetting has been regarded as a negative outcome; thus, any process that underlies forgetting must be a passive process that results from the decay of memory over time, the subsequent accumulation of similar experiences in memory, and changes in the environmental context, all of which make it more difficult to retrieve the past (Baddeley et al. 2009).

Even before researchers acknowledged the advantages of forgetting, more than a century ago, Freud proposed an active process for purging unpleasant memories from consciousness, which he referred to as "repression" (Freud 1966). It should be noted that his theory regarding repression changed over time. Freud originally referred to an intentional attempt to prevent unwanted experiences from entering conscious awareness. However, he subsequently used the term to refer to one of several defense mechanisms that operate automatically outside of an individual's awareness (Erdelyi 2006). Currently, his earlier idea of an intentional, voluntary inhibition of unpleasant memories is referred to as "suppression" of memory, and his subsequent idea of automatic, involuntary inhibition is referred to as repression of memory, both of which deserve attention. In the twenty-first century, memory studies have begun to regard forgetting not only as a passive process, but also as an active process that interrupts mnemonic processing, which Freud proposed long ago. Although some researchers once claimed that memory suppression was a clinical myth (Kihlstrom 2002), other researchers have provided evidence that memory suppression is an active forgetting process supported by concrete brain mechanisms (Anderson and Green 2001; Anderson et al. 2004; Depue et al. 2006, 2007). Moreover, memory repression has emerged as an existing process concomitant with altered patterns of brain activity (Kikuchi et al. 2010). Therefore, theories of forgetting that underestimate the active control of awareness of unpleasant memories should be considered to omit a powerful and fundamental force that shapes our retention of the past (Anderson and Hanslmayr 2014).

Here, we review neural evidence for the role of active inhibitory control processes in mnemonic processing. We review both voluntary and involuntary processes, namely, the suppression and repression of memory. Suppression of memory is mainly studied in the fields of psychology and cognitive neuroscience, whereas repression of memory is often investigated in clinical fields, including psychiatry. Evidence indicates that both voluntary and involuntary inhibitory control processes exist to exclude unpleasant memories from awareness, and common brain mechanisms may underlie these two inhibitory processes.

8.2 Brain Mechanisms of Voluntary Memory Suppression: Knowledge from Studies of Motivated Forgetting in Healthy Individuals

When we are reminded of unpleasant experiences, we become badly disposed toward them and attempt to limit their presence in our awareness. Once a reminder evokes a brief flash of the negative memory and feeling, attempts to exclude the unwanted memory from awareness rapidly follow. This process, which represents an intentional attempt to inhibit unwanted experiences from entering conscious awareness, is familiar to most individuals and is regarded as voluntary memory suppression. Here, we review the accumulated evidence regarding the neural mechanisms that underlie "motivated forgetting," which refers to the increased forgetting that arises from memory suppression (Anderson and Hanslmayr 2014). These inhibitory control processes have primarily been investigated via psychological and functional neuroimaging experiments with healthy individuals. It is important to note that these inhibitory processes may be targeted flexibly at different stages of mnemonic processing, namely, during encoding or retrieval. The inhibition of encoding may stop the consolidation of the targeted experience, which may result in a deterioration or disappearance of the memory trace of the unpleasant experience. The inhibition of retrieval may halt a series of relatively automatic mnemonic processes, from processing reminding cues to retrieving the unpleasant memory, which may subsequently make it difficult or impossible to access the previously consolidated memory trace. Through both these inhibitory processes, an experience may be excluded from conscious awareness and successfully forgotten.

8.2.1 Voluntary Suppression at Encoding

In the absence of the trace, we cannot access the memory. The most effective way of avoiding the retrieval of unwanted memories in conscious awareness is to not create the memory trace, that is, to not encode the unpleasant experience. The attenuation of encoding would also be effective and more pragmatic. Inhibitory control has been demonstrated to have a suppressive effect on encoding, which has been investigated with many experiments of directed forgetting paradigms (Bjork 1989). There are two types of directed forgetting paradigms, including the item method and the list method. In the item-method paradigm, participants study items one at a time, and each item

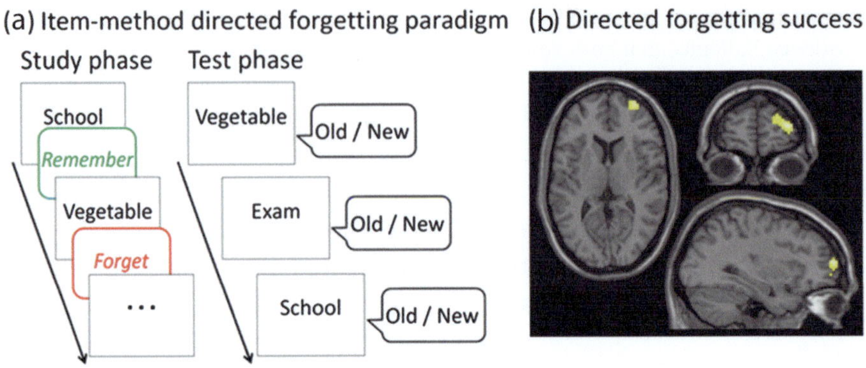

Fig. 8.1 An example of experimental directed forgetting paradigms and the underlying neural correlates. (**a**) The item-method-directed forgetting paradigm. In this paradigm, which measures voluntary suppression at encoding, participants study items one at a time, and each item is followed by an instruction to "forget" or "remember". Participants are subsequently tested regarding memory of all items (Muther 1965). (**b**) An activation map of a recent fMRI study of item-method-directed forgetting (modified with permission from Yang et al. 2016). Areas that are significantly more active regarding intentional forgetting compared with incidental forgetting (items to be forgotten that are successfully forgotten versus items to be remembered that are incidentally forgotten) are shown. Successful directed forgetting (compared with incidental forgetting) was associated with increased activation in the right middle frontal gyrus (i.e., the dorsolateral prefrontal cortex [DLPFC])

is followed by an instruction of "forget" or "remember"; participants are subsequently tested for memory of all items (Fig. 8.1a) (Muther 1965). The list-method paradigm comprises two lists of items. A forget or remember instruction is provided after participants have studied the first list. Participants subsequently study the second list, followed by a brief distracting task. At the end, participants are tested for memory of all items (Bäuml et al. 2010; Geiselman et al. 1983). The item method and list method differ with regard to the target of forgetting: the item method targets individual items, whereas the list method typically targets a set of items defined by the temporal context, namely, the former or latter list.

Many studies that have used the item-method paradigm have demonstrated a specific effect of directed forgetting. Items followed by a forget instruction are retrieved less often at the final test compared with items followed by a remember instruction. Moreover, the effect appears to be robust because it has been reported in a wide range of conditions, including both explicit and implicit memory tests (Basden and Basden 1998; MacLeod and Daniels 2000). Selective rehearsal, a passive and non-inhibitory process, has long accounted for item-method-directed forgetting; items to be remembered are actively and elaborately rehearsed, and items to be forgotten are isolated from further processing and are subject to passive forgetting (Bjork 1989). According to the selective rehearsal account, the "remember" condition should be more cognitively burdensome than the "forget" condition, which may be considered relatively passive and less demanding.

However, recent evidence suggests that an active and inhibitory process contributes to item-method-directed forgetting (Zacks et al. 1996). First, participants require a longer reaction time to perform a visual detection task after a forget instruction compared with after a remember instruction. This finding indicates that the "forget" condition is more cognitively burdensome and effortful than the "remember" condition, and the cognitive resource for the secondary task is therefore limited (Fawcett and Taylor 2008). Second, stopping a motor response is more successful in the "forget" condition than in the "remember" condition, which indicates that an intrinsically inhibitory process for regulating motor action is recruited by a forget instruction (Anderson 2003; Fawcett and Taylor 2010). Finally, other studies have directly demonstrated that participants are led to actively remove and expel items from working memory after a forget instruction, as opposed to losing them passively (Ecker et al. 2014; Williams et al. 2013). These findings indicate that item-method-directed forgetting occurs via an active, cognitively burdensome, and intrinsically inhibitory process. Studies that have used the list-method paradigm have also demonstrated a specific effect of directed forgetting via an active inhibitory process. Items in the first list are retrieved with greater difficulty when followed by a forget instruction than when they are followed by a remember instruction (Bäuml et al. 2010; Bjork 1989).

Recent neuroimaging studies using the item-method- or list-method-directed forgetting paradigm consistently support the view that directed forgetting is driven by an active inhibitory process, and all of these studies indicate that the process engages the dorsolateral prefrontal cortex (DLPFC; Fig. 8.1b) (Hanslmayr et al. 2012; Nowicka et al. 2011; Rizio and Dennis 2013; Wylie et al. 2008; Yang et al. 2016), a region implicated in cognitive control (Badre 2008; Miller and Cohen 2001). Functional magnetic resonance imaging (fMRI) studies using item-method-directed forgetting have demonstrated that the right DLPFC is consistently more active during intentional forgetting (forgetting items the individual is instructed to forget) compared with incidental forgetting (forgetting items the individual is instructed to remember) (Nowicka et al. 2011; Rizio and Dennis 2013; Wylie et al. 2008). Moreover, inter-regional connectivity analyses indicate that right DLPFC activation during forgetting trials predicts decreased activation in the left medial temporal lobe (MTL), which comprises a core structure in memory processing (Squire and Zola-Morgan 1991), especially during successful intentional forgetting (Rizio and Dennis 2013). These findings suggest that the right DLPFC might exert inhibitory control over the encoding activity in the MTL (Paller and Wagner 2002). Neuroimaging studies using list-method-directed forgetting have also demonstrated that an active process inhibits mnemonic activity. An fMRI study has demonstrated that left DLPFC activation is associated with intentional forgetting of the first list the individual is instructed to forget. Furthermore, this study attempted an intervention in the activity in the left DLPFC with repetitive transcranial magnetic stimulation (rTMS). The findings indicated that facilitatory stimulation of the left DLPFC with rTMS during a forget instruction magnified directed forgetting (Hanslmayr et al. 2012). Thus, these findings indicate a causal mechanism in which the left DLPFC drives the inhibitory control processes of encoding.

Taken together, these studies have consistently demonstrated that the DLPFC plays a key role in voluntary memory suppression at encoding, regardless of the experimental paradigms used. Similar to inhibitory control in the motor system, in which higher-order prefrontal control suppresses lower-order motor activity to terminate a movement (Aron et al. 2004), it appears that the DLPFC exerts inhibitory control over the activity of the MTL, which thereby down-regulates the encoding of targeted stimuli.

8.2.2 Voluntary Suppression at Retrieval

Although inhibitory processes occur during encoding, some unpleasant experiences are stored as memory traces, which may be accessed and retrieved in the presence of reminding cues. At this stage, the retrieval process determines whether the unwanted memory may be prevented from entering conscious awareness. As with the inhibitory control of encoding, it has been demonstrated that there is a specific suppressive effect in the inhibitory control of retrieval. This retrieval suppression effect has been investigated with experiments using the think/no-think (TNT) paradigm (Fig. 8.2a) (Anderson and Green 2001). In the TNT paradigm, participants first study pairs composed of a cue and target, such as "fruit–apple," to the extent that they retrieve the target "apple" when exposed to the reminding cue "fruit." The pairs are typically word pairs (Anderson and Green 2001); however, any stimuli, such as face–scene (Depue et al. 2006), face–word (Hanslmayr et al. 2009), or word and nonsense shape (Hart and Schooler 2012) pairs, are acceptable. Participants subsequently undergo the TNT phase, in which they are presented with reminding cues one at a time and asked to retrieve and think about the target (think trials) or to not retrieve the target, thereby preventing it from entering conscious awareness (no-think trials). Whether a trial represents a think or no-think trial is typically indicated by the color (e.g., green or red) of stimuli, such as the letters of the word or the picture frame. At the end, participants are tested regarding memory of all pairs; they are asked to retrieve the paired target given each reminding cue of the think trials, no-think trials, and trials that were not retrieved or prevented in the TNT phase (baseline trials).

Studies with the TNT paradigm have consistently demonstrated a reliable effect of retrieval suppression (Anderson and Green 2001; Anderson and Huddleston 2011; Anderson and Levy 2009; Depue et al. 2006; Hanslmayr et al. 2009; Hart and Schooler 2012). Items from no-think trials are retrieved significantly less frequently compared with items from think trials, which indicates that retrieval suppression does not enhance the rehearsal of memory contents, similar to other repeated learning processes. Moreover, items from no-think trials are retrieved even less often than items from baseline trials, which indicates that retrieval suppression operates as an active forgetting process and makes it difficult or impossible to access the trace of the unwanted memory. This effect is referred to as "suppression-induced forgetting," and detailed properties of this effect are now known. Suppression-induced

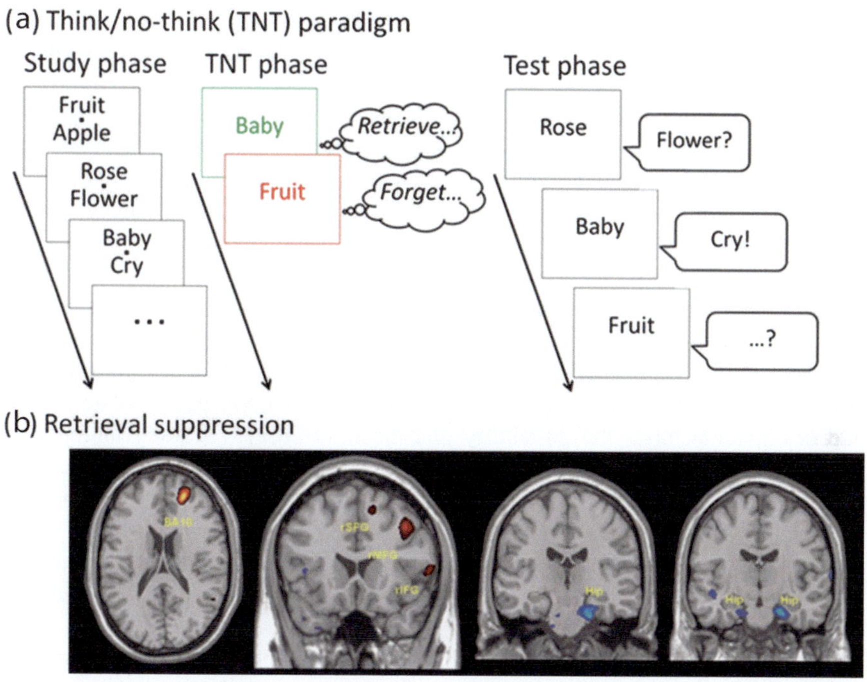

Fig. 8.2 An example of experimental retrieval suppression paradigms and the underlying neural correlates. (**a**) The think/no-think (TNT) paradigm. In this paradigm, which measures voluntary suppression at retrieval, participants first study cue–target pairs. Participants subsequently undergo the TNT phase, in which they are presented with reminding cues one at a time and asked to retrieve (think trial) or forget (no-think trial) the target, which is indicated by the colors *green* or *red*. At the end, participants are tested for their memory of all pairs (Anderson and Green 2001). (**b**) An activation map of a representative fMRI study of retrieval suppression using the TNT paradigm (modified with permission from Depue et al. 2007). Areas that are significantly more activated (*red*) or deactivated (*blue*) in no-think trials compared with think trials are depicted. Retrieval suppression was associated with increased activations in the right prefrontal regions (including the right superior, middle and inferior frontal gyri and Brodmann's area 10) and deactivations in the bilateral hippocampi

forgetting has cumulative effects, that is, as the number of times a memory is suppressed increases, the more frequently the memory tends to be forgotten (Anderson and Green 2001; Anderson and Huddleston 2011; Anderson et al. 2011; Hanslmayr et al. 2009). Suppression-induced forgetting not only occurs under the conditions of retrieving experimental pairs, but also with autobiographical experiences (Noreen and Macleod 2013, 2014). It is more important and suggestive that suppression-induced forgetting makes it difficult to retrieve the target from reminding cues other than the originally learned reminding cue (Anderson and Green 2001). This effect is referred to as "cue independence," which indicates that suppression-induced forgetting may not only inhibit the accessibility of the memory trace, but also, or even more so, disrupt the actual memory trace (Anderson and Huddleston

2011). Moreover, the same phenomenon is applied in situations without a reminding cue; individuals are less likely to retrieve the suppressed contents when subsequently tested, even in a free recall paradigm, and the suppression-induced forgetting effect appears more robust in these situations (Hertel et al. 2012). It should be noted that it is an unnatural situation for participants to be asked to retrieve contents that they were motivated to suppress. Therefore, it is plausible that the suppression-induced forgetting effect is more robust in real life (Anderson and Hanslmayr 2014).

Neuroimaging studies that have used the TNT paradigm support the view that retrieval suppression is driven and achieved by an active inhibitory process. Similar to the activation patterns associated with directed forgetting at encoding, the right DLPFC is more activated, and the MTL, including the hippocampus, is more deactivated during no-think trials compared with think trials (Fig. 8.2b) (Anderson et al. 2004; Benoit and Anderson 2012; Depue et al. 2007). The pattern of activation and deactivation raises the possibility that the right DLPFC actively inhibits the MTL, which results in suppression-induced forgetting. Cumulative findings support this possibility, and the most important and supportive findings comprise connectivity analyses that indicate that activation in the DLPFC during no-think trials predicts deactivation in the hippocampus, and that DLPFC activation that inhibits the hippocampus predicts the final amount of suppression-induced forgetting (Benoit and Anderson 2012; Gagnepain et al. 2014).

In summary, the neural mechanisms that underlie voluntary memory suppression at retrieval are similar to the mechanisms that operate at encoding; it appears that the DLPFC orchestrates the inhibition of memory retrieval through deactivation of the MTL memory system. This neural processing is likely to lead to limited access to memory traces and disruption of the stored memory content.

8.3 Brain Mechanisms of Involuntary Memory Repression: Knowledge from Studies of Patients with Dissociative Amnesia

Although it has been elucidated that Freud's early idea of intentional voluntary suppression of unpleasant memories exists as an active forgetting process supported by concrete brain mechanisms, the presence and nature of his subsequent idea of automatic involuntary repression had been neglected until recently. The reason for this may be quite simple: it is difficult to explicitly observe these processes. Even if the process exists, it goes unnoticed by the individual because it is an "involuntary" process, and by others without prior knowledge because both the memory and its concomitant unpleasant emotions are invisible. However, other individuals may notice and observe the forgetting induced by automatic involuntary repression in one clinical condition: "dissociative amnesia." Here, we introduce the concept and properties of dissociative amnesia and describe the recently suggested neural mechanisms that underlie involuntary memory repression. We mainly present our own research, which provided direct and powerful evidence regarding the neural mechanisms of dissociative amnesia. Dissociative amnesia is characterized by retrieval failure; thus, this section inevitably focuses on the neural mechanisms that underlie involuntary memory repression at retrieval. It is important to note that dissociative amnesia, which likely reflects involuntary memory repression, is associated with prefrontal activation and hippocampal deactivation, similar to the neural mechanisms of voluntary memory suppression.

8.3.1 Dissociative Amnesia: An Excessive Operation of Memory Repression

Dissociative amnesia is a psychiatric disorder in which patients have a continuous inability to retrieve specific memories. It typically follows a traumatic or stressful autobiographical event and is not the result of direct physiological effects of neurological or other general medical conditions (American Psychiatric Association 1994). Dissociative amnesia is therefore presumed to be psychogenic in origin; it is also referred to as "psychogenic amnesia" or "functional amnesia." Patients are typically not able to retrieve specific past memories, but are able to learn new episodes. Their symptoms are consistent with focal/isolated retrograde amnesia, which indicates retrograde amnesia without concomitant anterograde amnesia (Kapur 2000; Kopelman 2000, 2002). Dissociative amnesia often occurs soon after traumatic events; however, in some cases, it arises in association with continuous internal conflict or an ongoing intolerable situation (Kaplan and Sadock 1995). The triggers may include accidents, disasters, divorce, debt or poverty, and, in some cases, mild head trauma without organic brain injury accounting for amnesia. The impaired memories typically include unpleasant and unwelcome memories associated with the original problems. Therefore, this disorder is typically considered to be a way of managing the conflict or painful emotions of the problems, and may be regarded as involuntary memory repression. However, repression in dissociative amnesia appears to be exceptionally excessive.

Memory impairment in dissociative amnesia may be permanent, but it typically resolves spontaneously within several days or months once the original traumatic situation is ameliorated, and its recurrence is rare. Psychotherapy, including hypnotic therapy or an interview under anesthetic (Kopelman 2000), is conducted in some cases, and patients may restore their memories through therapy. Therefore, their impairment is limited to the process of retrieval. The experiences are ordinarily encoded and consolidated as the memory trace, and the memory trace is normally preserved. However, the accessibility of the memory trace is inhibited and disrupted in dissociative amnesia.

Here, we briefly present two patients with dissociative amnesia who exhibited typical symptoms and clinical courses and who also participated in our fMRI experiment (for details, see Kikuchi et al. 2010). Patient 1 was a 27-year-old man who had worked as a businessman for approximately 4 years following college graduation. He had retrograde amnesia that covered a period of 4.5 years preceding onset, that is, the last half year of his time at college and all 4 years of his time as a businessman (Table 8.1). His only symptom was retrograde amnesia; his anterograde memory was intact, and his symptoms were consistent with focal/isolated retrograde amnesia, which comprised retrograde amnesia for a limited period without concomitant anterograde amnesia (Kapur 2000; Kopelman 2000, 2002). He had been working continuously and extremely hard for approximately 6 months before the onset and had been worried about his impending marriage. Based on these findings and the criteria of the Diagnostic and Statistical Manual of Mental Disorders, Fourth Edition (DSM-IV) (American Psychiatric Association 1994), we diagnosed his disorder as dissociative amnesia. With the written informed consent of patient 1, therapeutic sessions with interviews using an intravenous injection of sodium thiopental were conducted to recover his memory (Kopelman 2000). After two sessions separated by a 10-day interval, his memories from the first 4 years of his amnesic period were nearly entirely recovered (Table 8.1). As a result of recovering his memory of his time as a businessman, he eventually returned to work.

Patient 2 was a 52-year-old man who had worked as a businessman for approximately 30 years following graduation from a university. More than a month after the car he was driving struck a guardrail, he suddenly became aware of his retrograde amnesia. He had retrograde amnesia for nearly all events after his high school graduation, that is, for a period of approximately 35 years (Table 8.1). He exhibited no other symptoms, which indicated that he had focal/isolated retrograde amnesia (Kapur 2000; Kopelman 2000, 2002) for a limited but extensive period. He had been troubled for several years regarding the divorce negotiations with his wife, and he was in debt, which had gradually increased. There was no evidence of neurological disease or traumatic brain injury. Based on these findings and the criteria of the DSM-IV (American Psychiatric Association 1994), we diagnosed his disorder as dissociative amnesia. Therapeutic sessions with interviews using an intravenous

Table 8.1 Amnesic period and changes with treatment in patients 1 and 2 with dissociative amnesia

Retrograde amnesia		Patient 1	Patient 2
Amnesic period	Pre-treatment	A period of 4.5 years preceding onset: the last half year of his time at college and all 4 years of his time as a businessman	A period of approximately 35 years preceding onset: entire period after high school graduation
(Treatment: interview under anesthetic)		↓	↓
	Post-treatment	All memories from the first 4 years of the amnesic period: nearly entirely recovered. The last half year of the amnesic period: not recovered	No memories were recovered

injection of sodium thiopental were conducted (Kopelman 2000) with the written informed consent of patient 2; however, he recovered only one memory, the model of a car he had driven in his twenties (Table 8.1). Although this retrieved memory was again impaired when the patient slept immediately after the interview, this therapeutic course supported the diagnosis of dissociative amnesia.

8.3.2 Neural Basis Underlying Dissociative Amnesia

Even if dissociative amnesia is originally a psychiatric disorder, it may be assumed that dissociative amnesia has a neural basis, and the memory impairment is caused by an altered pattern of activities in the regions responsible for mnemonic processing. Historically, two theoretical models have been proposed regarding the brain mechanisms that develop and maintain memory impairment in dissociative amnesia. One notable theoretical model, proposed by Kopelman (Kopelman 2002), is as follows: psychiatric stress initiates activities of cognitive control and the executive system in the frontal lobe, and these activities precipitate the inhibition of memory systems in the MTL and diencephalon, which results in memory retrieval inhibition. The other model is proposed by Markowitsch (2003), which he refers to as "mnestic block syndrome": an altered release or imbalance of stress-related hormones and neurotransmitters induced by traumatic events disconnects brain networks, especially within the right fronto-temporal regions. These regions are considered the critical junction areas for the initiation of autobiographical memory retrieval (Costello et al. 1998); therefore, the disconnection within these regions interrupts the retrieval of autobiographical memories.

To practically examine specific brain activity changes present in the background of dissociative amnesia and to test the competing theories previously described, approaches with functional imaging are useful. Studies regarding dissociative amnesia have used resting-state functional imaging, such as single photon emission computed tomography (SPECT) or positron emission tomography (PET). These studies have demonstrated activity changes in the broad temporal and frontal regions of the right hemisphere (Markowitsch et al. 1997), the left medial temporal region and the left thalamic region (Markowitsch et al. 2000), the bilateral border areas of the temporal and occipital lobes (Nakamura et al. 2002), and the right temporo-mesial brain region (Hennig-Fast et al. 2008); however, one study did not identify activity changes in the brain (De Renzi et al. 1995). These findings are diverse, and the interpretations of the data are controversial. Furthermore, these studies examined only one or two patients. It is therefore difficult to determine the specific abnormalities associated with dissociative amnesia with resting-state functional imaging in studies of a small number of patients.

One study examined brain metabolic changes with functional imaging in a resting state using PET in a group of 14 patients with dissociative amnesia (Brand et al. 2009). Compared with the healthy controls, the patients exhibited significantly decreased glycometabolism, i.e., decreased brain activity in the right inferior lateral prefrontal cortex. Decreased activity in the hippocampus was identified in only 2 of the 14 patients. These findings are consistent with the theory proposed by Markowitsch (2003). We further discuss the implications of this study later.

Other studies have measured task-related brain activation in dissociative amnesia. Brain activation has been assessed by PET or fMRI during a task in which information that the patients cannot retrieve because of amnesia is presented, and the patients are required to engage in memory retrieval. Of the fMRI experiments, Yang et al. (2005) investigated a patient with dissociative amnesia and identified deactivation in the MTL when the patient viewed faces that were unrecognizable to her because of her memory impairment. Botzung et al. (2007) reported a patient who exhibited reduced activity in the frontal networks and the left MTL, which was associated with the retrieval of autobiographical memory stemming from the forgotten period. Hennig-Fast et al. (2008) reported that during the patient's acute state, the temporo-occipital region, which is known to be related to autobiographical memory, was not activated even when the patient saw sentences from the forgotten autobiographical episodes.

These task-based neuroimaging findings indicate dysfunction of the MTL, which comprises the regions responsible for memory processing; however, the reason for why deactivation occurs has not been fully clarified. The results from studies of voluntary memory suppression in healthy participants are highly suggestive on this issue. It is presumed that voluntary suppression, which comprises an active inhibitory process used to consciously forget unpleasant memories, is a laboratory analogue of unconscious memory repression. Therefore, it may be hypothesized that, similar to voluntary memory suppression, the MTL, including the hippocampus, is deactivated by the inhibitory control of the activated DLPFC in the involuntary memory repression present in dissociative amnesia.

With this hypothesis, to clarify the brain mechanisms that underlie memory repression, we conducted fMRI experiments combined with face and name recognition tasks in patients 1 and 2, who both had dissociative amnesia (Kikuchi et al. 2010). Because each patient's period of retrograde amnesia was limited, each patient could easily retrieve their friends from high school; however, they could not retrieve colleagues from the company where they had recently worked. Therefore, in cooperation with their family and colleagues, we prepared three categories of face photographs and the names that corresponded to the photographs for each patient; these photographs consisted of "recognizable" high school friends, "unrecognizable" colleagues, and unfamiliar "control" strangers. The patients were visually presented with each stimulus and asked to indicate whether or not the presented individual was an acquaintance; the brain activation during the recognition task was measured via fMRI. Both patients participated in these fMRI experiments before and after their treatment with interviews under anesthesia.

The results are summarized in Fig. 8.3. Behaviorally, before the treatment, both patients judged all "recognizable" high school friends to be acquainted, and all "unrecognizable" colleagues and all "control" strangers to be unacquainted. After the treatment, patient 1, who recovered his memories from his amnesic period, judged all "recognizable" stimuli and all, with the exception of one, "unrecogniz-

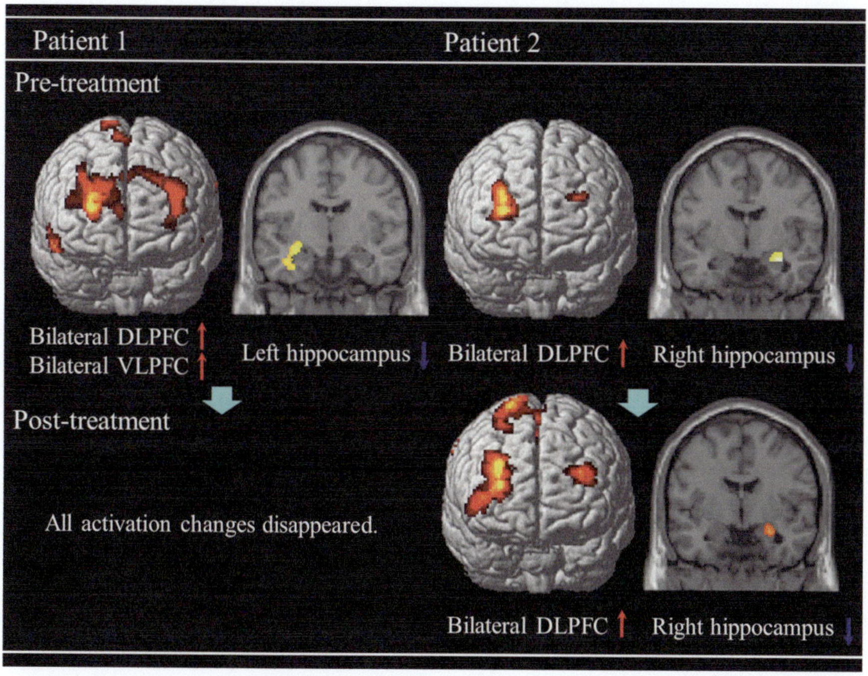

Fig. 8.3 Activation changes associated with dissociative amnesia from a task-related fMRI study (modified with permission from Kikuchi et al. 2010). Significant activation changes in response to "unrecognizable" stimuli versus "recognizable" stimuli, inclusively masked with significant activation changes in response to "unrecognizable" stimuli versus "control" stimuli, are shown. Before treatment, patients 1 and 2 with dissociative amnesia exhibited altered patterns of brain activations, with more activated prefrontal regions and more deactivated hippocampal regions. Following treatment, all activation changes disappeared in patient 1, who recovered nearly all his lost memories. In contrast, the altered pattern of brain activations remained virtually unchanged in patient 2, who did not recover his lost memories. *DLPFC* dorsolateral prefrontal cortex, *VLPFC* ventrolateral prefrontal cortex

able" stimuli to be acquainted. Patient 2, who did not recover his memories, exhibited the same behavioral performance as before the treatment. Both patients judged none of the "control" strangers to be acquainted. The pre-treatment fMRI results of patient 1 indicated that the bilateral DLPFC and ventrolateral prefrontal cortex (VLPFC) were activated, and the left hippocampus was deactivated during the presentation of the "unrecognizable" stimuli compared with both the "recognizable" ($p < 0.001$, uncorrected) and "control" ($p < 0.05$, uncorrected) stimuli. The pre-treatment fMRI results of patient 2 indicated bilateral DLPFC activation and right hippocampus deactivation in the same contrast as patient 1. Following treatment, these activation changes disappeared in patient 1, who had recovered his memories. However, the altered pattern of brain activations remained virtually unchanged in patient 2, who did not recover his memories.

These findings clearly demonstrate that dissociative amnesia is caused by activation changes in the brain. The pathological amnesic state in dissociative amnesia is characterized by activation of the prefrontal cortex and deactivation of the hippocampus. The potential interactions between the prefrontal cortex and the hippocampus identified in our study strongly support the theory proposed by Kopelman (2002): the activities in the frontal lobe precipitate inhibition of the MTL, which results in the inhibition of memory retrieval. It should be noted that our findings appear to be incompatible with the results of the previously described resting-state PET study (Brand et al. 2009), which fits with the theory proposed by Markowitsch (2003). Thus, the question arises as to how we may reconcile these two distinct theories. In general, the two theories appear contradictory; Kopelman claims that memory retrieval is actively inhibited by increased frontal control, and Markowitsch claims that the initiation of memory retrieval is altered and disrupted by decreased frontal function.

However, these two processes may theoretically be present in parallel, as suggested by Bell et al. (2011). These authors suggested the following two mechanisms: first, baseline activity in the right ventral fronto-temporal regions is decreased as a result of chronic frontal executive inhibition, which induces the interruption of the initiation of autobiographical memory retrieval, that is, "mnestic block syndrome." Second, when the "mnestic block syndrome" fails to prevent the initiation of memory retrieval, frontal executive inhibitory activations are excessively recruited to actively inhibit memory retrieval. Because there is evidence that supports each of the mechanisms, it is sufficiently plausible that both mechanisms operate in different time windows. Consistent with the proposition of Bell et al. (2011), Tramoni et al. (2009) examined a single patient using various methodologies, such as neuropsychology and psychophysiology. The authors suggested that dissociative amnesia might result from a combination of two processes, including a relatively passive process of "hypo-retrieval" (interruption of the initiation of memory retrieval) and an active process of "hyper-suppression" (retrieval inhibition driven by executive control) of pre-onset memories. Additional studies are needed to determine whether this theory is plausible for characterizing the clinical symptoms and the mechanisms of memory repression in dissociative amnesia.

8.4 Concluding Remarks

In this chapter, we reviewed the neural evidence for the role of active inhibitory control processes in mnemonic processing. We reviewed both voluntary and involuntary inhibitory processes, namely, the suppression and repression of memory. The findings from studies of voluntary memory suppression in healthy participants indicate that individuals may weaken or disrupt encoding and retrieval of the unwanted memory by voluntary suppression, and that the inhibitory control process is intrinsically active rather than passive. The neural basis of memory suppression is composed of the activated DLPFC, which operates as an inhibitor of the MTL, and the deactivated MTL memory system, which results in motivated forgetting.

Involuntary memory repression has been researched in patients with dissociative amnesia. Notably, both the voluntary and involuntary inhibitory control processes are thought to be driven by shared brain mechanisms to exclude unpleasant memories from awareness. The results from our own research indicate that the interactions between prefrontal cortex activation and MTL deactivation are closely linked with involuntary memory repression. These findings are consistent with the concept of "hyper-suppression"; however, the mechanism of "hypo-retrieval" is also noteworthy and is supported by several findings, in which the baseline activity in the regions responsible for the initiation of memory retrieval is decreased. A recently proposed model attempts to reconcile these two ideas, and we believe that this interpretation provides the most coherent account of the memory impairment present in dissociative amnesia.

Several issues should be considered for future studies of the neural mechanisms that underlie the active forgetting of unpleasant memories. First, in studies of voluntary memory suppression, directed forgetting paradigms and the TNT paradigm do not intrinsically capture an individual's natural motivation. The motivations for forgetting in these experiments are artificial and unnatural and do not represent true motivations, such as regulating an individual's emotional state or protecting their sense of self, which urges them to forget unpleasant memories in real life. Thus, it is necessary to develop customized experimental materials for participants that cause emotional reactions or motivations for forgetting to precisely determine the impact and mechanisms of motivated forgetting. Second, most studies have investigated correlations among specific brain regions, and the causal relationships among these regions have not been sufficiently established. Interventional techniques, such as rTMS, which may modulate brain activities, are thought to be very helpful in examining the causal relationships among brain activity changes. Finally, most studies, especially those on involuntary memory repression, are based on a relatively small number of patients with dissociative amnesia. Large numbers of patients with dissociative amnesia should be examined using functional neuroimaging with a unified experimental paradigm. These lines of research should be considered to establish and update the neurobiologically based models of voluntary suppression and involuntary repression of memory.

Although unsettled issues remain, the fundamental neural mechanisms of the active forgetting processes that control mnemonic processing have been gradually clarified. These observations provide information regarding our ability to control and change our memory. It is not uncommon for memories to be distorted by several personal factors, including emotions. This phenomenon is thought to be supported, in part, by the brain mechanisms for the adaptive aspects of forgetting unpleasant memories. The active processes of voluntary suppression and involuntary repression of memory, therefore, have the power to shape our memory and are essential and important in the context of human society. Thus, future studies of memory in a social context should focus further on the adaptive aspects of forgetting experiences and the neural mechanisms. We are hopeful that the field will make significant advances and promote our understanding of memory and forgetting in a social context in the future.

References

American Psychiatric Association (1994) Diagnostic and statistical manual of mental disorders, 4th edn. American Psychiatric Association, Washington, DC

Anderson MC (2003) Rethinking interference theory: executive control and the mechanisms of forgetting. J Mem Lang 49(4):415–445

Anderson MC, Green C (2001) Suppressing unwanted memories by executive control. Nature 410(6826):366–369

Anderson MC, Hanslmayr S (2014) Neural mechanisms of motivated forgetting. Trends Cogn Sci 18(6):279–292

Anderson MC, Huddleston E (2011) Towards a cognitive and neurobiological model of motivated forgetting. In: Belli RF (ed) True and false recovered memories: toward a reconciliation of the debate: Nebraska Symposium on Motivation. Springer, New York, pp 53–120

Anderson MC, Levy BJ (2009) Suppressing unwanted memories. Curr Dir Psychol Sci 18(4):189–194

Anderson MC, Ochsner KN, Kuhl B, Cooper J, Robertson E, Gabrieli SW, Glover GH, Gabrieli JDE (2004) Neural systems underlying the suppression of unwanted memories. Science 303(5655):232–235

Anderson MC, Reinholz J, Kuhl BA, Mayr U (2011) Intentional suppression of unwanted memories grows more difficult as we age. Psychol Aging 26(2):397–405

Aron AR, Robbins TW, Poldrack RA (2004) Inhibition and the right inferior frontal cortex. Trends Cogn Sci 8(4):170–177

Baddeley A, Eysenck MW, Anderson MC (2009) Memory. Psychology Press, Hove

Badre D (2008) Cognitive control, hierarchy, and the rostro-caudal organization of the frontal lobes. Trends Cogn Sci 12(5):193–200

Basden BH, Basden DR (1998) Directed forgetting: a contrast of methods and interpretations. In: Golding JM, MacLeod CM (eds) Intentional forgetting: interdisciplinary approache. Erlbaum, Mahwah, pp 139–172

Bäuml KH, Pastötter B, Hanslmayr S (2010) Binding and inhibition in episodic memory: cognitive, emotional, and neural processes. Neurosci Biobehav Rev 34(7):1047–1054

Bell V, Oakley DA, Halligan PW, Deeley Q (2011) Dissociation in hysteria and hypnosis: evidence from cognitive neuroscience. J Neurol Neurosurg Psychiatry 82(3):332–339

Benoit RG, Anderson MC (2012) Opposing mechanisms support the voluntary forgetting of unwanted memories. Neuron 76(2):450–460

Bjork RA (1989) Retrieval inhibition as an adaptive mechanism in human memory. Varieties of memory and consciousness: essays in honor of Endel Tulving. Erlbaum, Hillsdale

Botzung A, Denkova E, Manning L (2007) Psychogenic memory deficits associated with functional cerebral changes: an fMRI study. Neurocase 13(5–6):378–384

Brand M, Eggers C, Reinhold N, Fujiwara E, Kessler J, Heiss WD, Markowitsch HJ (2009) Functional brain imaging in 14 patients with dissociative amnesia reveals right inferolateral prefrontal hypometabolism. Psychiatry Res Neuroimaging 174(1):32–39

Costello A, Fletcher PC, Dolan RJ, Frith CD, Shallice T (1998) The origins of forgetting in a case of isolated retrograde amnesia following a haemorrhage: evidence from functional imaging. Neurocase 4(6):437–446

De Renzi E, Lucchelli F, Muggia S, Spinnler H (1995) Persistent retrograde amnesia following a minor trauma. Cortex 31(3):531–542

Depue BE, Banich MT, Curran T (2006) Suppression of emotional and nonemotional content in memory: effects of repetition on cognitive control. Psychol Sci 17(5):441–447

Depue BE, Curran T, Banich MT (2007) Prefrontal regions orchestrate suppression of emotional memories via a two-phase process. Science 317(5835):215–219

Ecker UKH, Lewandowsky S, Oberauer K (2014) Removal of information from working memory: a specific updating process. J Mem Lang 74:77–90

Erdelyi MH (2006) The unified theory of repression. Behav Brain Sci 29(5):499–551

Fawcett JM, Taylor TL (2008) Forgetting is effortful: evidence from reaction time probes in an item-method directed forgetting task. Mem Cogn 36(6):1168–1181

Fawcett JM, Taylor TL (2010) Directed forgetting shares mechanisms with attentional withdrawal but not with stop-signal inhibition. Mem Cogn 38(6):797–808

Freud S (1966) In: Strachey J (ed) The standard edition of the complete psychological works of Sigmund Freud 1. Hogarth, London, pp 117–128

Gagnepain P, Henson RN, Anderson MC (2014) Suppressing unwanted memories reduces their unconscious influence via targeted cortical inhibition. Proc Natl Acad Sci U S A 111(13):E1310–E1319

Geiselman RE, Bjork RA, Fishman DL (1983) Disrupted retrieval in directed forgetting: a link with posthypnotic amnesia. J Exp Psychol 112(1):58–72

Hanslmayr S, Leipold P, Pastötter B, Bäuml KH (2009) Anticipatory signatures of voluntary memory suppression. J Neurosci 29(9):2742–2747

Hanslmayr S, Volberg G, Wimber M, Oehler N, Staudigl T, Hartmann T, Raabe M, Greenlee MW, Bäuml KH (2012) Prefrontally driven downregulation of neural synchrony mediates goal-directed forgetting. J Neurosci 32(42):14742–14751

Hart RE, Schooler JW (2012) Suppression of novel stimuli: changes in accessibility of suppressed nonverbalizable shapes. Conscious Cogn 21(3):1541–1546

Hennig-Fast K, Meister F, Frodl T, Beraldi A, Padberg F, Engel RR, Reiser M, Möller HJ, Meindl T (2008) A case of persistent retrograde amnesia following a dissociative fugue: neuropsychological and neurofunctional underpinnings of loss of autobiographical memory and self-awareness. Neuropsychologia 46(12):2993–3005

Hertel PT, Large D, Stück ED, Levy A (2012) Suppression-induced forgetting on a free-association test. Memory 20(2):100–109

Kaplan HI, Sadock BJ (1995) Comprehensive textbook of psychiatry, 6th edn. William & Wilkins, New York, pp 650–653

Kapur N (2000) Focal retrograde amnesia and the attribution of causality: an exceptionally benign commentary. Cogn Neuropsychol 17(7):623–637

Kihlstrom JF (2002) No need for repression. Trends Cogn Sci 6(12):502

Kikuchi H, Fujii T, Abe N, Suzuki M, Takagi M, Mugikura S, Takahashi S, Mori E (2010) Memory repression: brain mechanisms underlying dissociative amnesia. J Cogn Neurosci 22(3):602–613

Kopelman MD (2000) Focal retrograde amnesia and the attribution of causality: an exceptionally critical review. Cogn Neuropsychol 17(7):585–621

Kopelman MD (2002) Disorders of memory. Brain 125(10):2152–2190

MacLeod CM, Daniels KA (2000) Direct versus indirect tests of memory: directed forgetting meets the generation effect. Psychon Bull Rev 7(2):354–359

Markowitsch HJ (2003) Psychogenic amnesia. NeuroImage 20(S1):S132–S138

Markowitsch HJ, Fink GR, Thöne A, Kessler J, Heiss WD (1997) A PET study of persistent psychogenic amnesia covering the whole life span. Cogn Neuropsychiatry 2(2):135–158

Markowitsch HJ, Kessler J, Weber-Luxenburger G, Van der Ven C, Albers M, Heiss WD (2000) Neuroimaging and behavioral correlates of recovery from mnestic block syndrome and other cognitive deteriorations. Neuropsychiatry Neuropsychol Behav Neurol 13(1):60–66

Miller EK, Cohen JD (2001) An integrative theory of prefrontal cortex function. Annu Rev Neurosci 24:167–202

Muther WS (1965) Erasure or partitioning in short-term memory. Psychon Sci 3(1):429–430

Nakamura H, Kunori Y, Mori K, Nakaaki S, Yoshida S, Hamanaka T (2002) Two cases of functional focal retrograde amnesia with impairment of object use. Cortex 38(4):613–622

Noreen S, MacLeod MD (2013) It's all in the detail: intentional forgetting of autobiographical memories using the autobiographical think/no-think task. J Exp Psychol Learn Mem Cogn 39(2):375–393

Noreen S, MacLeod MD (2014) To think or not to think, that is the question: individual differences in suppression and rebound effects in autobiographical memory. Acta Psychol 145:84–97

Nowicka A, Marchewka A, Jednoróg K, Tacikowski P, Brechmann A (2011) Forgetting of emotional information is hard: an fMRI study of directed forgetting. Cereb Cortex 21(3):539–549
Paller KA, Wagner AD (2002) Observing the transformation of experience into memory. Trends Cogn Sci 6(2):93–102
Rizio AA, Dennis NA (2013) The neural correlates of cognitive control: successful remembering and intentional forgetting. J Cogn Neurosci 25(2):297–312
Squire LR, Zola-Morgan S (1991) The medial temporal lobe memory system. Science 253(5026):1380–1386
Tramoni E, Aubert-Khalfa S, Guye M, Ranjeva JP, Felician O, Ceccaldi M (2009) Hypo-retrieval and hyper-suppression mechanisms in functional amnesia. Neuropsychologia 47(3):611–624
Williams M, Hong SW, Kang MS, Carlisle NB, Woodman GF (2013) The benefit of forgetting. Psychon Bull Rev 20(2):348–355
Wylie GR, Foxe JJ, Taylor TL (2008) Forgetting as an active process: an fMRI investigation of item-method-directed forgetting. Cereb Cortex 18(3):670–682
Yang JC, Jeong GW, Lee MS, Kang HK, Eun SJ, Kim YK, Lee YH (2005) Functional MR imaging of psychogenic amnesia: a case report. Korean J Radiol 6(3):196–199
Yang T, Lei X, Anderson M (2016) Decreased inhibitory control of negative information in directed forgetting. Int J Psychophysiol 100:44–51
Zacks RT, Radvansky G, Hasher L (1996) Studies of directed forgetting in older adults. J Exp Psychol Learn Mem Cogn 22(1):143–156

Chapter 9
The Impact of Interoception on Memory

Yuri Terasawa and Satoshi Umeda

Abstract Using interoception as a keyword, this chapter has summarized how changes in the internal physical state are encoded and how the processing of interoceptive information affects the memory of episodes and events. Interoceptive information processing is discussed frequently in the context of emotion in psychology. On the other hand, research studies and literature that have examined its relation to episodic memory are rare. However, changes in the internal physical state exist at the foundation of various concepts that have examined its relationship with memory. To understand the relationship more precisely, we focused on the following issues in this chapter: emotional arousal and bodily responses; effects of bodily responses on memory encoding; judgment related to memories and bodily response; and insular cortex and memory consolidation. Throughout the chapter, we presented the possibility that change in the physical state itself plays a role in latently supporting the encoding and recollection process.

Keywords Interoception • Arousal • Amygdala • SCR • Heart rate • Insula cortex • Emotion • Memory consolidation

9.1 Introduction

Investigating the relationship between emotions and memories has generated many beneficial studies by serving as the main topic for cognitive–emotional interaction. Comparing the memorization of events that have positive or negative emotional valence with that of neutral events has shown that the number of events that can be recalled, the accuracy of details, and vividness differ with emotionality (Kensinger and Schacter 2008). It has also been demonstrated that the emotional state during recollection affects the events that are recalled and their content. It has been implied that people suffering from affective disorders (i.e., depression or anxiety) demonstrate

Y. Terasawa (✉) • S. Umeda
Department of Psychology, Faculty of Letters, Keio University,
2-15-45 Mita, Minato-ku, Tokyo, Japan
e-mail: terasawa@flet.keio.ac.jp

T. Tsukiura, S. Umeda (eds.), *Memory in a Social Context*,
https://doi.org/10.1007/978-4-431-56591-8_9

certain characteristics when processing memories. According to these findings, few would cast doubt on the significance of studying the impact that emotions have on memories. With that in mind, what about the relationship between memories and changes in bodily responses? Most people may not have presumed the existence of any relationship between these two factors, and they may be taken aback by such a seemingly random association. However, are memories and bodily responses really two factors with a tenuous relationship?

One such example is classical fear conditioning. In animal experiments, whether or not fear conditioning has been achieved can be judged based on whether a fear reaction is observed when a conditioned stimulus is presented. With rats, reactions such as freezing can serve as indicators, whereas with humans, sympathetic nervous system activations are used as indicators. These can include an increase in heartbeat and respiration rate or a skin conductance response (SCR). According to the dimensional theory of emotions, "valence" and "arousal" are concepts that cannot go unmentioned. Of these two, arousal has been shown to have an extremely strong connection with bodily response. "Arousal," as a term, becomes an important keyword when discussing the relationship between memories and emotions. As such, it demonstrates that the instinctive judgment that the relationship between memories and bodily responses is only tenuous may be wrong.

Although interest in research into the impact an individual's body has on the mind is wide, reviewing previous research shows that few studies directly examined the impact of the physical state upon memories (in particular, declarative memory). In this chapter, we present various phenomena related to bodily responses and memories, and then discuss the roles of bodily responses during memory encoding and recollection.

9.2 Emotional Arousal and Bodily Responses

First, we would like to consider the relationship between the emotions that modulate memories and bodily responses. Although several studies have discussed the relationship between memories and emotions based on valence, this section focuses on "arousal," a term that needs extra attention, as an entry point. Arousal, along with "valence," is a concept that is used when defining emotions with a dimensional method. Although valence has relatively easy-to-understand characteristics such as pleasantness/unpleasantness, positive/negative, or approach/avoidance, the definition of arousal is slightly more complex. When Wundt discussed his three concepts used for defining the human emotional state, arousal was one of them. Here, it was used as a concept focusing on the contrast between excitement and tranquility. An indication that human emotions cannot be defined solely through pleasantness/unpleasantness and that they cannot be separated from arousal has served as the foundation for the circumplex theory of emotion (Russell and Lemay 2000), which is widely accepted today. The core affect theory (Barrett 2006), which originated from the circumplex theory, agrees with this assertion, and claims that affect is a fundamental factor within the human mind and is a central existence that affects various psychological processes. Discussions on whether these two axes intersect orthogonally or obliquely, or whether it is a structure that is common among individuals, are still ongoing. However, many studies on emotions that use this perspective as their foundation have been conducted (Kuppens et al. 2013).

Incidentally, how is arousal generated and adjusted? As mentioned above, this concept represents excitement and tranquility. As you read this, I would like you to imagine a state of excitement or a tranquil state in which you are about to fall asleep. What is the difference between these two situations? You would probably notice that the physical state is extremely different. During an excited state, a person would feel their heart beating or blood going to their head. On the other hand, a tranquil state conjures up an image of the body feeling heavy and listless or a quiet state in which the mind feels as if it has been cut off from the body. As you could see here, the state of your body strongly affects arousal, which in turn may strongly affect the emotional state. In reality, measuring the peripheral autonomic response (i.e., SCR, which reflects sympathetic nervous system activity) is mainly believed to be effective in evaluating arousal (Boucein 1992). As such, it is inseparable from the change in physical state. This is a fact that conforms highly to the peripheral theory of emotions, which proposes that an individual's physical state affects their emotional state.

An event with high arousal tends to be remembered later. In one study, the participants were asked to record the events that occurred each day, their content, and the arousal rating in a journal. The participants were later asked to recall the events using cues as clues. Such research studies have confirmed that events that attained a high arousal rating are likely to be recalled later (that said, whether the recalled content is correct when collated with the actual event is an enduring issue – in reality, the relationship between arousal and the accuracy of memory may be shaped like an inverted U, as with other cognitive functions).

The brain area that shows the activity that corresponds to arousal discussed here is known as the amygdala. This area has been identified as a brain area whose activity differs depending on the arousal level (LaBar and Cabeza 2006). The amygdala connects the medial temporal lobe with the memory system (hippocampus, entorhinal cortex, and perirhinal cortex) bi-directionally. This connection cannot be excluded when explaining the modulation of emotions related to memory. In reality, the tendency for the amygdala's activity to increase when encoding emotional events has been confirmed to be homologous with the increase in the activity in the entorhinal cortex, part of the memory system. Furthermore, the level of amygdala activity during encoding is a good predictor of whether or not the event can be recalled later (Kensinger and Schacter 2008).

As inferred from the close relationship with the memory system, case studies of bilateral amygdala lesions show that enhancements of memory due to emotions are unlikely to occur. However, judging from the fact that no significant decline in general declarative memory is seen, the modulation of the memory system by the amygdala is lost, rather than the memory system function itself. This supports the idea that an increase in arousal plays a key factor in memory. The fact that the change in arousal correlates with memory performance suggests that the mechanism that heightens the body's tension might be triggered because of experiencing a certain event, which may be involved in consolidating the memory so that this leads to the recollection of the event later. Arousal effects on memory would be implemented by the associated activity of amygdala and hippocampus corresponding to the arousal.

Focusing on the neural substrates of arousal shows how much the characteristics of arousal depend on an individual's physical state. Let us discuss functional magnetic resonance imaging (fMRI) studies that have examined the brain areas related to recognition of arousal. Looking at these studies highlights the insular cortex, in addition to the amygdala, as a key area. The insular cortex is located in the interior frontal operculum area and its function is related to the perception and modulation of information originating from the body, particularly visceral sensation (Damasio 1994). An fMRI study that used emotion-related words showed that some regions respond to both arousal level and valence, whereas areas that have been identified as showing activity that only corresponds to arousal were the anterior ventral insula, the ventral striatum, the globus pallidus, and the amygdala (Lewis et al. 2007). In terms of neuropsychological research, Berntson et al. (2011) presented images that elicit emotions to an insular cortex lesion group, amygdala lesion group, and a lesion control group, and evaluated the emotional valance and arousal toward the images. The results showed a pattern in which the insular cortex lesion group estimated the arousal to be lower regardless of the valence.

An investigation that we have conducted also obtained an experimental result in which the right insular cortex lesion group similarly suggested a lowered ability to perceive arousal. On the other hand, however, no significant difference compared with a healthy person was seen in SCR, a physiological indicator closely related to arousal (Terasawa et al. 2010). This suggests that the insular cortex might be deeply involved in the perception of a change in physical state, which serves as the basis of arousal, rather than generating such changes. In fact, earlier studies have shown that even if there is no salient bodily response, the insular cortex refers to the internal bodily state and integrates the state and environmental information for yielding emotional awareness based on interoception (Critchley et al. 2004; Terasawa et al. 2013). Activity in this region, in addition to neuropsychological findings, implies how accurately the internal physical state affects recognition of arousal, and subsequently how its impact on memory is by no means non-existent.

9.3 How Are Bodily Responses Encoded in Memory?

In the previous section, we discussed how bodily responses affect memory via arousal. However, "arousal" is a concept that psychologists have proposed as a dimension comprising states of emotion and awareness. When people go about their day-to-day lives, they are rarely aware of whether they have a high or low level of arousal. What people feel are the emotions that change from moment to moment depending on events. The change in an individual's body during an excited state (i.e., changes in heartbeat, breathing, and perspiration) may be perceived as a situational state or as an emotion. What is important here is that we experience such changes in emotional and physical states continuously along with the transitions between events. This integrated experience is indispensable when we evaluate events or objects that we faced. When people are asked to recall episodic memories, it is common for people to report the content of the events or their overall valence. However, do people remember the physical sensation that affects that arousal itself? Alternatively, do people interpret it as emotion and simply encode or recall the overall evaluation?

McCall et al. (2015) explored this issue using a virtual reality (VR) setting. They asked participants to experience a threatening scene in VR and continuously recorded their heart rate and SCR. Afterward, the participants were shown the same scene on a computer and were asked to provide continuous reports of how aroused they were in each scene. Even with this retrospective method, the reports by the participants and the changes in SCR and heart rate while watching the 3D images coincided extremely well. It was also shown that the level of congruence was high among subjects who had acute interoception – in other words, people who excelled in sensing the changes that were occurring within their bodies. This fact also indicates vividly how the arousal experienced during an event is memorized along with the details of the event. The authors proposed the arousal encoding hypothesis, stating that "memory spontaneously encodes an individual's physiological signals during an experience." It could be interpreted from this research that the encoded details and the accompanying emotional reactions based on the bodily response are much more strongly linked than predicted. Such a close relationship may regulate the reality or salience of the memory.

As mentioned previously, interoception not only generates the perception of physiological changes, but it is often processed latently as a part of evaluation by a subject of objects, situations, and events. The contents that are perceived explicitly are the values of the subject and not the change in the accompanying physical state. However, when we think of how such values are formed, we realize that interoception may play a significant part. There have been many studies on the effects that accompanying emotions have on the memory of the event's content. However, it is very interesting that at the root of these phenomena are the continuously changing bodily responses and their perception. When considering the recollection of an event that accompanies several vivid sensations in particular, as seen with post-traumatic stress disorder, we realize the importance of closely examining the significant, albeit latent, effect that interoception has on the consolidation of memories.

Many other research data imply that interoceptive information is strongly associated with episodic memory. Hirsh (1974) stated that interoception provides important cues for memory retrieval, whereas a recent study conducted by Araujo et al. (2015) indicates that hippocampus activity increases when instructed to pay attention to interoception rather than exteroception. This may signify that interoception triggers more memories. Furthermore, although it has been indicated that exteroceptive information regarding episodic memories is encoded by the associations between the interior and exterior of the hippocampus, recent research has also indicated that the hippocampus is involved in the encoding of interoceptive factors (Kassab and Alexandre 2015). If we were to assume that, with episodic memories, there is a system in the hippocampus that supports the encoding and recollection of interoceptive and exteroceptive information, this means that interoceptive information is also included in episodic memories, as McCall et al. (2015) have indicated. Thus, it would be a reasonable inference that exteroceptive cue information would be useful in recalling interoceptive sensation.

9.4 Judgment Related to Memories and Bodily Response

In the previous section, we indicated that episodic memory may also include interoceptive content and not just the exteroceptive aspects of events, as has been pointed out by previous research. In particular, somatic information was noted as being a part of the information to be encoded, and its relationship with arousal was discussed. However, the relationship between memory and somatic responses has been repeatedly shown, even in the field of relatively classical psychology.

Previous studies that use classical psychophysiological methods maintain the position that bodily responses that occur with information that has been experienced before are different from the responses to information that has never been experienced. The origin of this approach can be traced back to the theory that Sokolov (1963) proposed regarding the orientation response to the novelty of the stimuli. This theory hypothesizes that when the neuronal model of stimuli that exists in the cerebral cortex concerning the stimuli that the individual is currently in contact with is compared with the neural model of stimuli that the individual has come into contact with in the past, an orientation response occurs when there is no previously existing model that matches. Although there are aspects concerning the accuracy of the detailed process that cannot be agreed on in view of current research results, the existence of an implicit processing system that generates a bodily response differing from the explicit memory response is noteworthy.

In fact, in studies with amnesia (Rapcsak et al. 1998; Verfaellie et al. 1991), prosopagnosia (Tranel and Damasio 1985), and children (Newcombe and Fox 1994; Stormark 2004) as examples, it has been reported that the performance of recognition tasks and the autonomic responses of the task being performed were dissociated. In all these studies, specific changes in autonomic reactions (i.e., increases in SCR and heart rate) were seen toward stimuli that the subjects had learned, even if they were unable to report that they "knew" the stimuli. For example, a study on patients with prosopagnosia showed that when the correct name to the face presented was read aloud, or the face of a person with high familiarity was shown to the patients, a large SCR was observed, despite the subjects being unable to distinguish the face explicitly (Tranel and Damasio 1985). When pictures of current classmates, former classmates, or unknown children were shown to kindergarten pupils and schoolchildren, and they were asked if they could determine whether or not they knew the children in the pictures, they were not able to recognize previous classmates and showed the same reaction as when they saw children they did not know. However, the SCR during the presentation of facial stimuli showed a large response when the current and previous classmates were shown, whereas the response was smaller in the unknown children condition (Newcombe and Fox 1994; Stormark 2004).

Such processing implies the existence of latent processing for constructing familiarity; it also indicates that transitions in bodily responses are involved in such processing. It is not clear whether the observed change in autonomic response appears simply in association with the implicit processing or if it has some sort of impact on our cognitive processes. However, it has been pointed out that perceiving the generated autonomic response enables the effective use of limited cognitive resources, focusing on matters in which a person should develop deeper processing; in other words, it is a function for creating alertness (Oehman 1979). The perception of an increase in arousal is probably also included as one such process. For example, the feeling of knowing (FoK) judgment is a phenomenon that can detect the relationship between the interoception process and decision-making. This judgment seeks to instantly determine whether an individual feels like they know the item presented.

The accuracy of metacognition can be evaluated by the reaction and the actual known/unknown state. The accuracy of the FoK judgment has been reported to be significantly lower in examples involving ventromedial prefrontal area lesions (Schnyer et al. 2004). In other words, the percentage of times in which things that an individual claimed to know were actually unknown, or the percentage of times when things they felt they did not know were actually something they knew, increased in patients with ventromedial prefrontal cortex lesions. Generally, lesions to this area are not associated with severe amnesia. As such, this area is believed to be deeply involved in the feeling that an individual's own judgment is certain, rather than the recollection of memory itself. Our research has identified the ventromedial prefrontal area as a significant area in interoception (Terasawa et al. 2013).

Previously, Craig (2009) inferred that the insular cortex operates supportively with regard to FoK judgment. However, the existence of interoceptive processing in FoK judgment had not been mentioned explicitly. Nevertheless, an interesting hypothesis can be derived when the results from the preceding studies and this study are summed up. The hypothesis states that accurately perceiving the bodily responses generated in association with some sort of psychological process improves the accuracy of monitoring an individual's own self. In actuality, a study that used a questionnaire showed a result in which the accuracy of metamemory – a self-monitoring function of an individual's own memory – and the accuracy of interoception are correlated, supporting this hypothesis (Chua and Bliss-Moreau 2016). With judgment related to recognition and FoK, it is not clear if interoception improves the accuracy of judgment or has some effect on the generation of a bias. However, the results of this study have a possible application to such fields.

It is not easy to elucidate whether the bodily responses accompanying recognition have the function of increasing the accuracy of judgments related to memory or whether they are merely a reaction accompanying cognitive processing. We would like to introduce our study, in which we have used a framework known as prospective memory (PM) to conduct an experiment concerning this question (Umeda et al. 2016). PM refers to an ability in which an individual can recall an intention memorized in the past at the right time in the future and perform an action as they had intended. In daily life, people often remember things in anticipation of recalling them at a certain point in the future. For example, this could include remembering to e-mail a certain person tomorrow or asking a certain question when you meet your colleague. As seen here, PM is a memory function that is indispensable for supporting a smooth daily life. However, as an analogical inference from the Zeigarnik effect can show, psychological stress exists when maintaining an intention. This psychological stress can be evaluated by observing changes in autonomic activity, including the SCR and the heart rate. Past studies have observed an enhancement of SCR in conjunction with the presentation of a cue associated with an intention, which may promote the processing related to the recollection of the intention (Kliegel et al. 2007; Rothen and Meier 2014). In the experiment, we had our subjects conduct a 2-back task that employed alphabet letters. During this 2-back task, we asked the subjects to respond using specific buttons only when certain letters (a, i, u, e, and o) were presented and not to react to the 2-back task. These specific letters were the PM targets, and whether or not the subjects were able to press specific buttons in accordance with the PM targets served as the PM performance.

When observing changes in heart rate during PM performance and when the PM targets were presented, there was an interesting positive correlation between the two. Those who were able to perform without forgetting the PM had an accelerated heartbeat during PM target observation. Furthermore, their performance showed a positive relationship with individual interoceptive accuracy as measured during the heartbeat perception task (Fig. 9.1). This means that those who have accurate interoception show better PM performance. How should we understand the relationship among the three factors of PM performance, change in heart rate, and the interoceptive accuracy that was observed in our study? We believe that by perceiving the changes in the body that occur in relation to psychological processing associated with the presentation of the PM target, the recollection of intentions encoded in the past at an appropriate time becomes enhanced.

The idea that autonomic responses affect the success or failure of the PM may seem difficult to accept at a first glance. However, past research studies have also indicated that realization of PM targets is a factor that affects the success of recalling PM (West and Craik 1999). This realization is a phenomenon that has a large commonality with the so-called orientation response. Given the significant contribution of autonomic responses toward the orientation response, changes in the autonomic responses and their perception can be interpreted to underlie the realization of the PM targets. We believe that this result is not limited to PM but, in conjunction with the aforementioned findings of previous studies, is a significant piece of information that suggests a close relationship between memories and bodily responses. There is a hypothesis that states that a crucial role of the insular cortex, the neural correlate

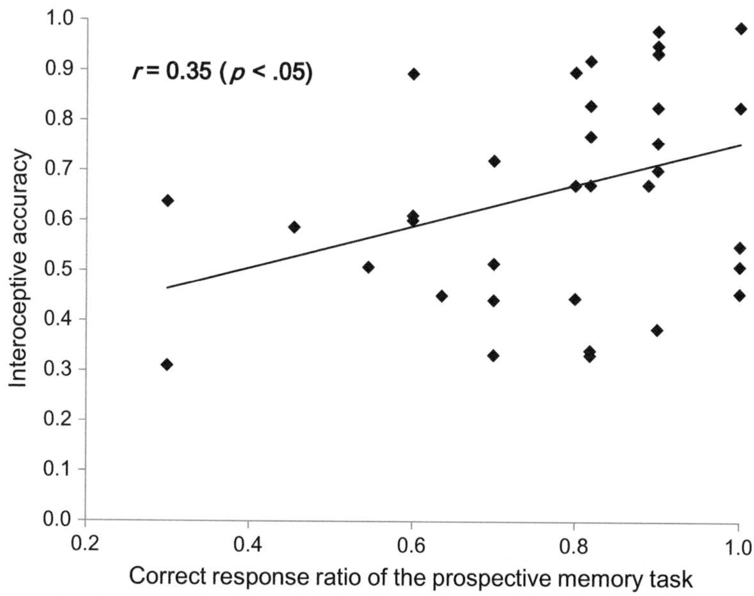

Fig. 9.1 Positive correlation between prospective memory (PM) performance and interoceptive accuracy. Participants who showed higher PM performance could also feel their heartbeats accurately. The authors of Umeda et al., 2016 allowed reproducing the figure based on thieir data

of interoception, is being responsible for switching attention toward the interior or exterior (Menon and Uddin 2010). An event that is the recognition of prospective or episodic memory involves detecting a trigger from abundant environmental information and accumulating further cognitive processing concerning that trigger. This then recalls a situation that has been experienced before or refines its content. Instead, it may incite a specific behavior that has been associated with the trigger. This system may be better explained if it is hypothesized that the change in the interior environment might function as a signal that promotes the detection of an important cue or deepening of the process, rather than explaining it by merely referencing attention and processing with regard to the external environment.

9.5 Insular Cortex and Memory Consolidation

We have been discussing performance, bodily responses, and interoception in relation to episodic memory. From the data so far, we can infer that memories and bodily response are correlated during a fundamental processing of episodic memory. We can also presume that the insular cortex as a key area for connecting interoception and episodic memory processing.

Detailed examinations of the relationship between the insular cortex and memory using humans as subjects have not been conducted very often. However, a meta-analysis of an fMRI study indicated that the anterior dorsal part of the insular cortex shows memory-related activity. Kurth et al. (2010) conducted an activation likelihood estimation (ALE) using 811 research papers and revealed that activity in the frontal insular cortex was often reported among studies that dealt with working memory and episodic memory. As the insular cortex was not a member of a so-called neural network related to memory, most memory researchers did not believe that the insular cortex was closely related to memory. As a result, activity in the insular cortex has been identified in many fMRI studies related to memory, but an in-depth examination was never conducted because the function of the insular cortex in memory was unclear. When examining the role of this area in memory, because its connections were revealed by a meta-analysis, it is possible to interpret it as associated with interoception processing. In actuality, recall of PM, which can accurately recall the intention, was revealed to be proportionate to the ability of the individual to perceive the change in their body (Umeda et al. 2016). This study did not conduct an examination by directly comparing activity in the insular cortex with the recall performance for PM. However, recent animal research has indicated interesting aspects of memory functions of the insular cortex.

In animal studies, the insular cortex was, for a long time, believed to be an area related to the sense of taste. Indeed, this area does show activity related to gustatory sensation. However, recent studies have suggested the possibility that this area might not be specifically active in relation to taste.

What is interesting is that there have been studies that indicate that the insular cortex is, in fact, related to memory consolidation. Memory consolidation is indispensable to the formation of a long-term declarative memory. Past studies have revealed that memory consolidation takes a little time after first experiencing an event. If an operation is added that prevents consolidation during this time, the memory will not be consolidated, decreasing the probability of this episode being recalled later. Bermudez-Rattoni et al. (2005) allowed rats to freely explore white glass light bulbs or transparent glass vials and injected scopolamine (a muscarinic cholinergic receptor antagonist) into the rats' insular cortex immediately after learning about the objects or after 160 min. In a retention test conducted 24 h later, rats that had had scopolamine injected into their insular cortex immediately after learning showed no difference in their exploratory behavior between the object they studied the day before and the new object. In other words, no effects of learning were observed, as blocking of memory consolidation was believed to have occurred. Based on the fact that such an effect was not observed in the rats that had had a similar injection 160 min after learning, it follows that the insular cortex plays a significant role in the memory consolidation immediately after learning. Bermudez-Rattoni published a review in 2014 related to the role of the insular cortex in the formation of recognition memory, indicating that acetylcholine and dopamine were released from the insular cortex in reaction to novel stimuli, promoting cortical plasticity (Bermudez-Rattoni 2014). Bermudez-Rattoni stated that these chemicals enhance cortical remodeling in the insula for the consolidation of recognition memory. Furthermore, he also stated his view that the area in which cortical organization occurs would differ depending on whether the target stimuli are hedonic or aversive.

Even in studies that employed drugs and conditioned place preferences, it was found that injecting a chemical known as anisomycin into the high-order rostral agranular insular cortex when reconsolidation is occurring eradicates the effects of conditioned place preferences (Contreras et al. 2012). This effect was also confirmed with injection of the chemical into the primary interoceptive posterior insular cortex (pIC). Anisomycin is believed to reduce the expression of zif268, a protein that plays a crucial role in memory consolidation. Further, pIC is an area that is known to be deeply involved with the sensation of cravings toward drugs. As such, this area is suggested to be involved in the formation and recollection of memories that connect interoception with a context. Casanova et al. (2016) later performed auditory fear conditioning on rats to examine the relationship between this effect and the activity of the insula. In this experiment, anisomycin was injected into the pIC immediately or 6 h after the auditory fear conditioning took place. If the conditioning was maintained on the following day, or even the day after, a freezing reaction would be seen when a conditioned stimulus was presented. However, such a reaction was not observed among the group in which anisomycin was injected immediately after conditioning. On the other hand, such eradication of conditioning was not seen among the 6-h post-training injection group either, and no similar effect was seen when anisomycin was simply injected anywhere in the brain. It is noteworthy that no such effect was seen when anisomycin was injected into the primary somatosensory cortex.

Bodily sensation underlies pleasure from drug administration, craving from the lack of it, and terrifying fear. Such experiences have the effect of establishing robust learning, even if the number of experiences is small. This is what triggers a state that cannot easily be controlled with reasoning, such as addiction or phobia. Although it could have been assumed that the insular cortex, which is believed to be the neural correlate of interoception, would display a connection in this context, its strong impact on the formation of memory is a result that goes beyond prediction. When familiar with memory research in humans, the likely belief is that memory is mainly formed by information input from five main senses (seeing, hearing, tasting, smelling, and touching), known as exteroception. This is because memory functions are studied by displaying sentences, words, and images in such studies. As mentioned previously, the trend in memory studies within the past 20 years has mainly focused on the relationship between memory and emotions, and these studies have discussed the topic as if emotion is related to a particular context. However, thinking about this matter carefully brings about the realization that it is not surprising for evaluating information input through the five senses (this could be explicit or implicit). Although the evaluation may vary it shares many commonalities with emotions, such as whether the subject is important or whether it is positive or negative. Results from animal studies suggest that, in the end, such information might be coded via the body. The memory of senses that occurred within an individual's body functions as a fast and clear signal to induce reactions that the organism should take, if they encounter a similar situation or event. Even for this purpose, the fact that the neural correlate of interoception supports such memories would be extremely reasonable.

What about such an association in human? Many studies have revealed the activation of the anterior cingulate cortex (ACC) and the insula when evaluating an individual's own emotional state toward the situation they are facing. Inferences from rat experiments state that this activity should contribute to the act of adding values to the environmental information and to events that brought about such an emotional state, and to the formation of memories. However, this system lacks much clarity, including how the memories are being boosted. Pais-Vieira et al. (2016) displayed pictures from the International Affective Picture System to subjects and asked them to evaluate what kind of impact they had through such stimuli to conduct an interoceptive self-orientation. fMRI was used for imaging of cerebral activity during this experiment. The authors found that anterior insula and ACC activity could predict later memory performance. Furthermore, it was revealed that when encoding memory that was later recalled with great confidence, the connectivity between the amygdala and dorsal ACC was enhanced. As we wrote at the beginning of this chapter, these regions are essential in the adjustment of arousal, and this result implies that modulation of emotion-related memory takes place via the body. Pais-Vieira has also suggested the possibility that insular cortex activity might promote memory consolidation after referencing various research studies utilizing rats. However, we need to wait for further updates concerning the examination of this hypothesis.

9.6 Conclusion

Using interoception as a keyword, this chapter has summarized how changes in the internal physical state are encoded and how the processing of interoceptive information affects the memory of episodes and events. Interoceptive information processing is discussed frequently in the context of emotion in psychology. On the other hand, research studies and literature that have examined its relation to episodic memory are rare. However, changes in the internal physical state exist at the foundation of various concepts that have examined its relationship with memory (i.e., arousal, emotion, and the weight of episodic values). We found that changing the perspective slightly could become an important first step in subsuming these concepts. In other words, the possibility that change in the physical state itself might play a role in latently supporting the encoding and recollection process, and not merely as content to be encoded, has been revealed. In actuality, studies that focused on the relationship between autonomic nerve activity and memory have commenced in very recent years. As an example, an interesting result regarding the relationship between the cardiac cycle and interoception has been reported. If the words presented during the systolic period and the diastolic period and the recognition score were calculated respectively, only the score during the systolic period showed a positive correlation with individuals' interoceptive accuracy (Garfinkel et al. 2013). This suggests that the fidelity of interoceptive representation might mitigate the interference of physiological arousal with memory. Significant progress would be made if it became clear what kind of neural network is involved in the background of this mechanism. Reviewing the newly discovered findings and data accumulated on memories in cognitive neuroscience can contribute to constructing the overall picture of the cognitive neural process of memory. At present, few studies have directly dealt with this issue, and we await future developments.

References

Araujo HF, Kaplan J, Damasio H, Damasio A (2015) Neural correlates of different self domains. Brain Behav 5(12):e00409

Barrett LF (2006) Emotions as natural kinds? Perspect Psychol Sci 1(1):28–58

Bermudez-Rattoni F (2014) The forgotten insular cortex: its role on recognition memory formation. Neurobiol Learn Mem 109:207–216

Bermudez-Rattoni F, Okuda S, Roozendaal B, McGaugh JL (2005) Insular cortex is involved in consolidation of object recognition memory. Learn Mem 12(5):447–449

Berntson GG, Norman GJ, Bechara A, Bruss J, Tranel D, Cacioppo JT (2011) The insula and evaluative processes. Psychol Sci 22(1):80–86

Boucein W (1992) Electrodermal activity. Plenum Press, New York

Casanova JP, Madrid C, Contreras M, Rodriguez M, Vasquez M, Torrealba F (2016) A role for the interoceptive insular cortex in the consolidation of learned fear. Behav Brain Res 296:70–77

Chua EF, Bliss-Moreau E (2016) Knowing your heart and your mind: the relationships between metamemory and interoception. Conscious Cogn 45:146–158

Contreras M, Billeke P, Vicencio S, Madrid C, Perdomo G, Gonzalez M, Torrealba F (2012) A role for the insular cortex in long-term memory for context-evoked drug craving in rats. Neuropsychopharmacology 37(9):2101–2108

Craig AD (2009) How do you feel – now? The anterior insula and human awareness. Nat Rev Neurosci 10(1):59–70

Critchley HD, Wiens S, Rotshtein P, Ohman A, Dolan RJ (2004) Neural systems supporting interoceptive awareness. Nat Neurosci 7(2):189–195

Damasio AR (1994) Descarte's error. Penguin Putnam, New York

Garfinkel SN, Barrett AB, Minati L, Dolan RJ, Seth AK, Critchley HD (2013) What the heart forgets: cardiac timing influences memory for words and is modulated by metacognition and interoceptive sensitivity. Psychophysiology 50(6):505–512

Hirsh R (1974) The hippocampus and contextual retrieval of information from memory: a theory. Behav Biol 12(4):421–444

Kassab R, Alexandre F (2015) Integration of exteroceptive and interoceptive information within the hippocampus: a computational study. Front Syst Neurosci 9:87

Kensinger EA, Schacter DL (2008) Memory and emotion. In: Lewis M, Haviland-Jones JM, Barrett LF (eds) Handbook of emotions. The Guilford Press, New York, pp 601–617

Kliegel M, Guynn MJ, Zimmer H (2007) The role of noticing in prospective memory forgetting. Int J Psychophysiol 64(3):226–232

Kuppens P, Tuerlinckx F, Russell JA, Barrett LF (2013) The relation between valence and arousal in subjective experience. Psychol Bull 139(4):917–940

Kurth F, Zilles K, Fox PT, Laird AR, Eickhoff SB (2010) A link between the systems: functional differentiation and integration within the human insula revealed by meta-analysis. Brain Struct Funct 214(5–6):519–534

LaBar KS, Cabeza R (2006) Cognitive neuroscience of emotional memory. Nat Rev Neurosci 7(1):54–64

Lewis PA, Critchley HD, Rotshtein P, Dolan RJ (2007) Neural correlates of processing valence and arousal in affective words. Cereb Cortex 17(3):742–748
McCall C, Hildebrandt LK, Bornemann B, Singer T (2015) Physiophenomenology in retrospect: memory reliably reflects physiological arousal during a prior threatening experience. Conscious Cogn 38:60–70
Menon V, Uddin LQ (2010) Saliency, switching, attention and control: a network model of insula function. Brain Struct Funct 214(5–6):655–667
Newcombe N, Fox NA (1994) Infantile amnesia: through a glass darkly. Child Dev 65(1):31–40
Oehman A (1979) The orienting response, attention and learning: an information-processing perspective. In: Kimmel HD, van Olst EH, Orlebeke JF (eds) The orienting reflex in humans. Erlbaum, Hillsdale
Pais-Vieira C, Wing EA, Cabeza R (2016) The influence of self-awareness on emotional memory formation: an fMRI study. Soc Cogn Affect Neurosci 11(4):580–592
Rapcsak SZ, Kaszniak AW, Reminger SL, Glisky ML, Glisky EL, Comer JF (1998) Dissociation between verbal and autonomic measures of memory following frontal lobe damage. Neurology 50(5):1259–1265
Rothen N, Meier B (2014) Psychophysiology of prospective memory. Memory 22(7):867–880
Russell JA, Lemay G (2000) A dimensional-contextual perspective on facial expressions. Jpn Psychol Rev 43:161–176
Schnyer DM, Verfaellie M, Alexander MP, LaFleche G, Nicholls L, Kaszniak AW (2004) A role for right medial prefrontal cortex in accurate feeling-of-knowing judgements: evidence from patients with lesions to frontal cortex. Neuropsychologia 42(7):957–966
Sokolov EN (1963) Perception and the conditioned reflex. Pergamon, New York
Stormark JK (2004) Skin conductance and heart-rate responses as indices of covert face recognition in preschool children. Infant Child Dev 13(5):423–433
Terasawa Y, Umeda S, Saito F, Kato M (2010) Loss of discrimination among negative facial expressions following right insula lesion. High Brain Funct Res 30(2):349
Terasawa Y, Fukushima H, Umeda S (2013) How does interoceptive awareness interact with the subjective experience of emotion? An fMRI study. Hum Brain Mapp 34(3):598–612
Tranel D, Damasio AR (1985) Knowledge without awareness: an autonomic index of facial recognition by prosopagnosics. Science 288(4706):145–1454
Umeda S, Tochizawa S, Shibata M, Terasawa Y (2016) Prospective memory mediated by interoceptive accuracy: a psychophysiological approach. Philos Trans R Soc B 371(1708):20160005
Verfaellie M, Bauer RM, Bowers D (1991) Autonomic and behavioral evidence of "implicit" memory in amnesia. Brain Cogn 15(1):10–25
West R, Craik FI (1999) Age-related decline in prospective memory: the roles of cue accessibility and cue sensitivity. Psychol Aging 14(2):264–272

Chapter 10
Functional Neuroimaging Studies of Autobiographical Memory Retrieval: Past, Present, and Future

Donna Rose Addis, Kristina Wiebels, and Aleea L. Devitt

Abstract Remembering events from our past – a form of memory known as episodic autobiographical memory (AM) – not only allows us to reminisce, but also to imagine the future, solve open-ended problems, and engage in creative thought. Two decades of neuroimaging research have established the core regions comprising the brain network supporting AM retrieval. Overlapping substantially with the default mode network, the AM retrieval network includes medial and lateral cortices in the prefrontal, temporal, and parietal lobes, the posterior cerebellum, and critically, the hippocampus. We take a historical perspective on *past* neuroimaging studies of AM to elucidate how the development of various neuroimaging methods have yielded increasing clarity on how the brain supports AM retrieval, including identification of the core nodes of the AM retrieval network and how activation may vary depending on the phase of retrieval (search versus elaboration), the recency of the memory and its recollective qualities, and the age of the rememberer. We discuss some *presently* emerging findings that the hippocampus is involved not only in AM retrieval, but also future simulation, in addition to new work using multivariate pattern analysis to uncover the brain patterns that represent the content of AMs and the processes of remembering and imagining. Finally, we end with some speculations about where memory research will take us in the *future*.

Keywords Aging • Autobiographical memory • Default mode network • Episodic memory • Future thinking • Hippocampus • Imagination • Multivariate pattern analysis • Neuroimaging • Recollection

---

D.R. Addis (✉) • K. Wiebels
School of Psychology and Centre for Brain Research, The University of Auckland, Auckland, New Zealand

Brain Research New Zealand, Auckland, New Zealand
e-mail: d.addis@auckland.ac.nz

A.L. Devitt
Department of Psychology, Harvard University, Cambridge, MA, USA

T. Tsukiura, S. Umeda (eds.), *Memory in a Social Context*,
https://doi.org/10.1007/978-4-431-56591-8_10

10.1 Introduction

The term "autobiographical memory" (AM) is used to refer to memory for a range of personal representations – from very abstract, factual information about ourselves and the happenings in our lives, to vivid re-experiencing of the self in a specific moment in the past. Over the last 20 years, neuroimaging has contributed much to our understanding of the neural regions and networks that underlie the ability to retrieve personal memories, with a particular emphasis on episodic autobiographical memories of past events (for reviews, see Addis et al. 2016a; Cabeza and St Jacques 2007; Maguire 2001; Svoboda and Levine 2009). More recently, this emphasis has broadened to consider the role that episodic AM plays in our ability to imagine specific events in the future (e.g., Addis et al. 2007; Schacter et al. 2012) in addition to means-end problem-solving (Madore and Schacter 2014; Sheldon et al. 2011) and creativity (Addis et al. 2016b; Madore et al. 2015).

The field has now accumulated more than two decades of research on the neural substrates of AM retrieval. The earliest study, by Andreasen et al. (1995), used positron emission tomography (PET) to examine the brain regions engaged when remembering episodic AMs relative to semantic memories. In doing so, they provided the first visualization of what would become known as the AM retrieval network – a distributed whole-brain network involving medial prefrontal, temporal and parietal cortices, in addition to lateral temporal and parietal regions and the cerebellum. Interestingly, Andreasen et al. also instructed participants to engage in a form of spontaneous mind-wandering they called random episodic silent thought (REST), and when contrasting both "focused" event retrieval and REST to a semantic control condition, an overlapping network of regions was evident. Importantly, this finding foreshadowed the later "discovery" of the default mode network (DMN) – the ubiquitous pattern of brain activity evident when one is not engaged with externally based cognitive tasks (e.g., during rest or fixation trials; see Raichle et al. 2001). The fact that the DMN was by-and-large the same network recruited by AM retrieval – likely because participants were thinking about their past during fixation tasks – was not lost on AM researchers of the time, although it would take many years for the notion that the DMN might be associated with spontaneous autobiographical cognition to take hold (for a review see Andrews-Hanna 2012).

As aptly described by Andreasen et al. (1995), the retrieval of specific episodic events is a "focused" version of the sort of remembering that occurs spontaneously when an individual is mind-wandering; and it is likely that "unfocused" AM retrieval includes not only the recollection of specific past events, but also retrieval of more semantic forms of AM. Thus, certain aspects of the DMN (as evident during mind-wandering) may be "ramped up" during AM tasks, depending on the particular requirements. Indeed, Andrews-Hanna reports that the DMN includes a medial temporal lobe (MTL) subsystem that is associated with self-reported instances of remembering (2012). Thus, in this chapter, we refer to the "AM retrieval network" when describing the aspects of the DMN engaged by AM retrieval tasks. Moreover, given that most neuroimaging studies in the literature have examined retrieval of specific autobiographical events, we restrict our review to episodic AM (and our use of the term "AM" refers to specific episodic events unless otherwise specified).

By taking a historical perspective on *past* neuroimaging studies of AM, we hope to elucidate how the development of various neuroimaging methods – in particular, AM paradigms and analytic techniques – have yielded increasing clarity on how the brain supports AM retrieval. Moreover, this provides a window into the major themes permeating the research, such as how the activation of AM network nodes may vary depending on the phase of retrieval, the recency of the memory and its recollective qualities, and the age of the rememberer. We also discuss some *presently* emerging work using multivariate pattern analysis to uncover the brain patterns that represent the content of AMs in addition to the process of remembering. We end our review with speculation about where memory research will take us in the *future*.

10.2 Autobiographical Memory Retrieval Network

Neuroimaging studies of AM retrieval have consistently revealed activation of a whole-brain network (Maguire 2001; Svoboda and Levine 2009). In a recent activation likelihood estimation (ALE) meta-analysis of 32 neuroimaging studies of specific AM retrieval, statistically reliable activity was evident in the left medial/lateral prefrontal cortex (PFC), the left lateral temporal cortex (including the temporal pole), the right thalamus, bilateral MTL (including the hippocampus and parahippocampal gyrus), bilateral medial parietal cortex (including precuneus, retrosplenial cortex, and posterior cingulate), bilateral angular gyrus (within the temporoparietal junction, TPJ), and the right posterior cerebellum (Fig. 10.1) (Addis et al. 2016a). Moreover, the results of a lateralization index (LI)[1] analysis on the ALE map indicates that the AM retrieval network is moderately left-lateralized, confirming Maguire's (2001) early qualitative description of AM retrieval network as a "medial and left-lateralized network" (p. 1445).

Although this meta-analysis only examined the neural correlates of specific AM retrieval relative to a control task, the AM retrieval network has been reported to be engaged during AM retrieval with a striking consistency, despite experimental variations. For instance, this network is engaged during different AM tasks, including recognition (Cabeza et al. 2004; Maguire and Mummery 1999), associative retrieval with direct personalized cues (e.g., Addis et al. 2012; Gilboa et al. 2004; Sheldon and Levine 2013; Sperduti et al. 2013), and strategic retrieval with impersonal cues (Addis et al. 2012; Conway et al. 2001; Daselaar et al. 2008; Ford et al. 2011; Holland et al. 2011; Rekkas and Constable 2005) or no cues (e.g., Andreasen et al. 1995). Moreover, although the particular control or baseline task may result in the

[1] LI values can range from −1, where activity is completely left-lateralized, to +1, where activity is completely right-lateralized. The result of this LI analysis was −0.66, reflecting 2,089 suprathreshold voxels in the left hemisphere vs 731 suprathreshold voxels in the right hemisphere.

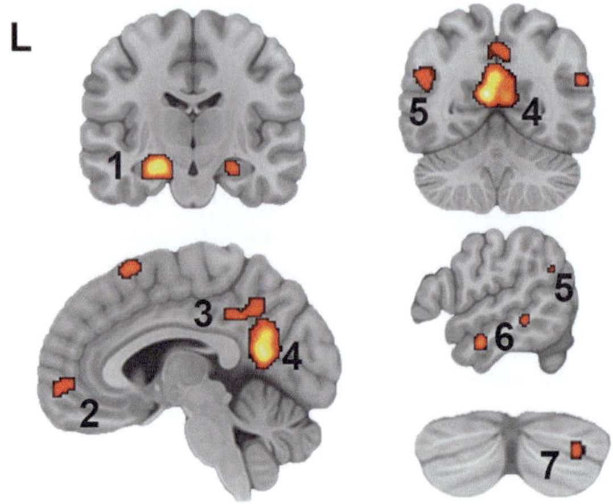

Fig. 10.1 Autobiographical memory (AM) retrieval network. Activation likelihood estimation meta-analysis of 32 studies revealed that the brain network reliably associated with specific AM retrieval. This network included (*1*) bilateral hippocampus, (*2*) medial prefrontal cortex, (*3*) posterior cingulate, (*4*) retrosplenial cortex and precuneus, (*5*) bilateral angular gyrus, (*6*) temporal pole, and (*7*) posterior right cerebellum (Crus I/II) (Adapted from Addis et al. 2016a)

obscuration of activity in certain regions (i.e., regions that are active during both the AM task and the baseline task), the network is largely evident when contrasting AM retrieval with episodic memory retrieval (Cabeza et al. 2004; Conway et al. 1999), semantic memory retrieval (Addis et al. 2012; Graham et al. 2003; Levine et al. 2004; Maguire and Mummery 1999; Ryan et al. 2001), imagined or impersonal events (e.g., Conway et al. 2003; Gilboa et al. 2004), perceptual decisions (e.g., Addis et al. 2012; Maguire and Mummery 1999), and rest (e.g., Andreasen et al. 1999; Piolino et al. 2004). Finally, the AM retrieval network is evident across a range of neuroimaging techniques, including surface electroencephalography (EEG; e.g., Conway et al. 2001), positron emission tomography (PET; e.g., Andreasen et al. 1999; Conway et al. 1999; Fink et al. 1996; Graham et al. 2003; Maguire and Mummery 1999; Piolino et al. 2004), and functional magnetic resonance imaging (fMRI; e.g., Addis et al. 2012; Gilboa et al. 2004; Levine et al. 2004; Maguire et al. 2000; Ryan et al. 2001). We briefly describe the known and hypothesized roles of the major nodes of the AM retrieval network. Note that some of the regions of the AM network, and the processes they mediate, overlap with those related to laboratory-based episodic memory (Cabeza et al..2004).

10.2.1 Prefrontal Cortex

The medial PFC is known to respond preferentially to, and predict memory for, self-referential stimuli (Gusnard et al. 2001; Macrae et al. 2004). Thus, during AM retrieval, activation of this region is thought to reflect a self-referential response to the personal nature of the cues and/or retrieved information. This response is a critical aspect of AM, providing an awareness that this memory is part of an individual's personal past (Prebble et al. 2013; Wheeler et al. 1997). More recently, it has been suggested that the medial PFC supports the reactivation of schemas during retrieval (Benoit et al. 2014; Ghosh et al. 2014), and for AMs, abstract autobiographical knowledge provides a critical framework for organizing retrieval of specific past events (Conway and Pleydell-Pearce 2000).

Lateral PFC regions are typically recruited during AM retrieval when the available cues are not sufficient to directly elicit an AM and iterative search-retrieval cycles are required to locate a relevant AM. This "strategic" (Moscovitch 1992; Moscovitch and Melo 1997) or "generative" (Conway and Pleydell-Pearce 2000) retrieval process requires first establishing the goals of the search, which are then held in working memory (supported by the dorsolateral PFC; e.g., D'Esposito et al. 2000). Available cues are then elaborated upon, or if there are none, cues are internally generated – a process thought to be supported by the mid-ventrolateral PFC (Moscovitch 1992). Once cue information is sufficiently specific, pattern completion of a memory trace ensues (Horner et al. 2015). The recovered information is then monitored and evaluated to determine whether it meets the search criteria and whether the memory is authentic and accurate (Johnson et al. 1993). In contrast to laboratory-based episodic memory tasks that recruit the dorsolateral PFC for monitoring purposes, AMs are thought to rely on a rapid, intuitive form of monitoring called *feeling-of-rightness* (Moscovitch and Winocur 2002) mediated by the ventromedial PFC. Indeed, damage to this region can result in confabulation (Gilboa et al. 2006). If the recovered AM is unsatisfactory, it is elaborated further and used as a cue for another search – a cycle that repeats until an appropriate AM is retrieved.

10.2.2 Lateral Temporal Cortex

Both the lateral temporal cortex and the temporal poles are consistently engaged during AM retrieval. It is notable that these regions also feature in the brain network mediating semantic processing (Binder and Desai 2011; Binder et al. 2009), and thus, within the context of AM retrieval, the lateral temporal cortex is thought to support the conceptual knowledge that is invariably accessed during the reconstruction of an AM (Graham et al. 2003; Irish and Piguet 2013). Indeed, when describing specific AMs, semantic information – including personal semantic information, general events, relevant concepts, and general knowledge – is intertwined with episodic details (Levine et al. 2002). The lateral temporal cortex may play a particularly important role during the retrieval of remote AMs (Hodges and Graham 2001; Irish et al. 2011) that have become semanticized owing to repeated rehearsal and the abstraction of gist over time (Irish and Piguet 2013; Winocur et al. 2007); indeed, more activity in lateral temporal regions is observed during the retrieval of remote AMs (Sheldon and Levine 2013).

10.2.3 Hippocampus

Considerable evidence exists for the critical role of the hippocampus in the retrieval of specific AMs. Its activation is ubiquitously associated with AM (Maguire 2001) and its damage is known to result in dense amnesia for AMs (the most famous case being Patient H.M., Steinvorth et al. 2005). Nevertheless, the nature of the hippocampal contribution to AM retrieval remains contentious. Although it has been argued to reflect consolidation of recent AMs (Squire and Alvarez 1995), the balance of neuroimaging evidence suggests that the hippocampus might play a lifelong role in the retrieval of episodic AMs (e.g., Addis et al. 2004b; Nadel et al. 2007; Rekkas and Constable 2005; Ryan et al. 2001; Sheldon and Levine 2013) in line with the predictions of the multiple trace theory (Nadel and Moscovitch 1997). This time-invariant role of the hippocampus is thought to involve pattern completion of the AM from cue information (Horner et al. 2015), including the reinstatement and reintegration of the episodic details comprising AM representations (Addis et al. 2004b; Moscovitch et al. 2005). This putative function of the hippocampus draws on evidence that the hippocampus is critical for the retrieval of relational information (Cohen and Eichenbaum 1993). Others propose that the fundamental computation of the hippocampus in AM retrieval is scene construction (Maguire and Mullally 2013), although integrating spatial information to form the spatial context of the event may be just one instantiation of relational processing during AM retrieval.

10.2.4 Parahippocampal Gyrus and Medial Parietal Cortex

Retrieval of AM is characterized by a swathe of continuous activity from the hippocampus, extending through the parahippocampal gyrus and retrosplenial cortex, and into the posterior cingulate and precuneus. These regions, particularly the parahippocampal and retrosplenial cortices, appear to function as a network supporting the processing of spatial and nonspatial context (Bar and Aminoff 2003; Burgess et al. 2001), in addition to integrating allocentric and egocentric spatial information (Epstein 2008; Vann et al. 2009). Indeed, these areas are more strongly activated when retrieving AMs versus imagining future events (Gilmore et al. 2016), reflecting the fact that AMs typically comprise familiar and vivid contextual information (Szpunar et al. 2009). Evidence suggests that the precuneus might be involved in both episodic memory and visuospatial imagery (Cavanna and Trimble 2006) and as such, this region has been described as "the mind's eye" (Fletcher et al. 1995). For instance, the precuneus supports memory not only for visual (Johnson and Johnson 2014) and imagined scenes (Hassabis et al. 2007), but also for other imageable stimuli, such as words (Fletcher et al. 1995). On the other hand, some have argued that cortical midline regions are sensitive to self-referential information (Northoff and Bermpohl 2004) and support the integration of emotion and memory during AM retrieval (Maddock 1999).

10.2.5 Lateral Parietal Cortex

The inferior portion of the lateral parietal cortex is also consistently activated during AM retrieval; known as the TPJ, this includes the angular gyrus and inferior parietal lobule.[2] A number of theories have been advanced to explain this recruitment during AM retrieval. Wheeler and Buckner (2004) argue that the left lateral parietal cortex supports the perceived "pastness" of a memory (i.e., that a memory is old) as distinct from the retrieval of mnemonic content supported by other posterior regions. This idea draws on evidence that the left parietal cortex is engaged during successful retrieval (i.e., an old/new effect; Henson et al. 1999) and during the presentation of new combinations of old stimuli (Donaldson et al. 2001). Others have posited that the lateral parietal cortex supports the orienting of attention to internally generated representations (i.e., retrieved AMs; Wagner et al. 2005) or the capture of attention by the retrieval of salient information (Cabeza et al. 2008). However, given evidence that activity in the inferior lateral parietal cortex modulates according to the amount of information recollected, it has been advanced that this region may act to hold-in-mind an integrated representation of retrieved information – possibly as part of the network supporting Baddeley's (2000) "episodic buffer" (Vilberg and Rugg 2008; see also the output-buffer hypothesis, Wagner et al. 2005). Indeed, damage to the lateral parietal cortex is associated with an impoverished ability to retrieve detailed AMs (including spatial, perceptual, emotional, and referential details) in the absence of general mental imagery impairment (Berryhill et al. 2007), suggesting that this area of the cortex might support the recollection of multimodal representations.

10.2.6 Cerebellum

Cerebellar activation is often observed during AM retrieval, but is either not reported or, when it is, it is not discussed or interpreted. One of the few studies to highlight cerebellar activity during AM retrieval, Andreasen et al. (1999) reported that the bilateral posterior cerebellum was activated, in line with a now growing body of evidence that the posterior cerebellum contributes to cognition (Buckner 2013). A meta-analysis by Svoboda et al. (2006) confirmed that the cerebellum was reported to be activated in approximately half of the available neuroimaging studies of AM retrieval (Svoboda et al. 2006), leading them to designate it a "core" AM region. More recently, an ALE meta-analysis by Addis et al. (2016a) identified reliable AM-related activity in the cerebellum, localized to the right Crus I (extending into Crus II; Fig. 10.1), and confirmed this that subregion exhibits both intrinsic and

[2]Although activity in the supramarginal gyrus is sometimes associated with AM retrieval, this region was not evident in the ALE meta-analysis by Addis et al. (2016a), consistent with evidence that it is more responsive to familiarity than to recollection (Vilberg and Rugg 2008).

AM-related connectivity with the AM retrieval network. The exact role this region plays in AM retrieval is yet to be determined, but on the basis of pervasive cognitive sequencing deficits in patients with pathological cerebellar conditions (Leggio et al. 2008, 2011), we suggested that the cerebellum might support the sequencing of events within the life narrative and the sequencing of details within an event (Addis et al. 2016a).

10.3 Early Neuroimaging Studies of Autobiographical Memory Retrieval

10.3.1 Autobiographical Memory Recognition

Around the turn of the century, an influential series of imaging studies were published by Maguire and colleagues using an AM recognition paradigm (Maguire and Frith 2003a, b; Maguire and Mummery 1999; Maguire et al. 2000, 2001a, b). Specifically, during a pre-scan interview conducted a few weeks before scanning, AMs were collected and knowledge of semantic facts and public events was assessed. This information was then used to create sets of verbal statements for four experimental conditions that varied according to temporal specificity or personal relevance, as illustrated in Fig. 10.2. During each 8-s trial, participants listened to a statement and indicated whether it was true or false.

In the first of these studies, Maguire and Mummery (1999) used the AM recognition paradigm in PET to test whether temporally-specific and personally-relevant memories (i.e., episodic AMs) preferentially engaged the hippocampus. The comparison of all memory conditions (irrespective of memory type) to the control conditions revealed activation of the AM retrieval network (Fig. 10.2). Although a number of these regions exhibited preferential activity for personally-relevant memories (e.g., medial PFC, left temporal pole, left TPJ), the interaction of temporal specificity and personal relevance supported their prediction, revealing significant activation *only* in the left hippocampus. Recognizing that, although the hippocampus plays a critical role in AM retrieval, it is situated within the context of a wider brain network, Maguire et al. (2000) examined whether the interactions between regions comprising this network differed according to the type of memory being retrieved. Path analyses revealed that left hippocampal–parahippocampal connections were stronger during episodic AM retrieval, whereas left middle temporal–temporopolar connectivity was stronger during forms of semantic retrieval.

10.3.2 Autobiographical Memory Recollection

One critique of the AM recognition paradigm used by Maguire and colleagues is that an episodic AM trial can be completed without fully re-experiencing the past event. For instance, an individual could "know" that the statement *"You were Mike's*

(a)

Memory Type	Temporally Specific?	Personally Relevant?	Example
Episodic AM	+	+	"You were Mike's best man at his wedding"
Semantic AM	-	+	"Ray is the youngest of your brothers"
Public event	+	-	"Windsor Castle was damaged by a fire"
Semantic facts	-	-	"The people of Holland are known as Dutch"
Control task	-	+	"While yet your therefore still about this"
Control task	-	-	"Ago rather he yet if because"

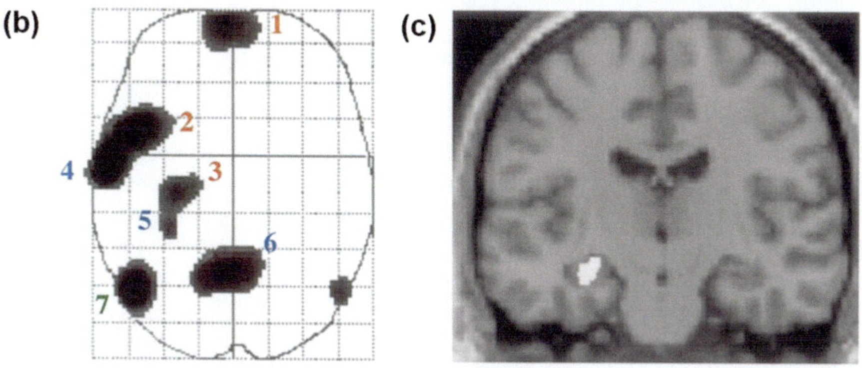

Fig. 10.2 Medial and left-lateralized AM retrieval network, as revealed by the autobiographical recognition task. (**a**) The autobiographical recognition task conditions vary according to temporal specificity and personal relevance. (**b**) The glass brain clearly demonstrates that activation during AM recognition is primarily medial and left-lateralized, as seen in: (*1*) medial prefrontal cortex, (*2*) left temporal pole, (*3*) left hippocampus, (*4*) left anterolateral middle temporal gyrus, (*5*) left parahippocampal gyrus, (*6*) medial parietal cortex/posterior cingulate, and (*7*) temporoparietal junction. (**c**) Superimposed over an MRI image, increased activity in the left hippocampus is evident when retrieving temporally-specific, personally-relevant AMs (i.e., episodic AM relative to semantic AM, public events, and general semantic memory) (Adapted from Maguire and Mummery 1999)

best man at his wedding" is true or false without having to recollect the wedding in detail. Moreover, the language-based nature of the stimuli and response task may also underpin the left-lateralized activation evident during AM recognition. In the early 2000s, an alternative paradigm emerged in the literature – participants were presented in the scanner with personally-tailored stimuli that directly cued AMs with the instruction to recollect the memory in as much detail as possible. In many cases, measures of re-experiencing (e.g., detail ratings; Addis et al. 2004b) were also collected, to ensure that participants were recollecting the events.

Ryan et al. (2001) used this approach in a theoretical debate that pervades the AM literature: whether the involvement of the hippocampus is limited to relatively recent memories (standard consolidation theory; Squire and Alvarez 1995) or whether it is ubiquitous as long as the AM remains episodic (as posited by the multiple trace theory and the transformation hypothesis; Nadel and Moscovitch 1997; Winocur et al. 2007). Specifically, they examined whether the activation of the hippocampus differed between retrieval of recent events (from the past 4 years) and remote events (from more than 20 years ago). Immediately before the scan, participants selected cues (e.g., "wedding day," "recent holiday") related to their recent and remote experiences. When retrieving AMs associated with these cues during scanning, the hippocampus was equally active for recent and remote events (relative to a semantic control task), supporting the multiple trace theory/transformation hypothesis.

Interestingly, Ryan et al. (2001) noted that the recent and remote events were similar with regard to the number of details and the intensity of emotion elicited, as assessed by participant ratings. This observation touches on another modulator of hippocampal activity during AM retrieval: the nature of the recollective experience. Piefke et al. (2003) reported that recent events were associated with stronger hippocampal activity than remote events, but critically, these recent events were also rated as higher in richness of detail, emotionality and re-experiencing. According to the multiple trace theory, if an AM is truly episodic in that the elements of the AM are reactivated and reintegrated upon retrieval, then the hippocampus should be involved, regardless of the age of the memory.

Addis et al. (2004b) designed a study that tested the modulatory effects of recency *and* the richness of the recollective experience on hippocampal activity. During scanning, participants retrieved AMs elicited by personalized cues, and rated retrieved AMs for the level of detail, emotionality, and personal significance; following scanning they dated each event. Parametric modulation analyses using these ratings revealed that increasing recency was associated with a linear increase in hippocampal activity – but when the level of recollective experience (detail, emotionality, and personal significance) was accounted for, the effect of recency was reduced or eliminated. In contrast, all three of these recollective qualities were associated with linear increases in hippocampal activity, even when controlling for the age of the AM. These findings suggest that recency effects likely reflect changes in the recollective experience with memory age, and that the hippocampus is more sensitive to processes supporting recollective experience (e.g., reactivating and reintegrating event components).

A general critique of these direct cueing paradigms, however, is that the reactivation of the AM during the prescan phase (when personalized cues are collected) alters the recency of the memory. Moreover, the hippocampal activity evident during AM retrieval could arguably reflect the re-encoding of the event. Gilboa et al. (2004) addressed both of these concerns: they eliminated the prescan interview by cueing participants with family photographs provided by a confederate, and they controlled for re-encoding by having participants imagine an event in response to non-personal photographs (i.e., the photographs of other participants). Despite these critical controls, Gilboa et al. still found that both the remote and the recent AMs elicited significant activation in the left hippocampus. Moreover, when comparing remote AMs that were vivid versus those that were impoverished, they found that the left hippocampus was one of a few regions exhibiting a significant effect. Sheldon and Levine (2013) took a different approach to this issue, and had participants record descriptions of AMs as the events happened (prospective design). When comparing recent (1-month-old) and relatively more remote (1.5-year-old)

AMs, they found that both conditions resulted in the activation of the left hippocampus and parahippocampal gyrus when the vividness of remote AMs matched that of recent AMs. Together, these findings support the idea that the hippocampus is associated with the recollection of richly detailed AMs, irrespective of age.

10.4 Neuroimaging of Constructive Retrieval Processes

10.4.1 Reconstructing Autobiographical Memories

Although studies using personally-relevant cues typically capture direct retrieval, a number of studies have examined the protracted generative search for an AM. Using EEG, Conway and his collaborators (Conway et al. 2001, 2003) reported AM-related activity during the initial search phase across a large area of the left PFC, including the dorsolateral and ventrolateral regions, followed by a shift of activity to posterior networks in the right temporal and occipital lobes when the AM is retrieved (as indicated by a response). Immediately following retrieval, activation shifted to bilateral frontal regions likely reflecting monitoring processes. Similarly, Steinvorth et al. (2006) found activity in both the dorsolateral and ventrolateral PFC during an initial search period, whereas following AM retrieval, only dorsolateral PFC remained active.

Drawing on the Crovitz cueing paradigm (Crovitz and Schiffman 1974), generic cues have been used during scanning to elicit the generative search for AMs. This approach necessitates the collection of responses during scanning (to confirm an AM was retrieved) as well as a post-scan interview to determine the type of AM retrieved (e.g., specific versus generic). Moreover, from an analytic perspective, this experimental design requires each trial to be broken into distinct retrieval phases; namely, AM access/construction and AM elaboration. Using this approach, Daselaar et al. (2008) found that the construction phase was characterized by activity in the hippocampus, retrosplenial cortex, medial and right lateral PFC, whereas the elaboration phase was associated with activity in the left lateral PFC, precuneus, and visual cortices. These findings fit with the idea that different aspects of the lateral PFC are recruited during the search and monitoring phases, and that the elaboration of an AM is associated with activity in regions supporting mental imagery.

In a creative version of the cueing paradigm, Ford et al. (2011) used non-personal musical cues to elicit the search for AMs, and the retrieved AMs were categorized by participants according to specificity (specific events, general events, and lifetime periods). It was found that the search for all three types of AMs engaged the MTL. Interestingly, the search phase also engaged the medial PFC and posterior cingulate, possibly reflecting the self-referential nature of cues that resulted in successful retrieval of autobiographical knowledge.

Addis et al. (2012) compared the neural correlates of direct and generative retrieval by instructing participants to retrieve AMs in response to generic word cues and personalized cues referring to specific AMs (obtained during a session approximately 1 month before scanning). To equate previous exposure to the cues for the generative retrieval condition with that of the direct retrieval condition, the generic nouns were presented during the prescan session in the context of a future imagination task. Although both tasks commonly recruited the AM retrieval network relative to semantic and visuospatial control tasks, there were some important temporal and spatial distinctions between the two retrieval processes. In particular, generative retrieval recruited the left anterolateral PFC and the right inferior frontal gyrus 4–6 s after cue presentation, possibly reflecting strategic search processes, and the left medial PFC and left inferior frontal gyrus 16–18 s into the trial, likely reflecting the monitoring of retrieved AMs. Interestingly, the laterality of lateral PFC activity (right during the search and left during monitoring) is consistent with that reported by Daselaar et al. (2008).

10.4.2 Constructing Future Events

In recent years, neuroimaging has begun to examine the neural correlates of another form of autobiographical construction: imagining future events. In the first neuroimaging study on this question, Okuda et al. (2003) used PET to measure brain activity while participants freely talked about the past or the future for a minute at a time. Although there was considerable overlap in the regions engaged by both of these tasks, there were some crucial differences: future thinking engaged the medial PFC (frontal pole) and MTL to a *greater* degree than AM retrieval.

In a follow-up study, Addis et al. (2007) used an event-related fMRI design to enable analysis of different task phases (construction and elaboration) and the restriction of analysis to specific events only (by excluding unsuccessful trials). During the construction phase, the left hippocampus and right inferior parietal lobule (in addition to posterior visual regions such as superior/middle occipital gyrus and cuneus) were activated for both the AM task and the future event task. In contrast, elaboration was associated with activation of the AM retrieval network, including medial and lateral PFC, hippocampus and parahippocampal gyrus, lateral temporal cortex (including temporal poles), medial parietal and inferior lateral parietal cortices (Fig. 10.3a). The striking overlap of the regions engaged by the elaboration of past and future events led to the proposal of the constructive episodic simulation hypothesis, and the idea that accessing AM details is not only critical for remembering, but also for imagining episodes in the future (Addis and Schacter 2012; Schacter and Addis 2007) and the past (Addis et al. 2009; De Brigard et al. 2013).

One of the most intriguing findings to come out of the study by Addis et al. (2007) was that the right anterior hippocampus was more active during the imagination of future events relative to retrieval AMs of past events; this finding was replicated at the same peak voxel (Talairach coordinates: *xyz* 40-22-11) in a follow-up study by Addis et al. (Fig. 10.3b) (2011a). Although the hippocampus has traditionally been thought of as a canonical "memory structure," particularly because the

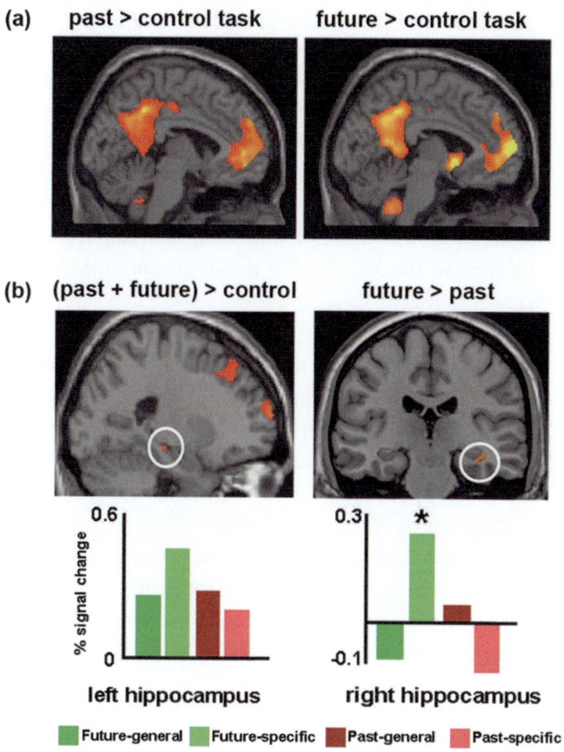

Fig. 10.3 Activation associated with remembering the past and imagining the future. (**a**) Common engagement of the AM retrieval network when remembering past events and imagining future events (relative to a semantic/visuospatial control task). (**b**) Common (*left*) and differential (*right*) hippocampal activity during the simulation of specific and general past and future events. Differential hippocampal activity was evident for the simulation of specific future events (Adapted from Addis et al. 2007 and 2011a)

studies on Patient H.M. (Milner 1972; Scoville and Milner 1957), a recent wave of findings have uncovered its involvement in other cognitive functions, including future thinking (Addis and Schacter 2012), problem solving (Sheldon et al. 2011), creativity (Duff et al. 2013), and visual perception (Lee et al. 2012).

The question of why the right anterior hippocampus is more intensely activated during the construction of specific future events has received much attention. One possibility is related to the fact that future events (and other imagined events, e.g., imagined past events; Addis et al. 2009) differ from AMs in that they are representations of *novel* episodes that have not yet been experienced (Addis and Schacter 2012). Although it has been shown that the first instance of constructing a novel future event is associated with higher levels of right anterior hippocampal activity, which then drops off with repeated imaginings of the same event (van Mulukom et al. 2013), the results of other studies suggest that the novelty of the imagined episode might not be critical. For instance, imagining general future events (e.g., routines) was not associated with greater anterior hippocampal activity than past AMs, even though these imagined routines were novel (Fig. 10.3b) (Addis et al. 2011a).

It is likely, then, that the increased anterior hippocampal activity during future simulation reflects a more intensive constructive process than when *re*-constructing AMs or imagining generic information (e.g., routines). According to the constructive episodic simulation hypothesis, simulation of specific events involves not only the access to episodic details, but also the recombination and integration of these details into a coherent event. This constructive process includes not only assembling the spatial elements of the imagined scene (see the scene construction hypothesis; Hassabis and Maguire 2007; Maguire and Mullally 2013), but also the people, objects, thoughts, emotions and actions (Addis and Schacter 2012; Schacter et al. 2012). Given the role of the hippocampus in relational processing (Cohen and Eichenbaum 1993) it is not surprising that activity in the anterior hippocampus has been linked to the amounts of detail comprising future events (Addis and Schacter 2008).

Addis and Schacter (2012) propose that future simulation comprises at least three components: access to episodic details; recombination of details into a coherent simulation, and encoding the future event into memory. Indeed, if a future simulation is to be used to enhance our behavior in future situations, then we must be able to later recall the simulated event when faced with that situation. This encoding process also appears to be associated with anterior hippocampal activity, even when controlling for the amount of detail accessed and recombined (Martin et al. 2011). However, more detailed and plausible simulations are easier to remember (McLelland et al. 2015); thus, it is likely that these processes are closely intertwined in terms of recruitment of the hippocampus during event construction and their effects on the memorability of future simulations.

10.5 Multivariate Approaches to Examining AM Processes and Content

As described above, activation-based fMRI analyses have provided insight into many aspects of AM; however, the ways in which AM processes and representations map onto patterns of brain activity cannot be fully captured by standard analyses of mean activity. Multivoxel pattern analysis (MVPA) (Haxby et al. 2001) enables researchers to address these questions using pattern classification algorithms applied to brain activity (either within a region of interest or across the whole brain). Using this technique, mental states or processes are "decoded" from fMRI data and related to behavior on a trial-by-trial basis. This makes MVPA particularly suited to both testing theories about specific AM *processes*, and examining where the *content* of different AMs is represented.

The first study to examine whether the content of single memories could be decoded using MVPA was conducted by Maguire and colleagues (Chadwick et al. 2010). Before scanning, participants viewed three short video clips, each showing a different everyday event. During fMRI scanning, participants recalled each of these events several times. Using MVPA in the hippocampus, the authors could classify which of the three events participants were remembering. Information about the different memories was also present in the two other regions of interest (entorhinal cortex and posterior parahippocampal gyrus), showing that these structures also contained some episodic representations, but to a lesser degree than the hippocampus. The bilateral anterior and right posterior hippocampus carried the most episodic information across participants.

To further investigate exactly which aspects of the memories were decodable by MVPA, a follow-up study was conducted, looking at highly overlapping memories (Chadwick et al. 2011). Specifically, two different events were overlaid on two different spatial contexts, resulting in four overlapping video clips. As in the earlier study, participants viewed the video clips before scanning and recalled the events from the videos repeatedly during scanning. Using a four-way classifier, all four memories could be discriminated on the basis of hippocampal activity, demonstrating that this region held a representation of each memory.

Although these studies showed for the first time that the content of single episodic memories can be decoded from hippocampal activity, the memories were laboratory-based instead of truly autobiographical events. Indeed, there are critical differences between these forms of memory, with the recollection of AMs engaging the medial PFC, MTL, and visual cortices more than memories of laboratory-based events, and the latter activating the right dorsolateral PFC more than the former (Cabeza et al. 2004; Gilboa 2004).

To address this issue, Bonnici et al. (2012) examined whether single events can be decoded when a paradigm is used that taps into *autobiographical* events. A week before scanning, participants provided details of three recent and three remote AMs. During scanning, participants recalled these AMs repeatedly in response to verbal cues. Results indicated that classifiers in the MTL (hippocampus, entorhinal/perirhinal cortex, parahippocampal gyrus), retrosplenial cortex, temporal pole, lateral temporal cortex, and medial PFC could all distinguish among all three recent and all three remote memories. Moreover, a comparison of recent and remote memories revealed higher classifier accuracies for remote than recent AMs in the medial PFC, but even so, medial PFC patterns representing recent and remote memories overlapped. Information about both types of memories was also present in the hippocampus, but here the representations did not overlap. Specifically, although both recent and remote memories were represented in the anterior hippocampus, the posterior hippocampus showed higher classification accuracies for remote than for recent memories. Importantly, however, these results show that remote and recent AMs are represented in the hippocampus, providing further support for the transformation hypothesis/multiple trace theory (Nadel and Moscovitch 1997; Winocur et al. 2007).

In addition to decoding the *content* of memories – in which case information is decoded across individual AMs – MVPA can also be used to examine *processes* involved in AM – where information is decoded across different conditions. For example, Kirwan et al. (2014) examined whether neural patterns can distinguish between remembering past AMs and imagining future events. Although univariate analyses did not reveal any hippocampal differences between the past and future conditions in these data, an MVPA classifier trained in hippocampal data could reliably distinguish between these two conditions. Moreover, classifiers trained using data from the MTL and the whole brain were also successful at differentiating remembering from imagining. Together, these findings indicate that the closely related processes of remembering and imagining are distinguishable on the basis of activity patterns in the MTL and across the whole brain.

Rissman et al. (2016) used MVPA to examine whether subjective memory states (i.e., AMs versus novel events) are associated with specific neural patterns. After wearing a portable camera for 3 weeks, participants were shown event cues (a sequence of four photos) in the scanner that either used their own photos (AMs) or photos from other participants (novel). Whole-brain MVPA could distinguish between AM and novel events, and could also determine the degree to which participants recollected the event (i.e., recollection/familiarity/novelty). Interestingly, the neural features distinguishing the degree of recollection were consistent across participants, such that a classifier trained on one participant's data could successfully decode recollection in another participant, indicating the robustness of the patterns related to these subjective memory states.

10.6 Neuroimaging of Autobiographical Memory in Aging

The episodic richness charactertistic of AMs tends to decrease as we age, as evident on objective measures of richness. For instance, Levine et al. (2002) used an objective scoring technique (the Autobiographical Interview) to parse the content of AMs into episodic (e.g., details about percepts, place, time, emotions, thoughts and action) and non-episodic (e.g., general conceptual information, semantic facts and metacognitive statements) categories. This scoring revealed that older adults' AMs contained fewer episodic details compared with those of younger adults, instead comprising more semantic information (Piolino et al. 2002; St. Jacques and Levine 2007). Similar age effects are seen when imagining both future (Addis et al. 2010; Addis et al. 2008; Cole et al. 2013; De Beni et al. 2013; Rendell et al. 2012) and past events (Devitt et al. 2016).

Declines in the episodic aspects of AM are not surprising given evidence that the MTL and PFC decline structurally and functionally with age, whereas the lateral temporal regions involved in semantic processing tend to be preserved (Buckner et al. 2006; Dennis et al. 2008b; Prull et al. 2000; Raz et al. 2005). Indeed, there is a disproportionate effect of aging on episodic memory for laboratory events, whereas semantic memory is relatively unaffected (Spaniol et al. 2006). Medial temporal – and particularly hippocampal – dysfunction is thought to result in deficits in retrieving associative information (Dennis et al. 2008a; Mitchell et al. 2000; Naveh-Benjamin 2000), which likely contributes to age-related difficulties in generating episodically-rich past and future events. Indeed, older adults are less able than younger adults to integrate key details when imagining future episodes (Addis et al. 2010). Moreover, providing older adults with retrieval support boosts retrieval of episodic detail for both past and future autobiographical events, suggesting that dysfunction in PFC-mediated strategic retrieval mechanisms also contributes to the reduced effectiveness of generative AM retrieval (Levine et al. 2002; Madore and Schacter 2014).

Neuroimaging studies have confirmed these age-related changes in PFC and MTL activity during AM and future thinking tasks. Although older adults recruit the AM retrieval network when thinking about past (St Jacques et al. 2012) and future autobiographical events (Addis et al. 2011b; Viard et al. 2011), they do so to a lesser extent than younger adults. In effect, the neural signatures of autobiographical, episodic, and semantic memory are less distinguishable with age (St-Laurent et al. 2011), consistent with a decline in the episodic content of autobiographical events (for a review, see Schacter et al. 2013). St Jacques et al. (2012) further demonstrated that age-related activation changes during AM retrieval are most prominent during the elaboration phase (which depends on retrieval of specific episodic details) as opposed to the search phase (which tends to be guided by preserved semantic processes). In that study, when retrieving AMs low in episodic richness, older adults activated regions supporting episodic construction, such as the hippocampus and ventrolateral PFC, to a lesser extent than younger adults. In contrast, when retrieving episodically rich memories, fewer age differences were evident, although older adults still exhibited a reduction in the coupling of the ventrolateral PFC and the hippocampus. These results further implicate difficulties in strategic retrieval processes, mediated by the PFC, in the decline in episodic detail of AMs in older age (see Piolino et al. 2010, for further evidence of a link between age-related reductions in episodic AM and executive functioning).

These findings stand in contrast to previous reports of an age-associated increase in right hippocampal activity during AM retrieval (Maguire and Frith 2003a). This increased activation of the bilateral hippocampus with age is thought to reflect recruitment of additional neural resources, or increased use of spatial context, during AM retrieval. However, methodological differences may account for this discrepancy in hippocampal findings; it is possible that future research differentiating between different phases of retrieval (search versus elaboration), the retrieval support given (e.g., single cue word versus memory description), and the type of retrieval process required (e.g., recall versus recognition) will yield different age effects on the AM retrieval network.

Age differences in brain activation are also observed when thinking about the future. Addis et al. (2011b) explored age-related changes in whole-brain patterns of activity during AM retrieval and future imagination. Although younger adults exhibited increased recruitment of the AM retrieval network during the past and future tasks relative to the control task, older adults did not exhibit this differentiation across conditions in regions associated with the retrieval of episodic detail, such as the hippocampus and the precuneus. Instead, during the elaboration phase of the autobiographical tasks, older adults engaged regions supporting semantic and conceptual retrieval (including the right lateral temporal cortex; Addis et al. 2004a; Graham et al. 2003) more so than younger adults. This pattern of activation is in line with an age-related increased role of conceptual information when thinking about autobiographical events. Furthermore, the age groups exhibited correlations with detail ratings in different brain areas. For younger adults, increasing detail ratings were associated with the engagement of areas involved in mediating episodic imagery, such as the MTL (Addis and Schacter 2008; Fletcher et al. 1995; Hassabis et al.

2007; Szpunar et al. 2009). Yet, for older adults, the regions that correlated with detail were limited to those involved in conceptual information, further suggesting that the balance of episodic to semantic content in AMs might change dramatically with age.

10.7 Conclusions: The Future of AM Research

Nearly twenty-five years of neuroimaging AM has yielded a considerable body of findings that has informed our theoretical understandings of AM, especially with regard to the role of particular neural structures (such as the hippocampus) and processes such as consolidation and transformation. The AM retrieval network has been reliably mapped, and its involvement in other processes such as future simulation has been established. Indeed, this latter finding has been particularly influential, as it indicates that AM retrieval is but one form of mental simulation – of (re)creating a dynamic, specific episode within the confines of the mind.

One of the ongoing challenges in this field of research is that we are reliant on subjective reports of memory content and phenomenological experience. The current state-of-the-art in terms of objectively assessing the characteristics of the AM retrieved is to examine event descriptions using scoring methods such as the autobiographical interview (Levine et al. 2002). Although these methods have yielded considerable advances in terms of understanding the types of details comprising an AM, these methods are inherently reliant on other processes unrelated to memory. The most obvious of these is language and narrative processes: in forming a coherent story, the individual chooses the features of their memory to tell us about and the words with which to tell us. These processes may have a considerable impact on our measure of the episodicity of memory, particularly in aging and patient populations. For instance, older adults typically describe events with more conceptual detail, and whether this increase in non-episodic detail reflects a narrative process (e.g., providing a broad context for the event) or a mnemonic process (e.g., compensation for a reduction in episodic details) remains to be determined. Interestingly, although subjective ratings of detail or vividness by older adults do not divulge the apparent reductions in their episodic detail (with their scores being on a par with the ratings of young adults), patterns of brain activity have revealed differences in what these ratings reflect in the young and the old (Addis et al. 2011b).

To further advance our understanding of AM, we need to "get inside" the mind of the rememberer so that we can characterize an AM in a *fully* objective fashion. Perhaps decoding techniques such as MVPA will provide us with the necessary tools. Indeed, MVPA is providing us with a "window" of sorts into those mental representations – not just the processes recruited to construct them, but also the content that the rememberer is experiencing. These techniques, and the promise of how they will be developed and leveraged to characterize AM, provide an exciting landscape for the future of AM research.

References

Addis DR, Schacter DL (2008) Effects of detail and temporal distance of past and future events on the engagement of a common neural network. Hippocampus 18(2):227–237

Addis DR, Schacter DL (2012) The hippocampus and imagining the future: where do we stand? Front Hum Neurosci 5:173

Addis DR, McIntosh AR, Moscovitch M, Crawley AP, McAndrews MP (2004a) Characterizing spatial and temporal features of autobiographical memory retrieval networks: a partial least squares approach. NeuroImage 23(4):1460–1471

Addis DR, Moscovitch M, Crawley AP, McAndrews MP (2004b) Recollective qualities modulate hippocampal activation during autobiographical memory retrieval. Hippocampus 14(6):752–762

Addis DR, Wong AT, Schacter DL (2007) Remembering the past and imagining the future: common and distinct neural substrates during event construction and elaboration. Neuropsychologia 45(7):1363–1377

Addis DR, Wong AT, Schacter DL (2008) Age-related changes in the episodic simulation of future events. Psychol Sci 19(1):33–41

Addis DR, Pan L, Vu M-A, Laiser N, Schacter DL (2009) Constructive episodic simulation of the future and the past: distinct subsystems of a core brain network mediate imagining and remembering. Neuropsychologia 47(11):2222–2238

Addis DR, Musicaro R, Pan L, Schacter DL (2010) Episodic simulation of past and future events in older adults: evidence from an experimental recombination task. Psychol Aging 25(2):369–376

Addis DR, Cheng T, Roberts RP, Schacter DL (2011a) Hippocampal contributions to the episodic simulation of specific and general future events. Hippocampus 21(10):1045–1052

Addis DR, Roberts RP, Schacter DL (2011b) Age-related neural changes in autobiographical remembering and imagining. Neuropsychologia 49(13):3656–3669

Addis DR, Knapp K, Roberts RP, Schacter DL (2012) Routes to the past: neural substrates of direct and generative autobiographical memory retrieval. NeuroImage 59(3):2908–2922

Addis DR, Moloney EE, Tippett LJ, Robert RP, Hach S (2016a) Characterizing cerebellar activity during autobiographical memory retrieval: ALE and functional connectivity investigations. Neuropsychologia 90:80–93

Addis DR, Pan L, Musicaro R, Schacter DL (2016b) Divergent thinking and constructing episodic simulations. Memory 24(1):89–97

Andreasen NC, O'Leary DS, Cizadlo T, Arndt S, Rezai K, Watkins GL, Ponto LL, Hichwa RD (1995) Remembering the past: two facets of episodic memory explored with positron emission tomography. Am J Psychiatr 152(11):1576–1585

Andreasen NC, O'Leary D, Paradiso S, Cizadlo T, Arndt S, Watkins GL, Boles Ponto LL, Hichwa RD (1999) The cerebellum plays a role in conscious episodic memory retrieval. Hum Brain Mapp 8(4):226–234

Andrews-Hanna JR (2012) The brain's default network and its adaptive role in internal mentation. Neuroscientist 18(3):251–270

Baddeley AD (2000) The episodic buffer: a new component of working memory? Trends Cogn Sci 4(11):417–423

Bar M, Aminoff E (2003) Cortical analysis of visual context. Neuron 38(2):347–358

Benoit RG, Szpunar KK, Schacter DL (2014) Ventromedial prefrontal cortex supports affective future simulation by integrating distributed knowledge. Proc Natl Acad Sci U S A 111(46):16550–16555

Berryhill ME, Phuong L, Picasso L, Cabeza R, Olson IR (2007) Parietal lobe and episodic memory: bilateral damage causes impaired free recall of autobiographical memory. J Neurosci 27(52):14415–14423

Binder JR, Desai RH (2011) The neurobiology of semantic memory. Trends Cogn Sci 15(11):527–536

Binder JR, Desai RH, Graves WW, Conant LL (2009) Where is the semantic system? A critical review and meta-analysis of 120 functional neuroimaging studies. Cereb Cortex 19(12):2767–2796

Bonnici HM, Chadwick MJ, Lutti A, Hassabis D, Weiskopf N, Maguire EA (2012) Detecting representations of recent and remote autobiographical memories in vmPFC and hippocampus. J Neurosci 32(47):16982–16991

Buckner RL (2013) The cerebellum and cognitive function: 25 years of insight from anatomy and neuroimaging. Neuron 80(3):807–815

Buckner RL, Head D, Lustig C (2006) Brain changes in aging: a lifespan perspective. In: Bialystok E, Craik FIM (eds) Lifespan cognition: mechanisms of change. Oxford University Press, New York, pp 27–42

Burgess N, Maguire EA, Spiers HJ, O'Keefe J (2001) A temporoparietal and prefrontal network for retrieving the spatial context of lifelike events. NeuroImage 14(2):439–453

Cabeza R, St Jacques P (2007) Functional neuroimaging of autobiographical memory. Trends Cogn Sci 11(5):219–227

Cabeza R, Prince SE, Daselaar SM, Greenberg DL, Budde M, Dolcos F, LaBar KS, Rubin DC (2004) Brain activity during episodic retrieval of autobiographical and laboratory events: an fMRI study using a novel photo paradigm. J Cogn Neurosci 16(9):1583–1594

Cabeza R, Ciaramelli E, Olson IR, Moscovitch M (2008) The parietal cortex and episodic memory: an attentional account. Nat Rev Neurosci 9(8):613–625

Cavanna AE, Trimble MR (2006) The precuneus: a review of its functional anatomy and behavioural correlates. Brain 129(3):564–583

Chadwick MJ, Hassabis D, Weiskopf N, Maguire EA (2010) Decoding individual episodic memory traces in the human hippocampus. Curr Biol 20(6):544–547

Chadwick MJ, Hassabis D, Maguire EA (2011) Decoding overlapping memories in the medial temporal lobes using high-resolution fMRI. Learn Mem 18(12):742–746

Cohen NJ, Eichenbaum H (1993) Memory, amnesia, and the hippocampal system. The MIT Press, Cambridge, MA

Cole SN, Morrison CM, Conway MA (2013) Episodic future thinking: linking neuropsychological performance with episodic detail in young and old adults. Q J Exp Psychol 66(9):1687–1706

Conway MA, Pleydell-Pearce CW (2000) The construction of autobiographical memories in the self-memory system. Psychol Rev 107(2):261–288

Conway MA, Turk DJ, Miller SL, Logan J, Nebes RD, Meltzer CC, Becker JT (1999) A positron emission tomography (PET) study of autobiographical memory retrieval. Memory 7(5–6):679–703

Conway MA, Pleydell-Pearce CW, Whitecross SE (2001) The neuroanatomy of autobiographical memory: a slow cortical potential study of autobiographical memory retrieval. J Mem Lang 45(3):493–524

Conway MA, Pleydell-Pearce CW, Whitecross SE, Sharpe H (2003) Neurophysiological correlates of memory for experienced and imagined events. Neuropsychologia 41(3):334–340

Crovitz HF, Schiffman H (1974) Frequency of episodic memories as a function of their age. Bull Psychon Soc 4(5):517–518

Daselaar SM, Rice HJ, Greenberg DL, Cabeza R, LaBar KS, Rubin DC (2008) The spatiotemporal dynamics of autobiographical memory: neural correlates of recall, emotional intensity, and reliving. Cereb Cortex 18(1):217–229

De Beni R, Borella E, Carretti B, Zavagnin M, Lazzarini L, Milojevi G (2013) Remembering the past and imagining the future: age-related differences between young, young-old and old-old. Aging Clin Exp Res 25(1):89–97

De Brigard F, Addis DR, Ford JH, Schacter DL, Giovanello KS (2013) Remembering what could have happened: neural correlates of episodic counterfactual thinking. Neuropsychologia 51(12):2401–2414

Dennis NA, Hayes SM, Prince SE, Madden DJ, Huettel SA, Cabeza R (2008a) Effects of aging on the neural correlations of successful item and source memory encoding. J Exp Psychol: Learn Mem Cogn 34(4):791–808

Dennis NA, Kim H, Cabeza R (2008b) Age-related differences in brain activity during true and false memory retrieval. J Cogn Neurosci 20(8):1390–1402

D'Esposito M, Postle BR, Rypma B (2000) Prefrontal cortical contributions to working memory: evidence from event-related fMRI studies. Exp Brain Res 133(1):3–11

Devitt AL, Tippett LJ, Schacter DL, Addis DR (2016) Autobiographical memory conjunction errors in younger and older adults: evidence for a role of inhibitory ability. Psychol Aging 31(8):927–942

Donaldson DI, Petersen SE, Buckner RL (2001) Dissociating memory retrieval processes using fMRI: evidence that priming does not support recognition memory. Neuron 31(6):1047–1059

Duff MC, Kurczek J, Rubin R, Cohen NJ, Tranel D (2013) Hippocampal amnesia disrupts creative thinking. Hippocampus 23(12):1143–1149

Epstein RA (2008) Parahippocampal and retrosplenial contributions to human spatial navigation. Trends Cogn Sci 12(10):388–396

Fink GR, Markowitsch HJ, Reinkemeier M, Bruckbauer T, Kessler J, Heiss W (1996) Cerebral representation of one's own past: neural networks involved in autobiographical memory. J Neurosci 16(13):4275–4282

Fletcher P, Frith C, Baker SC, Shallice T, Frackowiak RS, Dolan R (1995) The mind's eye – precuneus activation in memory-related imagery. NeuroImage 2(3):195–200

Ford JH, Addis DR, Giovanello KS (2011) Differential neural activity during search of specific and general autobiographical memories elicited by musical cues. Neuropsychologia 49(9):2514–2526

Ghosh VE, Moscovitch M, Melo Colella B, Gilboa A (2014) Schema representation in patients with ventromedial PFC lesions. J Neurosci 34(36):12057–12070

Gilboa A (2004) Autobiographical and episodic memory--one and the same? Evidence from prefrontal activation in neuroimaging studies. Neuropsychologia 42(10):1336–1349

Gilboa A, Winocur G, Grady CL, Hevenor SJ, Moscovitch M (2004) Remembering our past: functional neuroanatomy of recollection of recent and very remote personal events. Cereb Cortex 14(11):1214–1225

Gilboa A, Alain C, Stuss DT, Melo B, Miller S, Moscovitch M (2006) Mechanisms of spontaneous confabulations: a strategic retrieval account. Brain 129(6):1399–1414

Gilmore AW, Nelson SM, McDermott KB (2016) The contextual association network activates more for remembered than for imagined events. Cereb Cortex 26(2):611–617

Graham KS, Lee AC, Brett M, Patterson K (2003) The neural basis of autobiographical and semantic memory: new evidence from three PET studies. Cogn Affect Behav Neurosci 3(3):234–254

Gusnard DA, Akbudak E, Shulman GL, Raichle ME (2001) Medial prefrontal cortex and self-referential mental activity: relation to a default mode of brain function. Proc Natl Acad Sci U S A 98(7):4259–4264

Hassabis D, Maguire EA (2007) Deconstructing episodic memory with construction. Trends Cogn Sci 11(7):299–306

Hassabis D, Kumaran D, Maguire EA (2007) Using imagination to understand the neural basis of episodic memory. J Neurosci 27(52):14365–14374

Haxby J, Gobbini MI, Furey M, Ishai A, Schouten J, Pietrini P (2001) Distributed and overlapping representations of faces and objects in ventral temporal cortex. Science 293(5539):2425–2430

Henson RN, Rugg MD, Shallice T, Josephs O, Dolan RJ (1999) Recollection and familiarity in recognition memory: an event-related functional magnetic resonance imaging study. J Neurosci 19(10):3962–3972

Hodges JR, Graham KS (2001) Episodic memory: insights from semantic dementia. Philos Trans R Soc Lond B Biol Sci 356(1413):1423–1434

Holland AC, Addis DR, Kensinger EA (2011) The neural correlates of specific versus general autobiographical memory construction and elaboration. Neuropsychologia 49(12):3164–3177

Horner AJ, Bisby JA, Bush D, Lin W-J, Burgess N (2015) Evidence for holistic episodic recollection via hippocampal pattern completion. Nat Commun 6:7462
Irish M, Piguet O (2013) The pivotal role of semantic memory in remembering the past and imagining the future. Front Behav Neurosci 7:27
Irish M, Hornberger M, Lah S, Miller L, Pengas G, Nestor PJ, Hodges JR, Piguet O (2011) Profiles of recent autobiographical memory retrieval in semantic dementia, behavioural- variant frontotemporal dementia, and Alzheimer's disease. Neuropsychologia 49(9):2694–2702
Johnson MR, Johnson MK (2014) Decoding individual natural scene representations during perception and imagery. Front Hum Neurosci 8:59
Johnson MK, Hashtroudi S, Lindsay DS (1993) Source monitoring. Psychol Bull 114(1):3–28
Kirwan CB, Ashby SR, Nash MI (2014) Remembering and imagining differentially engage the hippocampus: a multivariate fMRI investigation. Cogn Neurosci 5(3–4):177–185
Lee ACH, Yeung LK, Barense MD (2012) The hippocampus and visual perception. Front Hum Neurosci 6:91
Leggio MG, Tedesco AM, Chiricozzi FR, Clausi S, Orsini A, Molinari M (2008) Cognitive sequencing impairment in patients with focal or atrophic cerebellar damage. Brain 131(5):1332–1343
Leggio MG, Chiricozzi FR, Clausi S, Tedesco AM, Molinari M (2011) The neuropsychological profile of cerebellar damage: the sequencing hypothesis. Cortex 47(1):137–144
Levine B, Svoboda E, Hay JF, Winocur G, Moscovitch M (2002) Aging and autobiographical memory: dissociating episodic from semantic retrieval. Psychol Aging 17(4):677–689
Levine B, Turner GR, Tisserand D, Hevenor SJ, Graham SJ, McIntosh AR (2004) The functional neuroanatomy of episodic and semantic autobiographical remembering: a prospective functional MRI study. J Cogn Neurosci 16(9):1633–1646
Macrae CN, Moran JM, Heatherton TF, Banfield JF, Kelley WM (2004) Medial prefrontal activity predicts memory for self. Cereb Cortex 14(6):647–654
Maddock RJ (1999) Retrosplenial cortex and emotion: new insights from functional imaging studies of the human brain. Trends Neurosci 22(7):310–316
Madore KP, Schacter DL (2014) An episodic specificity induction enhances means-end problem solving in young and older adults. Psychol Aging 29(4):913–924
Madore KP, Addis DR, Schacter DL (2015) Creativity and memory: effects of an episodic-specificity induction on divergent thinking. Psychol Sci 26(9):1461–1468
Maguire EA (2001) Neuroimaging studies of autobiographical event memory. Philos Trans R Soc Lond Ser B Biol Sci 356(1413):1441–1451
Maguire EA, Frith CD (2003a) Aging affects the engagement of the hippocampus during autobiographical memory retrieval. Brain 126(7):1511–1523
Maguire EA, Frith CD (2003b) Lateral asymmetry in the hippocampal response to the remoteness of autobiographical memories. J Neurosci 23(12):5302–5307
Maguire EA, Mullally SL (2013) The hippocampus: a manifesto for change. J Exp Psychol Gen 142(4):1180–1189
Maguire EA, Mummery CJ (1999) Differential modulation of a common memory retrieval network revealed by positron emission tomography. Hippocampus 9(1):54–61
Maguire EA, Mummery CJ, Buchel C (2000) Patterns of hippocampal-cortical interaction dissociate temporal lobe memory subsystems. Hippocampus 10(4):475–482
Maguire EA, Henson RN, Mummery CJ, Frith CD (2001a) Activity in prefrontal cortex, not hippocampus, varies parametrically with the increasing remoteness of memories. Neuroreport 12(3):441–444
Maguire EA, Vargha-Khadem F, Mishkin M (2001b) The effects of bilateral hippocampal damage on fMRI regional activations and interactions during memory retrieval. Brain 124(6):1156–1170
Martin VC, Schacter DL, Corballis MC, Addis DR (2011) A role for the hippocampus in encoding simulations of future events. Proc Natl Acad Sci U S A 108(33):13858–13863
McLelland VC, Devitt AL, Schacter DL, Addis DR (2015) Making the future memorable: phenomenology of remembered future events. Memory 23(8):1255–1263

Milner B (1972) Disorders of learning and memory after temporal lobe lesions in man. Clin Neurosurg 19:421–446

Mitchell KJ, Johnson MK, Raye CL, Mather M, D'Esposito M (2000) Aging and reflective processes of working memory: binding and test load deficits. Psychol Aging 15(3):527–541

Moscovitch M (1992) Memory and working-with-memory: a component process model based on modules and central systems. J Cogn Neurosci 4(3):257–267

Moscovitch M, Melo B (1997) Strategic retrieval and the frontal lobes: evidence from confabulation and amnesia. Neuropsychologia 35(7):1017–1034

Moscovitch M, Winocur G (2002) The frontal cortex and working with memory. In: Stuss DT, Knight RT (eds) Principles of frontal lobe function. Oxford University Press, Oxford, pp 188–209

Moscovitch M, Rosenbaum RS, Gilboa A, Addis DR, Westmacott R, Grady C, McAndrews MP, Levine B, Black S, Wincour G, Nadel L (2005) Functional neuroanatomy of remote episodic, semantic and spatial memory: a unified account based on multiple trace theory. J Anat 207(1):35–66

Nadel L, Moscovitch M (1997) Memory consolidation, retrograde amnesia and the hippocampal complex. Curr Opin Neurobiol 7(2):217–227

Nadel L, Campbell J, Ryan L (2007) Autobiographical memory retrieval and hippocampal activation as a function of repetition and the passage of time. Neural Plast 2007:90472

Naveh-Benjamin M (2000) Adult age differences in memory performance: tests of an associative deficit hypothesis. J Exp Psychol Learn Mem Cogn 26(5):1170–1187

Northoff G, Bermpohl F (2004) Cortical midline structures and the self. Trends Cogn Sci 8(3):102–107

Okuda J, Fujii T, Ohtake H, Tsukiura T, Tanji K, Suzuki K, Kawashima R, Fukuda H, Itoh M, Yamadori A (2003) Thinking of the future and the past: the roles of the frontal pole and the medial temporal lobes. NeuroImage 19(4):1369–1380

Piefke M, Weiss PH, Zilles K, Markowitsch HJ, Fink GR (2003) Differential remoteness and emotional tone modulate the neural correlates of autobiographical memory. Brain 126(3):650–668

Piolino P, Desgranges B, Benali K, Eustache F (2002) Episodic and semantic remote autobiographical memory in ageing. Memory 10(4):239–257

Piolino P, Giffard-Quillon G, Desgranges B, Chetelat G, Baron JC, Eustache F (2004) Re-experiencing old memories via hippocampus: a PET study of autobiographical memory. NeuroImage 22(3):1371–1383

Piolino P, Coste C, Martinelli P, Macé AL, Quinette P, Guillery-Girard B, Belleville S (2010) Reduced specificity of autobiographical memory and aging: do the executive and feature binding functions of working memory have a role? Neuropsychologia 48(2):429–440

Prebble SC, Addis DR, Tippett LJ (2013) Autobiographical memory and sense of self. Psychol Bull 139(4):815–840

Prull MW, Gabrieli JDE, Bunge SA (2000) Age-related changes in memory: a cognitive neuroscience perspective. In: Craik FIM, Salthouse TA (eds) The handbook of aging and cognition. Lawrence Erlbaum Associates Publishers, Mahwah, pp 91–153

Raichle ME, MacLeod AM, Snyder AZ, Powers WJ, Gusnard DA, Shulman GL (2001) A default mode of brain function. Proc Natl Acad Sci U S A 98(2):676–682

Raz N, Lindenberger U, Rodrigue KM, Kennedy KM, Head D, Williamson A, Dahle C, Gerstorf D, Acker JD (2005) Regional brain changes in aging healthy adults: general trends, individual differences and modifiers. Cereb Cortex 15(11):1676–1689

Rekkas PV, Constable RT (2005) Evidence that autobiographical memory retrieval does not become independent of the hippocampus: an fMRI study contrasting very recent with remote events. J Cogn Neurosci 17(12):1950–1961

Rendell PG, Bailey PE, Henry JD, Phillips LH, Gaskin S, Kliegel M (2012) Older adults have greater difficulty imagining future rather than atemporal experiences. Psychol Aging 27(4):1089–1098

Rissman J, Chow TE, Reggente N, Wagner AD (2016) Decoding fMRI signatures of real-world autobiographical memory retrieval. J Cogn Neurosci 28(4):604–620

Ryan L, Nadel L, Keil K, Putnam K, Schnyer D, Trouard T, Moscovitch M (2001) Hippocampal complex and retrieval of recent and very remote autobiographical memories: evidence from functional magnetic resonance imaging in neurologically intact people. Hippocampus 11(6):707–714

Schacter DL, Addis DR (2007) The cognitive neuroscience of constructive memory: remembering the past and imagining the future. Philos Trans R Soc Lond Ser B Biol Sci 362(1481):773–786

Schacter DL, Addis DR, Hassabis D, Martin VC, Spreng RN, Szpunar KK (2012) The future of memory: remembering, imagining, and the brain. Neuron 76(4):677–694

Schacter DL, Gaesser B, Addis DR (2013) Remembering the past and imagining the future in the elderly. Gerontology 59(2):143–151

Scoville WB, Milner B (1957) Loss of recent memory after bilateral hippocampal lesions. J Neurol Neurosurg Psychiatry 20(1):11–21

Sheldon S, Levine B (2013) Same as it ever was: vividness modulates the similarities and differences between the neural networks that support retrieving remote and recent autobiographical memories. NeuroImage 83:880–891

Sheldon S, McAndrews MP, Moscovitch M (2011) Episodic memory processes mediated by the medial temporal lobes contribute to open-ended problem solving. Neuropsychologia 49(9):2439–2447

Spaniol J, Madden DJ, Voss A (2006) A diffusion model analysis of adult age differences in episodic and semantic long-term memory retrieval. J Exp Psychol Learn Mem Cogn 32(1):101–117

Sperduti M, Martinelli P, Kalenzaga S, Devauchelle AD, Lion S, Malherbe C, Gallarda T, Amado I, Krebs MO, Oppenheim C, Piolino P (2013) Don't be too strict with yourself! Rigid negative self-representation in healthy subjects mimics the neurocognitive profile of depression for autobiographical memory. Front Behav Neurosci 7:41

Squire LR, Alvarez P (1995) Retrograde amnesia and memory consolidation: a neurobiological perspective. Curr Opin Neurobiol 5(2):169–177

St Jacques P, Rubin DC, Cabeza R (2012) Age-related effects on the neural correlates of autobiographical memory retrieval. Neurobiol Aging 33(7):1298–1310

St Jacques PL, Levine B (2007) Ageing and autobiographical memory for emotional and neutral events. Memory 15(2):129–144

Steinvorth S, Levine B, Corkin S (2005) Medial temporal lobe structures are needed to re-experience remote autobiographical memories: evidence from H.M. and W.R. Neuropsychologia 43(4):479–496

Steinvorth S, Corkin S, Halgren E (2006) Ecphory of autobiographical memories: an fMRI study of recent and remote memory retrieval. NeuroImage 30(1):285–298

St-Laurent M, Abdi H, Burianová H, Grady CL (2011) Influence of aging on the neural correlates of autobiographical, episodic, and semantic memory retrieval. J Cogn Neurosci 23(12):4150–4163

Svoboda E, Levine B (2009) The effects of rehearsal on the functional neuroanatomy of episodic autobiographical and semantic remembering: a functional magnetic resonance imaging study. J Neurosci 29(10):3073–3082

Svoboda E, McKinnon MC, Levine B (2006) The functional neuroanatomy of autobiographical memory: a meta-analysis. Neuropsychologia 44(12):2189–2208

Szpunar KK, Chan JCK, McDermott KB (2009) Contextual processing in episodic future thought. Cereb Cortex 19(7):1539–1548

Van Mulukom V, Schacter DL, Corballis MC, Addis DR (2013) Re-imagining the future: repetition decreases hippocampal involvement in future simulation. PLoS One 8(7):e69596

Vann SD, Aggleton JP, Maguire EA (2009) What does the retrosplenial cortex do? Nat Rev Neurosci 10(11):792–802

Viard A, Chetelat G, Lebreton K, Desgranges B, Landeau B, De La Sayette V, Eustache F, Piolino P (2011) Mental time travel into the past and the future in healthy aged adults: an fMRI study. Brain Cogn 75(1):1–9

Vilberg KL, Rugg MD (2008) Memory retrieval and the parietal cortex: a review of evidence from a dual-process perspective. Neuropsychologia 46(7):1787–1799
Wagner AD, Shannon BJ, Kahn I, Buckner RL (2005) Parietal lobe contributions to episodic memory retrieval. Trends Cogn Sci 9(9):445–453
Wheeler M, Buckner RL (2004) Functional-anatomic correlates of remembering and knowing. NeuroImage 21(4):1337–1349
Wheeler M, Stuss DT, Tulving E (1997) Toward a theory of episodic memory: the frontal lobes and autonoetic consciousness. Psychol Bull 121(3):331–354
Winocur G, Moscovitch M, Sekeres M (2007) Memory consolidation or transformation: context manipulation and hippocampal representations of memory. Nat Neurosci 10(5):555–557

Part III
Lifelong Effects on Memory and Social Cognition

Chapter 11
Development of Recognition Memory for Faces During Infancy

Yumiko Otsuka

Abstract An increasing number of studies have demonstrated that even newborn infants have the ability to recognize faces they have seen previously. Here, findings on recognition memory for faces during infancy were reviewed. First, the findings in newborn infants were summarized, including memory both for personally familiar faces (e.g., their own mother's face) and unfamiliar faces. Recent findings suggest that the differential pattern of preference between mother's face recognition and unfamiliar face recognition is due to the difference in the stimulus characteristics during the familiarization phase. Second, the total looking time during the habituation/familiarization phase were examined, which enabled infants to recognize faces in various previous studies. This examination revealed a clear developmental trend that the looking time during the learning phase decreased within a few months after birth for studies using the familiarization method. On the other hand, there was no clear trend across the studies using the infant-controlled habituation method. Although studies using looking time measures suggest exposure to a face for at least several seconds is required for successful face recognition, studies using nonbehavioral measures suggest the possibility that infants recognize face seen only for less than a second. Finally, the findings on long-term memory for faces during infancy were summarized. Although studies on this topic are relatively scarce, a few existing findings suggest that facial memory acquired in the latter half of the first year could be retained for a year or longer.

Keywords Face recognition • Infant • Memory • Habituation • Familiarization • Long-term memory

Y. Otsuka (✉)
Faculty of Law and Letters, Ehime University, Matsuyama, Ehime, Japan
e-mail: otsuka.yumiko.wf@ehime-u.ac.jp

T. Tsukiura, S. Umeda (eds.), *Memory in a Social Context*,
https://doi.org/10.1007/978-4-431-56591-8_11

11.1 Introduction

Adults hardly remember what they have experienced during infancy, a phenomenon called infantile amnesia or childhood amnesia (for reviews, see Bauer 2013; Hayne and Jack 2011). However, studies of infant cognition have revealed that infants are capable of forming memories of stimuli to which they have been exposed. Among various objects infants encounter in their daily lives, faces are perhaps the most socially significant visual stimuli. Reflecting the growing interest in infants' face recognition and face processing, several reviews on how the ability to process faces develops during infancy are available (e.g., Lee et al. 2011; Otsuka 2014; Simion and Di Giorgio 2015). Here, the available findings on the development of recognition memory for facial identity from birth through infancy are reviewed.

11.2 Recognition Memory for Mother's Face at Birth

Visual recognition memory for faces and the ability to discriminate between faces in newborns have been reported in the form of preferential looking toward the mother's face over an unfamiliar female face (Bushnell et al. 1989; Bushnell 2001, 2003; Field et al. 1984; Pascalis et al. 1995; Sai 2005). Most of the studies reporting the preference of newborns for their mother's face have used real face-to-face settings. Such experimental settings could potentially introduce various confounding factors such as olfactory cues and subtle facial gestures by the mother and strangers while they pose in front of the infants. However, several previous studies, as discussed below, clarified that newborns' preference for their mother's face could not be attributed to such confounding factors.

To control for olfactory cues, several studies used an air freshener during the experiment to mask the scent cue, and still found the newborns' preference for their mother's face over a stranger's face (Bushnell et al. 1989; Bushnell 2001; Pascalis et al. 1995; Sai 2005). By covering up the faces of both the mother and stranger, Bushnell et al. (1989) examined whether scent cues alone could induce preferential looking toward the mother over a stranger. Their results demonstrated that scent cues alone were not sufficient to elicit newborns' preferential looking at their mothers. Bushnell et al. (1989) also reported that when naïve adult observers were presented with the video recording of the mother's and stranger's faces during the preference experiment, they could not distinguish the mothers from the strangers (mothers of other newborns). This rules out the possibility that the mothers acted differently from the strangers when they were presented with their own newborn during the experiment.

Most of the studies that examined infants' preference for their mothers' faces have used mother–stranger face pairs that were matched in skin-tone and hair color (Barrera and Maurer 1981; Field et al. 1984; Sai 2005) and had similar hair styles (Bushnell et al. 2003; Pascalis et al. 1995). However, it was reported that newborns' preference for their mother's face disappeared when the hairline was masked by a scarf (Pascalis et al. 1995) or when the hairstyle was changed by a wig (Bushnell et al. 2003). These findings indicate the importance of the large outer hairline information in newborns' ability to recognize their mother's face. Whatever the visual information involved, these studies demonstrated that that even newborns can acquire visual memory for their mother's face through experience in real-life situations.

In most cases, newborns' naturally acquired memory for their mother's face has been demonstrated as a familiarity preference. However, in their study, Field et al. (1984) suggested that the pattern of preference could be experimentally reversed. Field et al. used a real face-to-face setting to measure the infants' spontaneous preference and the preference after habituation between the mother's face and an unfamiliar female face. Field et al. reported that newborns spontaneously preferred their mother's face over an unfamiliar female face in an initial preference test. After habituation to the mother's face, however, the newborns showed a preference toward the novel, unfamiliar female face over their mother's face. Barrera and Maurer (1981) reported a similar pattern of results in 3-month-old infants using photographs of faces.

This reversibility of the preference seems to contrast with the preference for female faces that develops a few months after birth. Several studies have reported that infants who were raised mainly by a female caregiver preferentially look at female faces over male faces at around 3–4 months of age (Liu et al. 2015; Quinn et al. 2002, 2008), although no such preference was observed in newborns (Quinn et al. 2008). Interestingly, familiarizing 3–4-month-old infants with multiple female faces could not reverse the preference for female over male faces (Quinn et al. 2002). In any case, the pattern in which natural interactions lead to a familiarity preference for the mother's face whereas experimental habituation trials lead to a novelty preference for an unfamiliar female face over the mother's face suggests that the interactions between mother and newborn that occur in natural situations do not induce sufficient habituation to elicit a novelty preference.

11.3 Recognition Memory for Unfamiliar Faces at Birth

Several research groups reported successful recognition memory for unfamiliar faces in newborns. Most of these studies observed a novelty preference after experimentally exposing newborns to images of unfamiliar female faces (de Heering et al. 2008; Pascalis and de Schonen 1994; Turati et al. 2006, 2008). For example, Pascalis and de Schonen (1994) assessed the recognition memory of newborns by using an infant-controlled habituation procedure. In this procedure, the duration of the habituation phase is determined by the infant's looking behavior. Specifically, the habituation trials are repeated until the infant's looking time decreases to 50% of initial trials. After habituation to a photograph of an unfamiliar face, the infants were shown the familiar habituated face and a novel unfamiliar face, side by side. Pascalis & de

Schonen reported that the newborns showed a novelty preference both when they were tested immediately after the habituation and after a 2-min delay.

Turati et al. (2006) reported that newborns successfully discriminated between upright faces regardless of whether whole-face images, inner features-only images, or outer features-only images were used, provided that the image condition was kept constant between the habituation and the test period. Even when the presence/absence of outer facial features was changed between the habituation and test phase, newborns successfully recognized the habituated face based on the external facial features. However, the newborns failed to discriminate between the novel and familiar faces based on the internal features in the internal features-only image after habituation to the whole-face image. The finding that the alteration of external features between the habituation phase and the test phase diminishes newborns' face recognition seems to be consistent with studies reporting infants' failure to recognize their mother's face when the hair line was occluded (Pascalis et al. 1995) or altered (Bushnell et al. 2003).

Newborns' recognition of unfamiliar faces seems to tolerate a certain degree of facial view change between the habituation and test phase. Turati et al. (2008) reported that newborns showed a novelty preference when they were habituated to an unfamiliar face in the frontal view and tested with a three-quarters view and vice versa. However, newborns showed no preference when their recognition was tested in a profile-view image after habituation to either the frontal view or a three-quarters view. Bulf and Turati (2010) further examined newborns' face recognition across views and the role of rigid rotational apparent motion. During the habituation phase, newborns were shown images of an unfamiliar face at various view angles (frontal face, ±30°, ±60°, ±90°). They were then shown the habituated and novel face in a profile-view angle. When the images of different views were shown in a dynamic manner that depicted the apparent motion of smooth facial rotation during the habituation, the newborns showed a novelty preference. In contrast, when the same images were shown in a dynamic manner, but in a random-order sequence, the newborns showed a familiarity preference. Bulf and Turati discussed that the familiarity preference reflects partial face recognition, and that this occurred because the random-order sequence was not as effective as the correct-order sequence in providing infants with a dynamic cue for facial structural information.

As in the above study by Bulf and Turati (2010) and Gava et al. (2008) reported that both familiarity and novelty preferences were shown by newborns depending on the transformation of the facial image introduced in the test phase. Gava et al. (2008) compared the effect of partial occlusion of the eye area and the occlusion of the mouth and nose area on newborns' face recognition. After habituation to a fully visible face, newborns' face recognition of partially occluded face images was tested. Gava et al. reported that the newborns showed a novelty preference for the images in which the nose and mouth, but not the eyes, were occluded with three vertical bars. In contrast, the newborns showed a familiarity preference for images in which the eyes were occluded by two vertical bars. Gava et al. discussed that the familiarity preference in the latter case suggest a weaker recognition than does a novelty preference, and suggest the importance of eyes in the face recognition of newborns.

With the exceptions of the studies by Bulf and Turati (2010) and Gava et al. (2008), the studies on unfamiliar face recognition mentioned above using an infant-controlled habituation method consistently found a novelty preference. On the other hand, several recent studies consistently reported that newborns showed a familiarity preference when they were *familiarized* with dynamic talking-face images. In a study by Coulon et al. (2011), 2- to 3-day-old newborns were familiarized with video recordings of unfamiliar female faces talking, with or without an accompanying voice. After familiarization, the infants were tested for visual preference between the photographs of novel and familiar female faces. They found that only infants who were exposed to a talking face video with a voice showed a familiarity preference. Similarly, Guellai et al. (2011) found that after familiarization with an unfamiliar talking face, newborns showed a familiarity preference both when they were tested using photographs of faces and when they were tested using the talking face videos. On the other hand, when the newborns were familiarized and tested with photographs of faces, they showed a novelty preference. The latter novelty preference result is consistent with the studies on unfamiliar face recognition in newborns using an infant-controlled habituation method (de Heering et al. 2008; Pascalis and de Schonen 1994; Turati et al. 2006, 2008).

Coulon et al. (2011) and Guellai et al. (2011) showed that exposure to a face and a voice led to a familiarity preference, whereas exposure to a static and silent face led to a novelty preference. These findings suggest that the presence or absence of a voice plays an important role in determining the direction of the preference in the subsequent test period. Such a factor may account for the different pattern of preference between the study on infants' recognition of their mothers' faces and unfamiliar face recognition. In most studies on the recognition of unfamiliar faces that reported a novelty preference, the infants were habituated to silent photographs of unfamiliar faces. In contrast, before infants showed a preference for their mother's face, they had had natural interactions with their mothers, which was very likely to involve exposure to both their mothers' voice and face. In fact, Sai (2005) reported that newborns who only had silent interactions with their mother before the experiment showed no looking preference between the mother's face and an unfamiliar female face.

A preference for the familiar stimulus is considered to reflect an early phase of stimulus processing in which the internal representation of a stimulus does not match well with the actual stimulus. Several studies have shown that infants' preference shifted from the familiar to the novel with increased exposure to repeated stimuli (Richards 1997; Roder et al. 2000; Rose et al. 1982). The findings of familiarity preference for the mother's face and an unfamiliar talking face in newborns may also be understood as partial recognition, which occurs because of an interference effect of complex multimodal stimulation on face processing in very young infants (Bahrick et al. 2013).

11.4 How Long Does It Take for Infants to Remember a Face?

As reviewed above, it has been demonstrated that infants are capable of forming memories for faces even at birth. A further question arises: how much exposure to faces is required for infants to form a memory that can lead to successful face recognition? To answer this question, the existing literature reporting successful face recognition in infants was surveyed and the total looking time during the learning phase was examined. It should be acknowledged that this is a very rough measure, but it could provide an intuitive understanding of how long an infant requires to memorize faces. There are two main ways of testing face recognition in infants: infant controlled habituation and familiarization. These are considered in turn below.

An infant-controlled habituation procedure has been used extensively in face recognition studies in infants. Table 11.1 summarizes the findings from the studies in which the value of the total looking time during the habituation phase could be found. Only the conditions that led to a significant preference are included and only the studies that examined the discrimination of faces among same-gender faces are included. Unless otherwise noted in Table 11.1, the same habituation stimulus was used for the test phase and the habituation phase.

To determine whether there is a developmental change in the duration of habituation, the total looking time during habituation in the studies listed in Table 11.1 was plotted against the ages of infants shown in Fig. 11.1. Figure 11.1 shows no obvious trend of the looking time systematically varying along with the infants' ages. Across the infants' age progression, the total looking time tended to be 50 s or longer. The shortest total habituation looking time in newborns among the studies that led to novelty preference was 40 s (Pascalis and de Schonen 1994), which is comparable to that for older infants (6 months: 42.07 s; 9 months: 38.88 s, Kelly et al. 2007).

At around 3–4 months of age, there is a large variation in the total looking time during habituation. The shortest total looking time around this age is 19.52 s, as reported by Turati et al. (2010), whereas the longest looking time is 452 s, reported by Barrera and Maurer (1981). This large variation can be at least partly explained by the characteristics of the habituation stimuli. Although most studies used whole faces as habituation stimuli, Turati et al. showed only the upper half of a face during habituation. Unlike most other studies, which used unfamiliar faces as stimuli, the habituation stimulus in Barrera and Maurer was a photograph of the mother's face. The facial image limited to top half would have attracted less attention than the whole face, whereas the mother's face in Barrera and Maurer's study would have attracted more attention than unfamiliar faces.

A close look at Fig. 11.1 shows some developmental trends. Although most of the studies targeted a particular age group, some of the studies tested infants of different ages. In Fig. 11.1, the data points from different age groups in a given study are connected by dashed lines. These connected data points across age groups show

Table 11.1 Summaries of the studies reporting successful face recognition memory in infants using an infant-controlled habituation method

Authors	*n*	Age (days)	Habituation stimuli	TL (s)	Preference test results
Field et al. (1984)	48	2	Mother's face in-person (talking or silent)	73.2	Novelty preference
Turati et al. (2006)	18	2	Female face with hairline	100.51	Novelty preference
	30	2	Female face without hairline	103.56	Novelty preference
	18	2	Female face including hairline without internal features	59.82	Novelty preference
	13	2	Female face with hairline	79.84	Novelty preference in images without internal features
	13	2	Female face with hairline without internal features	68.28	Novelty preference in images with internal features
	20	2	Inverted female face with hairline	50.57	Novelty preference
	20	2	Inverted female face including hairline without internal features	65.59	Novelty preference
Gava et al. (2008)	16	2	Female face with hairline	81.03	Novelty preference in nose and mouth occluded images
	19	2	Female face with hairline	74.09	Familiarity preference in eyes occluded images
De Heering et al. (2008)	14	2	Low-pass filtered (< 1 cpd) female face	70.15	Novelty preference
	12	2	Low-pass filtered (< 0.5 cpd) female face	48.2	Novelty preference
Turati et al. (2008)	9	2	Female face in frontal view	54.54	Novelty preference in ¾ view images
	9	2	Female face in ¾ view images	69.23	Novelty preference in frontal view images
Leo and Simion (2009)	16	2	Female face	43.94	Novelty preference (versus same face with inverted eyes and mouth)

(continued)

Table 11.1 (continued)

Authors	*n*	Age (days)	Habituation stimuli	TL (s)	Preference test results
	16	2	Female face with inverted eyes and mouth	46.91	Novelty preference (vs. same face with normal feature orientation)
Bulf and Turati (2010)	17	2	Apparent motion of smoothly rotating female face	59.38	Novelty preference in static profile view images
	17	2	Random motion female face from different views	52.77	Familiarity preference in static profile view images
Pascalis and de Schonen (1994)	12	4	Female face	53.72	Novelty preference
	12	4	Female face	40.60	Novelty preference after 2-min delay
Barrera and Maurer (1981)	12	88	Photograph of mother's face	452.00	Novelty preference
Sangrigoli and de Schonen (2004)	18	92	Female face (own race)	95.48	Novelty preference
Pascalis et al. (1998)	14	98	Female face	200.00	Novelty preference after 2-min delay
	15	186	Female face in various poses	65.90	Novelty preference in new expression images after 2-min delay
	11	189	Female face in various poses	63.00	Novelty preference in new expression images after 24-h delay
Turati et al. (2010)	10	99	Top half of a female face	19.62	Familiarity preference in misaligned top and bottom halves of face
Bulf et al. (2013)	11	100	Female face with hat	53.00	Novelty preference in images with hat
	17	99	Female face without hat	46.06	Novelty preference in images with hat
Farroni et al. (2007)	12	115	Female face (direct gaze)	72.36	Novelty preference
Turati et al. (2004)	14	124	Female face in ¾ view images	56.10	Novelty preference

(continued)

Table 11.1 (continued)

Authors	*n*	Age (days)	Habituation stimuli	TL (s)	Preference test results
	14	124	Female face in ¾ view images in inverted orientation	49.79	Novelty preference
	17	128	Multiple view images of a male/female face	74.77	Novelty preference in inverted images
Kelly et al. (2007)	16 per race face condition	94	Male/female face of own race or other race in frontal view	70.74	Novelty preference in new view for all race faces examined
	16 per race face condition	187	Male/female face of own race or other race in frontal view	42.67	Novelty preference in new view only for own race and one other race
	16 per race face condition	279	Male/female face of own race or other race in frontal view	38.88	Novelty preference in new view only for own race faces
Kelly et al. (2009)	14–15 per race face condition	94	Male/female face of own race or other race	105.82	Novelty preference in new view for all races examined
	16 per race face condition	190	Male/female face of own race or other race	59.86	Novelty preference in new view only for own race faces
	10–15 per race face condition	282	Male/female face of own race or other race	47.39	Novelty preference in new view only for own race faces
Gross and Schwarzer (2010)	38	224	Female face in frontal or ¾ view with positive expression	108.41	Novelty preference in new pose
	33	224	Female face in frontal or ¾ view with negative expression	100.15	Novelty preference in new pose
	38	284	Female face in frontal or ¾ view	85.26	Novelty preference in new pose
Macchi Cassia et al. (2014)	14	106	Adult female face	72.79	Familiarity preference
	14	106	Face of a newborn infant	78.79	Novelty preference
	14	290	Adult female face	66.90	Novelty preference
	14	291	Adult female face	62.18	Novelty preference

TL total looking time during the habituation phase

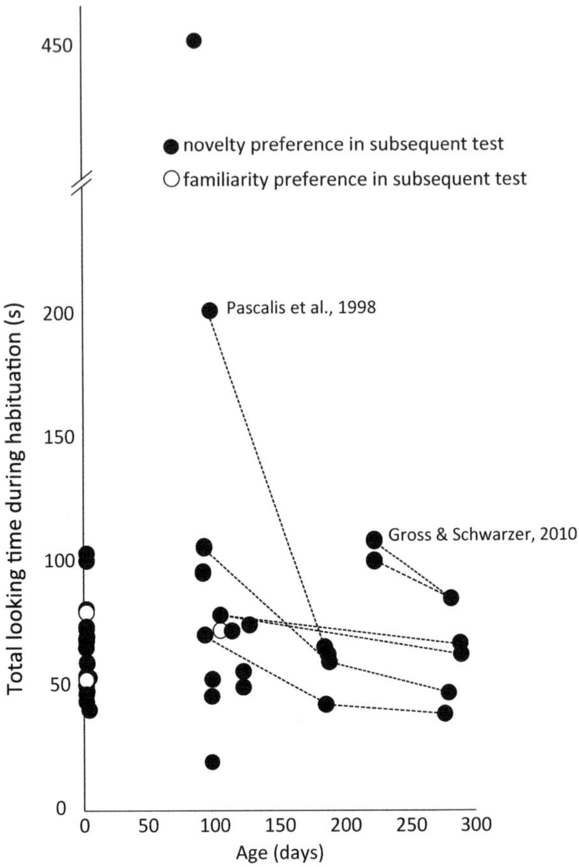

Fig. 11.1 The total looking times during the habituation phase for each age group for the studies listed in Table 11.1. Filled circle represents a group data point that elicited a novelty preference in the subsequent test phase. Open circle represents a group data point that elicited a familiarity preference in the subsequent test phase. The data points connected by dashed lines represent data from different age groups within the same study

that the looking time tended to decrease for the older age groups. The decline in looking time during the habituation phase in older infants is consistent with findings from studies that examined the developmental course of habituation to faces in infants (e.g., Colombo et al. 2004).

In a study by Pascalis et al. (1998), there is an especially sharp decline in the total looking time during the habituation phase between the 3-month-olds (200 s) and the 6-month-olds (65.9 s in a 2-min delay condition, 63.0 s in a 24-h delay condition). This difference occurred even though the habituation stimulus for the 3-month-olds was a photograph of a single female face, whereas that for the 6-month-olds was multiple photographs of a female face in various poses and views. When the 3-month-olds were habituated to the same stimuli as those used for the 6-month-olds, the total looking times during habituation were 175.5 s (n = 16, 2-min delay condition) and 128.6 s (n = 20, 24-h delay condition; not listed in Table 11.1 owing to the lack of preference in the subsequent test), which were again, much longer than the looking times in the 6-month-olds. Even with greater stimulus complexity, older infants seem to require a shorter time to habituate compared with younger infants.

The total looking times in the study by Gross and Schwarzer (2010)) are relatively longer compared with other studies examining similar age groups (Kelly et al. 2007, 2009; Pascalis et al. 1998). The stimulus used by Gross and Schwarzer during the habituation phase was a single color photograph of a face, which is similar to that in many other studies. Thus, stimulus complexity is unlikely to account for the longer looking times in this study. The relatively long total looking time in the study by Gross and Schwarzer may be related to the use of different habituation criteria from other studies. Except for Kelly et al. (2007, 2009) and Gross and Schwarzer, the studies listed in Table 11.1 defined the habituation criterion as a decline of at least 50% in the sum of looking times across three consecutive fixations compared with the sum of looking times across the first three fixations. Kelly et al. defined that the habituation criterion was at leaset 50% decline in the looking time witin a trial compared with the average looking time during the first two trials. To reach the habituation criterion in Gross and Schwarzer, however, the infants had to maintain at least a 50% decrease in looking time in comparison with the peak fixation level (the average of the two longest looking times across three consecutive trials). This habituation criterion of Gross and Schwarzer may have been more conservative than the criteria used in other studies.

Many studies have also used a familiarization method to examine face recognition in infants. In those studies, face recognition was tested after the presentation of facial images for a fixed duration, or after a fixed amount of accumulated looking time at facial images had been attained. Table 11.2 summarizes the findings of the studies using this familiarization method. As in Table 11.1, only the conditions that led to a significant preference are included, and only the studies examining the discrimination of faces among same-gender faces are included. Unless otherwise noted in Table 11.2, the same habituation stimulus was used for the test phase and familiarization phase.

To determine whether developmental changes affected the duration of familiarization required for face recognition, in Fig. 11.2, the total number of looking times during the familiarization period listed in Table 11.2 were plotted against the age of infants. As in Fig. 11.1, the data points from different age groups in a given study are connected by dashed lines. A tendency for the total looking time to decrease along with the infants' age is readily apparent in Fig. 11.2 without looking at connected data points. Studies that tested newborns reported the familiarization looking time to be around 70 s, whereas many studies reported a novelty preference after the familiarization looking time of less than 50 s for infants over 3 months of age.

The shortest looking time during familiarization that led to a successful novelty preference was reported by Chien et al. (2016) (the minimum of 7.63). However, caution is needed in interpreting this short looking time. Chain et al. tested infants repeatedly using three different types of face recognition tests, and each test followed a familiarization period to the same face across the tests. Therefore, the total looking time during familiarization before each test was not constant, but varied between the two values plotted as gray data points for each age in Fig. 11.2. The next shortest familiarization duration is 20 s, obtained by Pascalis et al. (2002) and Righi et al. (2014) by using a fixed accumulative looking time procedure.

Table 11.2 Summaries of the studies reporting successful face recognition memory in infants using the familiarization method

Authors	*n*	Age (days)	Habituation stimuli	TL (s)	Preference test results
Coulon et al. (2011)	16	2	Talking video of a female face	70	Familiarity preference
Guellai and Streri (2011)	12	2	Female face	69.7	Novelty preference
	12	2	Talking video of a female face	68	Familiarity preference in static silent images of faces
	12	2	Talking video of a female face	74.2	Familiarity preference
Guellai et al. (2015)	16	2	Talking video of a female face	71.9	Familiarity preference in static silent images of faces
	16	2	Talking video of a female face with a low-pass voice	75.6	Novelty preference in static silent images of faces
	16	2	Talking video of a female face with low-pass voice (unfamiliar language)	71	Novelty preference in static silent images of faces
De Haan et al. (2001)	13	29	Female face	52	Novelty preference
	14	85	Female face	22	Novelty preference
Sangrigoli and de Schonen (2004)	28	97	Female face shown together with another female face (own race)	68.63	Novelty preference
	28	97	Female face shown together with other female face (other race)	72.12	Novelty preference
Quinn et al. (2002)	8	104	Eight female faces, shown in pairs	32.96	Novelty preference
Otsuka et al. (2009)	12	103	Dynamic smiling female face image	27.8	Novelty preference in static images with new pose
	12	106	Smiling female face	80.89	Novelty preference
Quinn and Tanaka (2009)	32	108	Female face	49.68	Novelty preference for change in spacing between features, and change in eye size
	32	191	Female face	32.04	Novelty preference for change in spacing between features, and change in eye size

(continued)

Table 11.2 (continued)

Authors	n	Age (days)	Habituation stimuli	TL (s)	Preference test results
Chien et al. (2016)	16	120	Female face of own race or other race	9.92 to 59.52	Novelty preference only in own race faces
	19	184	Female face of own race or other race	9.88 to 59.32	Novelty preference in own race and in faces from one other ethnic group
	16	267	Female face of own race or other race	7.63 to 45.82	Novelty preference in all race faces examined
Bahrick et al. (2002)	12	164	Photograph of female face performing action	101.23	Novelty preference after 1-min delay
Fagan (1976)	96	204	Male face (frontal/¾ profile view)	21.6	Novelty preference in the same pose
	52	204	Male face (frontal/¾ profile view)	24	Novelty preference in the new pose
Pascalis et al. (2002)	15	186	Male or female face	20	Novelty preference
	15	274	Male or female face	20	Novelty preference
Righi et al. (2014)	23	218	Male or female face	20	Novelty preference
Tyrrell et al. (1987)	24	210	Male or female face	40	Novelty preference
	24	210	Caricature image of male or female face	40	Novelty preference
	40	231	Male or female face	40	Novelty preference in caricature images
Otsuka et al. (2013)	12	227	Female face	72.37	Novelty preference
	12	227	Contrast reversed female face with positive eyes	74.59	Novelty preference

Some of the TL values were estimated based on the reported mean looking time value across familiarization trials

TL total looking time during the familiarization phase

It is notable that some of the familiarization studies presented faces of multiple identities during the familiarization phase, then tested the infants' recognition of individual faces (de Haan et al. 2001; Quinn et al. 2002; Sangrigoli and de Schonen 2004). Among these studies, it is notable that Quinn et al. (2002) showed eight female faces to 4-month-olds during the familiarization period. Dividing the total looking time of 33 s by the number of the faces gives a looking time of only 4.04 s per face. The results described by Quinn et al. showed that such short exposure to faces was sufficient for 4-month-olds to maintain a memory for each face, at least for the period of two 10-s test trials that immediately followed the familiarization period.

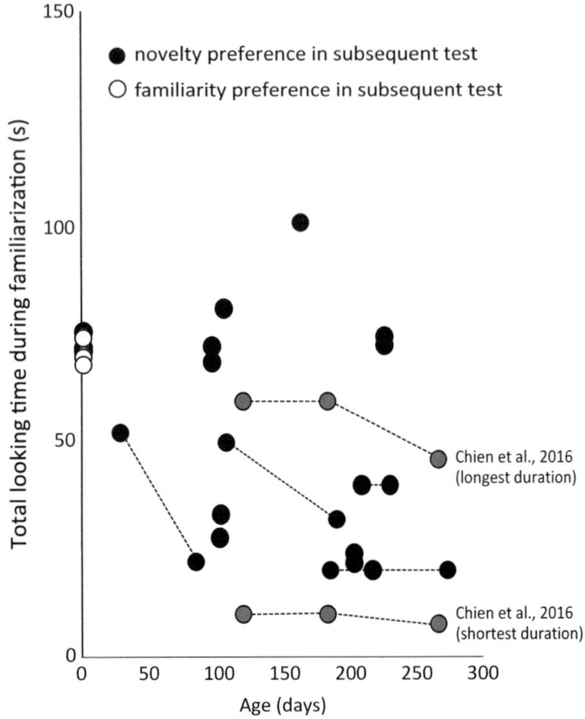

Fig. 11.2 The total looking times during the familiarization phase for each age group in the studies listed in Table 11.2. Filled circle represents a group data point that elicited a novelty preference in the subsequent test phase. Open circle represents a group data point that elicited a familiarity preference in the subsequent test phase. The data points connected by dashed lines represent data from different age groups within the same study. The data points in gray circles in each age group represent the shortest to the longest familiarization looking time in Chien et al. (2016)

Another line of studies attempted to determine the duration of familiarization required for infants to show a novelty preference by using a continuous familiarization technique. For example, in Rose et al. (2002a, b), infants received multiple test trials in which two photographs of infants' faces were shown side by side. One of the faces was repeatedly shown across all the trials, whereas the other face was a novel face that varied for each trial. The test trials were repeated until the infants consistently showed a novelty preference. The findings from this line of studies are summarized in Table 11.3. The reported total looking time at the familiarization face to reach the criterion tended to decline from about 30 s in 5-month-olds to about 20 s in 7-month-olds to about 15 s in 12-month-olds. These durations seem to be more conservative compared with the studies that presented multiple faces during the fixed duration familiarization trials as discussed above (Quinn et al. 2002; Sangrigoli and de Schonen 2004). Although Rose et al. (2002a, b) used faces of infants as the stimuli, the latter two studies used adult female faces as the stimuli. The difference between the required durations may be partly attributed to greater daily exposure to and resultant expertise for adult female faces compared with infant faces during infancy.

Table 11.3 The accumulated looking time to the familiarization face before novelty preference was shown in studies using the continuous familiarization technique

Authors	*n*	Age (days)	Familiar and novel stimuli	TL (s)
Rose et al. (2002a)	36	225	Infant face in frontal orientation	23.91
	24	221	Infant face in frontal orientation shown upside down	21.83
	36	376	Infant face in frontal orientation	13.34
	24	370	Infant face in frontal orientation shown upside down	13.79
Rose et al. (2002b)	134	Mean age in days not reported (5 months)	Infant face in frontal orientation	29.69
	128	Mean age in days not reported (7 months)	Infant face in frontal orientation	23.49
	117	Mean age in days not reported (12 months)	Infant face in frontal orientation	15.22

TL total looking time to the repeated (familiar) face during the multiple test trials
Although Rose et al. (2002b) examined both full-term and preterm infant populations, here only the results from full-term infants are considered

11.5 Is Habituation or Familiarization Necessary for Face Recognition in Infants?

The survey of the studies employing habituation or familiarization methods as discussed above showed that exposure to a face for several seconds to several tens of seconds is necessary to elicit the behavioral evidence of face recognition in infants. However, several neuroimaging studies suggest that infants might be capable of maintaining a representation of a face that has been shown for a shorter period. For example, using functional near-infrared spectroscopy, Kobayashi et al. (2011, 2012) reported that the repeated presentation of a single-identity face elicited a decreased hemodynamic response compared with multiple-identity faces in the temporal regions of the brain in 5- to 8-month-olds. Each face was shown for 800 ms, with a 200-ms interval between faces. In the different-identity condition, faces of five identities were shown during the 10-s test period with each repeated once. In the single-identity condition, a single-identity face was shown repeatedly throughout the 10-s test period. The differing hemodynamic responses between the single- and multiple-identity face conditions suggest that the infants differentiated multiple faces in a rapid image sequence in the different-identity condition, whereas they recognized repetition in identity in the single-identity condition.

Peykrjou et al. (2015) used a rapid repetition event-related potential (ERP) paradigm. In their study of 9-month-old infants, a face was shown for 1500 ms either following the presentation of the same face or following the presentation of a different face. The interval between the faces was 500–700 ms. The results revealed that the repeated face elicited a faster latency in N290 compared with the non-repeated novel face. Righi et al. (2014) measured ERP responses to male and female faces in 7-month-old infants. Using a 1-back paradigm, Righi et al. reported that a novel female face elicited a greater N290 amplitude compared with female faces that were repeated, whereas no such difference was found for novel and repeated male faces. In their study, each face was shown for only 500 ms, and the inter-stimulus-interval was 2 s. Thus, the results reported by Righi et al. suggest that a representation for a female face seen for only 500 ms might be maintained for at least a 4-s duration. These findings from neuroimaging studies indicate that infants over 5 months of age may be capable of discriminating between faces with only quick glances at them, and of recognizing faces that they have seen briefly before.

11.6 How Long Does an Infant's Memory for a Face Last?

Several reviews of the findings regarding long-term memory during infancy (Hayne 2007; Lukowski and Bauer 2014; Rose et al. 2007) consistently suggested that infants retain memories for longer periods as they grow older. There are several studies that included a memory retention period between the face-learning phase and the preference test phase. Findings from such previous studies are not contradictory to the general developmental trend of long-term memory. As already mentioned in the Sect. 11.3, Pascalis and de Schonen (1994) reported that newborn infants successfully recognized unfamiliar faces by showing novelty preference even when a 2-min delay period was inserted before the recognition test. Bushnell (2001) reported that newborn infants preferred their own mother's face over an unfamiliar female face after a delay of 15 min since they last saw their mother's face. Bahrick et al. (2002) reported that following familiarization to unfamiliar faces, 5-month-old infants showed a novelty preference after a 1-min delay, but not after a 7-week delay. Pascalis et al. (1998) reported that 6-month-olds showed a novelty preference following habituation to an unfamiliar face in various poses after a 2-min delay and after a 24-hour delay.

Studies that used real face-to-face interaction during the memorization phase suggest that the memory for faces acquired in infancy could be retained for a year or longer. Bornstein et al. (2004) examined whether infants could retain a memory for a face across a 15-month delay. The infants in that study participated in a still-face procedure experiment at 5 months of age. At that time, each infant had an initial normal interaction with a female experimenter for 1 min, followed by a still period in which the experimenter remained silent and expressionless for 2 min. The infant then participated in normal interaction with the experimenter again for 1 min. When they reached the age of 20 months, the same infants were shown a talking-face video recording of the familiar experimenter and that of two novel females. The infants looked significantly longer at the two novel female faces over the familiar experimenter's face. More recently, Kingo et al. (2014) reported that 3-year-olds showed recognition memory for a person with whom they had interacted once when they were 12 months old. In that study, 12-month-old infants had a face-to-face interaction for 48 min with one of two male experimenters. During the interaction, the experimenter manipulated objects in front of the infant and let the infant act on the object. Twenty-eight months later, when the infants were 3 years old, they were shown the video recording of the familiar experimenter and a novel experimenter in the original experimental setting, side by side. The children demonstrated recognition memory for the familiar experimenter by showing a novelty preference, although their response to an explicit verbal question about their memory was at the chance level. Unlike most other studies (Bushnell 2001; Bornstein et al. 2004; Pascalis et al. 1998), the two faces contrasted in Kingo et al. differed in race. The large appearance difference between the two faces may have contributed to the finding of very long-term face recognition memory.

11.7 Conclusion

The review of the findings on recognition memory for facial identity during infancy confirmed that although infants exhibit recognition memory for faces even at birth, there is considerable improvement in the efficacy of forming memories for faces and the durability of facial memory throughout infancy.

References

Bahrick LE, Gogate LJ, Ruiz I (2002) Attention and memory for faces and actions in infancy: the salience of actions over faces in dynamic events. Child Dev 73(6):1629–1643

Bahrick LE, Lickliter R, Castellanos I (2013) The development of face perception in infancy: intersensory interference and Unimodal visual facilitation. Dev Psychol 49(10):1919–1930

Barrera ME, Maurer D (1981) Recognition of mother's photographed face by the three-month-old infant. Child Dev 52(2):714–716

Bauer PJ (2013) The development of forgetting: childhood amnesia. In: Bauer PJ, Fivush (eds) The Wiley-Blackwell handbook on the development of children's memory. Wiley-Blackwell, West Sussex, pp 519–544

Bornstein MH, Arterberry ME, Mash C (2004) Long-term memory for an emotional interpersonal interaction occurring at 5 months of age. Infancy 6(3):407–416

Bulf H, Turati C (2010) The role of rigid motion in newborns' face recognition. Vis Cogn 18(4):504–512

Bulf H, Valenza E, Turati C (2013) How a hat may affect 3-month-olds' recognition of a face: an eye-tracking study. PLoS One 8(12):e82839

Bushnell IWR (2001) Mother's face recognition in newborn infants: learning and memory. Infant Child Dev 10(1–2):67–74

Bushnell IWR (2003) Newborn face recognition. In: Pascalis O, Slater A (eds) The development of face processing in infancy and early childhood current perspectives. Nova Science Pub Inc, New York, pp 41–53

Bushnell IWR, Sai F, Mullin JT (1989) Neonatal recognition of the mother's face. Br J Dev Psychobiol 7(1):3–15

Chien SH, Wang JF, Huang TR (2016) Developing the own-race advantage in 4-, 6-, and 9-month-old Taiwanese infants: a perceptual learning perspective. Front Psychol 7:1606
Colombo J, Shaddy DJ, Richman WA, Maikranz JM, Blaga OM (2004) The developmental course of habituation in infancy and preschool outcome. Infancy 5(1):1–38
Coulon M, Guellai B, Streri A (2011) Recognition of unfamiliar talking faces at birth. Int J Behav Dev 35(3):282–287
De Haan M, Johnson MH, Maurer D, Perrett DI (2001) Recognition of individual faces and average face prototypes by 1- and 3-month-old infants. Cogn Dev 16(2):659–678
De Heering A, Turati C, Rossion B, Bulf H, Goffaux V, Simion F (2008) Newborns' face recognition is based on spatial frequencies below 0.5 cycles per degree. Cognition 106(1):444–454
Fagan JF (1976) Infants' recognition of invariant features of faces. Child Dev 47(3):627–638
Farroni T, Massaccesic S, Menona E, Johnson MH (2007) Direct gaze modulates face recognition in young infants. Cognition 102(3):396–404
Field TM, Cohen D, Gracia R, Greenberg R (1984) Mother-stranger face discrimination by the newborn. Infant Behav Dev 7(1):19–25
Gava L, Valenza E, Turati C, de Schonen S (2008) Effect of partial occlusion on newborns' face preference and recognition. Dev Sci 11(4):563–574
Gross C, Schwarzer G (2010) Face recognition across varying poses in 7-and 9-month-old infants: I role of facial expression. Int J Behav Dev 34(5):417–426
Guellai B, Coulon M, Streri A (2011) The role of motion and speech in face recognition at birth. Vis Cogn 19(9):1212–1233
Guellai B, Mersad K, Streri A (2015) Suprasegmental information affects processing of talking faces at birth. Infant Behav Dev 38:11–19
Hayne H (2007) Infant memory development. In: Oakes LM, Bauer PJ (eds) Short- and long-term memory in infancy and early childhood: first steps toward remembering. Oxford University Press, Oxford, pp 209–239
Hayne H, Jack F (2011) Childhood amnesia. Wiley Interdiscip Rev Cogn Sci 2(2):136–145
Kelly DJ, Quinn PC, Slater AM, Lee K, Ge L, Pascalis O (2007) The other-race effect develops during infancy: evidence of perceptual narrowing. Psychol Sci 18(12):1084–1089
Kelly DJ, Liu SY, Lee K, Quinn PC, Pascalis O, Slater AM, Ge LZ (2009) Development of the other-race effect during infancy: evidence toward universality? J Exp Child Psychol 104(1):105–114
Kingo OS, Staugaard SR, Krojgaard P (2014) Three-year-olds' memory for a person met only once at the age of 12 months: very long-term memory revealed by a late-manifesting novelty preference. Conscious Cogn 24:49–56
Kobayashi M, Otsuka Y, Nakato E, Kanazawa S, Yamaguchi MK, Kakigi R (2011) Do infants represent the face in a viewpoint-invariant manner? Neural adaptation study as measured by near-infrared spectroscopy. Front Hum Neurosci 5:153
Kobayashi M, Otsuka Y, Kanazawa S, Yamaguchi MK, Kakigi R (2012) Size-invariant representation of face in infant brain: an fNIRS-adaptation study. Neuroreport 23(17):984–988
Lee K, Anzures G, Quinn PC, Pascalis O, Slater A (2011) Development of face processing expertise. In: Calder GRAJ, Johnson MH, Haxby JV (eds) The Oxford handbook of face perception. Oxford University Press, Oxford, pp 753–778
Leo I, Simion F (2009) Face processing at birth: a Thatcher illusion study. Dev Sci 12(3):492–498
Liu S, Xiao NG, Quinn PC, Zhu D, Ge L, Pascalis O, Lee K (2015) Asian infants show preference for own-race but not other-race female faces: the role of infant caregiving arrangements. Front Psychol 6:593
Lukowski AF, Bauer PJ (2014) Long-term memory in infancy and early childhood. In: Bauer PJ, Fivush R (eds) The Wiley handbook on the development of children's memory. Wiley, Chichester, pp 230–254
Macchi Cassia V, Bulf H, Quadrelli E, Proietti V (2014) Age-related face processing bias in infancy: evidence of perceptual narrowing for adult faces. Dev Psychobiol 56(2):238–248
Otsuka Y (2014) Face recognition in infants: a review of behavioral and near-infrared spectroscopic studies. Jpn Psychol Res 56(1):76–90

Otsuka Y, Konishi Y, Yamagushi MK, Kanazawa S, Abdi H, O'Toole AJ (2009) Recognition of moving and static faces by young infants. Child Dev 80(4):1259–1271
Otsuka Y, Motoyoshi I, Hill H, Kobayashi M, Kanazawa S, Yamagushi MK (2013) Eye contrast polarity is critical for face recognition by infants. J Exp Child Psychol 115(3):598–606
Pascalis O, de Schonen S (1994) Recognition memory in 3- to 4-day-old human neonates. Neuroreport 5(14):1721–1724
Pascalis O, de Schonen S, Morton J, Deruelle C, Fabregrenet M (1995) Mothers face recognition by neonates – a replication and an extension. Infant Behav Dev 18(1):79–85
Pascalis O, de Haan M, Nelson CA, de Schonen S (1998) Long-term recognition memory for faces assessed by visual paired comparison in 3- and 6-month-old infants. J Exp Psychol-Learn Mem Cogn 24(1):249–260
Pascalis O, de Haan M, Nelson CA (2002) Is face processing species-specific during the first year of life? Science 296(5571):1321–1323
Peykrjou S, Pauen S, Hoehl S (2015) 9-month-old infants recognize individual unfamiliar faces in a rapid repetition ERP paradigm. Infancy 21(3):288–311
Quinn PC, Tanaka JW (2009) Infants' processing of featural and configural information in the upper and lower halves of the face. Infancy 14(4):474–487
Quinn PC, Yahr J, Kuhn A, Slater AM, Pascalis O (2002) Representation of the gender of human faces by infants: a preference for female. Perception 31(9):1109–1121
Quinn PC, Uttley L, Lee K, Gibson A, Smith M, Slater AM, Pascalis O (2008) Infant preference for female faces occurs for same- but not other-race faces. J Neuropsychol 2(1):15–26
Richards JE (1997) Effects of attention on infants' preference for briefly exposed visual stimuli in the paired comparison recognition-memory paradigm. Dev Psychol 33(1):22–31
Righi G, Westerlund A, Congdon EL, Troller-Renfree S, Nelson CA (2014) Infants' experience-dependent processing of male and female faces: insights from eye tracking and event-related potentials. Dev Cogn Neurosci 8:144–152
Roder BJ, Bushnell EW, Sasseville AM (2000) Infants' preferences for familiarity and novelty during the course of visual processing. Infancy 1(4):491–507
Rose SA, Gottfried AW, Melloy-Carminar P, Bridger WH (1982) Familiarity and novelty preferences in infant recognition memory: implications for information processing. Dev Psychol 18(5):704–713
Rose SA, Jankowski JJ, Feldman JF (2002a) Speed of processing and face recognition at 7 and 12 months. Infancy 3(4):435–455
Rose SA, Feldman JF, Jankowski JJ (2002b) Processing speed in the 1st year of life: a longitudinal study of preterm and full-term infants. Dev Psychol 38(6):895–902
Rose SA, Feldman J, Jankowski JJ (2007) Developmental aspects of visual recognition memory in infancy. In: Bauer PJ, Oakes LM (eds) Short- and long-term memory in infancy and early childhood: taking the first steps toward remembering. Oxford University Press, Oxford, pp 153–178
Sai FZ (2005) The role of the mother's voice in developing mother's face preference: evidence for intermodal perception at birth. Infant Child Dev 14(1):29–50
Sangrigoli S, de Schonen S (2004) Recognition of own-race and other-race faces by three-month-old infants. J Child Psychol Psychiatry 45(7):1219–1227
Simion F, Di Giorgio E (2015) Face perception and processing in early infancy: inborn predispositions and developmental changes. Front Psychol 6:969
Turati C, Sangrigoli S, Ruel J, de Schonen S (2004) Evidence of the face inversion effect in 4-month-old infants. Infancy 6(2):275–297
Turati C, Macchi Cassia V, Simion F, Leo I (2006) Newborns' face recognition: role of inner and outer facial features. Child Dev 77(2):297–311
Turati C, Bulf H, Simion F (2008) Newborns' face recognition over changes in viewpoint. Cognition 106(3):1300–1321
Turati C, Di Giorgio E, Bardi L, Simion F (2010) Holistic face processing in newborns, 3-month-old infants, and adults: evidence from the composite face effect. Child Dev 81(6):1894–1905
Tyrrell DJ, Anderson JT, Clubb M, Bradbury A (1987) Infant's recognition of the correspondence between photographs and caricatures of human faces. Bull Psychon Soc 25(1):41–43

Chapter 12
Brain Maturation and Aging: Large Brain MRI Database

Yasuyuki Taki

Abstract Japan has a super-aging population; thus, it is important to preserve cognitive function for the entire lifespan by preventing pathological brain aging. To achieve the aim, we have compiled a large brain magnetic resonance imaging (MRI) database of around 3000 subjects aged from 5 to 80 years to reveal how the brain develops and ages. We have also collected data on several cognitive functions, lifestyle factors such as eating and sleeping habits, and genetic data. Using the database, we have revealed normal brain development and aging, and have revealed what factors affect brain development and aging. For example, there were significant negative correlations between alcohol consumption and the gray-matter volume of the fronto-parietal region, and between the body mass index and the gray-matter volume of the hippocampus in a cross-sectional analysis. In addition, having intellectual curiosity showed significant negative correlation with regional gray-matter volume decline rate in the temporo-parietal region. These findings help to understand the mechanism of brain development and aging, and to perform differential diagnosis or diagnosis at an early stage of several diseases/disorders such as autism and Alzheimer's. In addition, we introduce the Tohoku Medical Megabank Project, in which we will are compiling a brain MRI database of around 30,000 healthy subjects. By performing the project, we aim to build a system of preventive medicine for several diseases/disorders such as Alzheimer's.

Keywords Brain development • Brain aging • Magnetic resonance imaging • Database • Preventive medicine • Normal subject

Y. Taki (✉)
Department of Nuclear Medicine & Radiology, Institute of Development, Aging and Cancer (IDAC), Tohoku University, Sendai, Japan

Division of Medical Neuroimage Analysis, Department of Community Medical Supports, Tohoku Medical Megabank Organization, Tohoku University, Sendai, Japan
e-mail: yasuyuki.taki.c7@tohoku.ac.jp

T. Tsukiura, S. Umeda (eds.), *Memory in a Social Context*,
https://doi.org/10.1007/978-4-431-56591-8_12

12.1 Introduction

Japan has a super-aging population; thus, it is important to preserve cognitive function for the entire lifespan by preventing pathological brain aging. Recently, the importance of a human neuroimaging database was recognized. The database of normal brain structure and function can be used as the reference not only for neuroimaging study for humans, but also for the early diagnosis and computer-aided automated diagnosis of brain diseases. The most remarkable recently developed method of brain image analysis is voxel-based morphometry (VBM). It includes the anatomical standardization of the brain to a standard brain, brain tissue segmentation, and finally voxel-based statistical analysis based on a general linear model. This technique enables us to extract brain regions that show correlations between tissue volume and variables, such as age, sex and other characteristics of the subject. We can analyze not only age-related normal changes, but also diseased brain, suffering from disorders such as dementia and schizophrenia. It has been believed that functional imaging precedes structural imaging to detect early pathological findings of diseases. However, recent development of high-resolution structural imaging and a sophisticated analytical technique has enabled us to detect brain disease at a very early stage. Now, we have collected over 3000 brain MRIs of healthy Japanese individuals aged from 5 to 80 years and constructed an MRI database together with their characteristics such as age, sex, blood pressure, present and past disease history, and cognitive functions. To our knowledge, this is the largest brain MRI database in Japan and one of the largest in the world.

12.2 Imaging Studies of Brain Maturation

12.2.1 Correlation Between Gray-Matter Density-Adjusted Brain Perfusion and Age

In understanding brain aging, the knowledge of brain maturation is very important, for the relationship between brain maturation and brain aging is regarded as a "mirror image." In detail, brain regions that mature earlier such as occipital regions are robust in brain aging, whereas brain regions that mature rather late such as the prefrontal regions are vulnerable to aging. Brain development continues throughout childhood and adolescence. Recently, it has been revealed that human brain development is a structurally and functionally a nonlinear process. However, despite this growing wealth of knowledge about maturational changes in brain structure in children, the trajectory of brain perfusion with age in healthy children is not yet well documented.

Recently, arterial spin-labeling (ASL) perfusion magnetic resonance imaging (MRI) has been developed for evaluating brain perfusion. We examined the

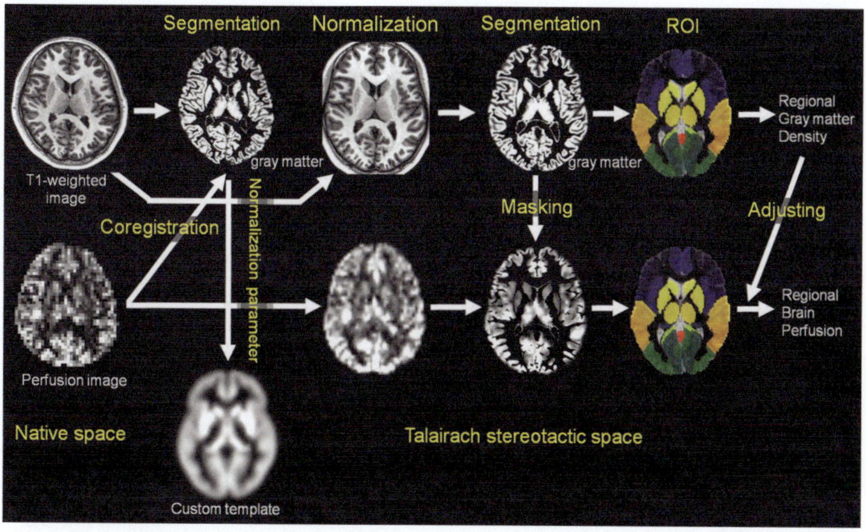

Fig. 12.1 Schematic of the image analysis

correlation between brain perfusion and age using pulsed ASL MRI in a large number of healthy children.

We collected data on brain structural and ASL perfusion MRI in 202 healthy children aged 5–18 years (Taki et al. 2011a). Structural MRI data were segmented and normalized, applying a voxel-based morphometric analysis. Perfusion MRI was normalized using the normalization parameter of the corresponding structural MRI. We calculated brain perfusion with an adjustment for gray-matter density (BP-GMD) by dividing normalized ASL MRI by normalized gray-matter segments in 22 regions. Next, we analyzed the correlation between BP-GMD and age in each region by estimating linear, quadratic, and cubic polynomial functions, using the Akaike information criterion (Fig. 12.1).

As a result, the correlation between BP-GMD and age showed an inverted U shape followed by a U-shaped trajectory in most regions. In addition, age at which BP-GMD was highest was different among the lobes and gray-matter regions, and the BP-GMD association with age increased from the occipital to the frontal lobe via the temporal and parietal lobes (Fig. 12.2).

In the frontal lobe, all gray-matter regions showed an inverted U-shaped trajectory for the correlation between BP-GMD and age, and the best fit was a negative quadratic or positive cubic polynomial function. The estimated age at which BP-GMD was highest was earlier in the precentral gyrus, cingulate gyrus, and anterior cingulate cortex than in the superior, middle, and inferior frontal gyri (Fig. 12.3).

We demonstrated a correlation between BP-GMD and age using ASL brain perfusion MRI in a large number of healthy children over a wide age range. As a result, the trajectory of the correlation between BP-GMD and age showed an inverted U-shaped second-order polynomial function in most regions in the frontal lobe, a

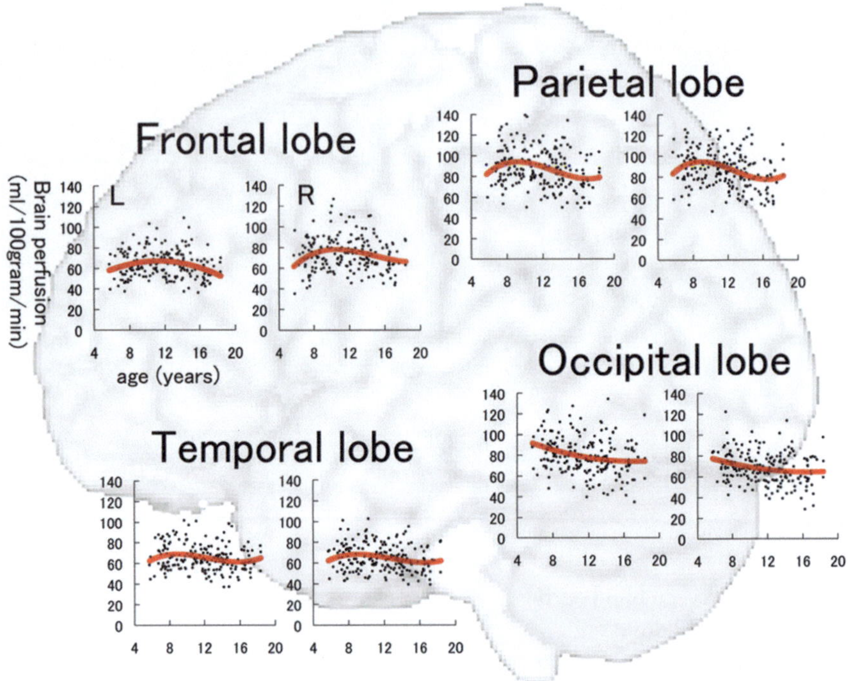

Fig. 12.2 Correlation between brain perfusion, adjusted for gray-matter density, and age in the frontal lobe, parietal lobe, occipital lobe, and temporal lobe in each hemisphere

third-order polynomial function in the parietal and temporal lobes, and a U-shaped second-order and negative linear correlation in the occipital lobe. Our results indicate that higher-order association cortices mature after the lower-order cortices in terms of brain perfusion. As a result, the trajectory of the correlation between BP-GMD and age showed an inverted U shape followed by a U-shaped trajectory in most regions. In addition, the age at which BP-GMD was highest was different among the lobes and gray-matter regions, showing a progression from the occipital lobe to the frontal lobe, via the temporal and parietal lobes. Our results indicate that higher-order association cortices mature later than the lower-order cortices. This may help not only to clarify the mechanisms of normal brain maturation from the viewpoint of brain perfusion, but also to distinguish normal from developmental disorders that show abnormal brain perfusion patterns.

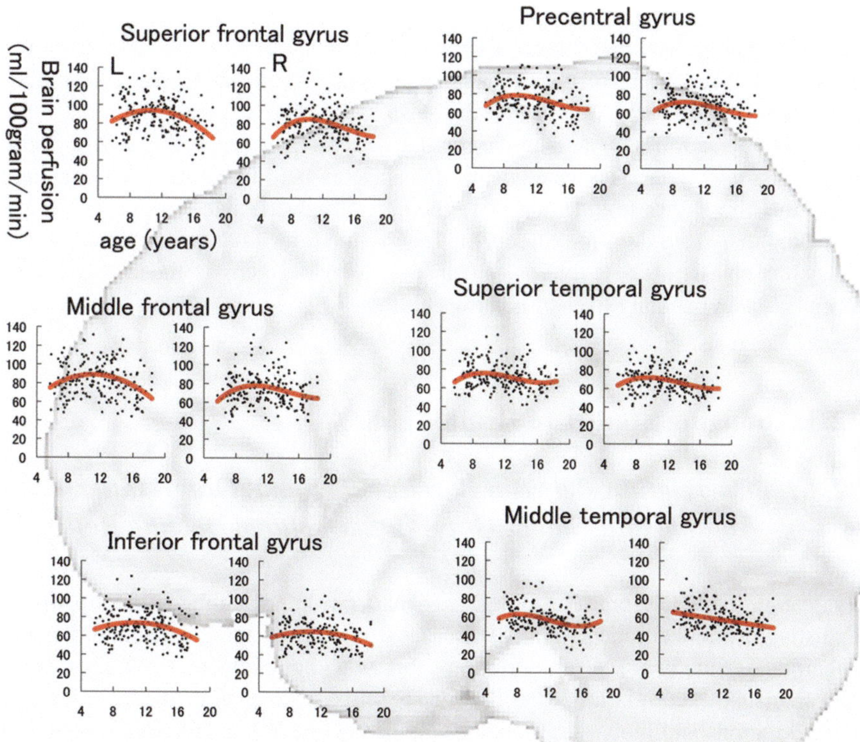

Fig. 12.3 Correlation between brain perfusion, adjusted for gray-matter density, and age in the precentral gyrus, superior frontal gyrus, middle frontal gyrus, inferior frontal gyrus, superior temporal gyrus, and middle temporal gyrus

12.2.2 Correlation Between Hippocampal Gray-Matter Volume and Sleep Duration in Healthy Children

Sleep is essential for living beings, and sleep loss has been shown to affect hippocampal structure and function in rats by inhibiting cell proliferation and neurogenesis in this region of the brain. We aimed to analyze the correlation between sleep duration and the hippocampal volume using brain magnetic resonance images of 290 healthy children aged 5–18 years (Taki et al. 2012). We examined the volume of gray matter, white matter, and the cerebrospinal fluid (CSF) space in the brain using a fully automated and established neuroimaging technique, VBM, which enabled global analysis of brain structure without bias toward any specific brain region while permitting the identification of potential differences or abnormalities in brain structures. We found that the volume of the bilateral hippocampal body was significantly positively correlated with the duration of sleep during weekdays after adjusting for age, sex, and intracranial volume, and after performing small-volume correction of the hippocampal region of interest (left: $t = 3.59$, $p = 0.014$, family-wise error, corrected; right: $t = 3.81$, $p = 0.007$, family-wise error, corrected). The whole-brain analysis showed that the duration of sleep during weekdays was substantially positively correlated with the regional gray-matter volume of the bilateral hippocampal body (left: $t = 3.59$, $p < 0.001$, uncorrected; right: $t = 3.81$, $p < 0.001$, uncorrected) and the right dorsolateral prefrontal cortex ($t = 3.95$, $p < 0.001$, uncorrected) after adjusting for age, sex, and intracranial volume, and using the liberal threshold ($p < 0.001$, uncorrected; cluster size >100). Our results indicated that sleep duration affects the hippocampal regional gray-matter volume of healthy children. These findings advance our understanding of the importance of sleep habits in the daily lives of healthy children.

12.3 Imaging Studies of Brain Aging

12.3.1 Correlation Between Baseline Regional Gray-Matter Volume and Global Gray-Matter Volume Decline Rate

Evaluating the whole-brain or global gray-matter volume decline rate is important in distinguishing neurodegenerative diseases from normal aging and in anticipating cognitive decline over a given period in subjects without dementia. Whether a significant negative correlation exists between baseline regional gray-matter volume of several regions and global gray-matter volume decline in the subsequent period in healthy subjects has not yet been clarified. Therefore, we analyzed the correlation between baseline regional gray-matter volumes and the rate of global gray-matter volume decline in the period following baseline using magnetic resonance images of the brains of 381 healthy subjects by applying a longitudinal design over 6 years using VBM (Taki et al. 2011b).

All subjects were Japanese individuals recruited from our previous brain-imaging project. From an initial 1,604 eligible persons, we selected participants who had lived in Sendai City at the time of the previous study, whose collected data had no missing values, and who had no serious medical problems. All participants were screened with a mail-in health questionnaire and underwent telephone and personal interviews. Persons who reported a history of any malignant tumor, head trauma with loss of consciousness for >5 min, cerebrovascular disease, epilepsy, any psychiatric disease, or claustrophobia were excluded from the study. All subjects were screened for dementia using the Mini-Mental State Examination (MMSE), and an experienced neuroradiologist examined the MR scans for any tumors and cerebrovascular disease. The final sample consisted of 381 participants (40.1% of the eligible cohort: 158 men, 223 women). All images were collected using the same 0.5-T MR scanner, including baseline images created using MP-RAGE pulse sequences. After the image acquisition, all MR images were analyzed using Statistical Parametric Mapping 2 in MATLAB. We calculated gray-matter volume and white-matter volume using fully automated techniques. To normalize the head

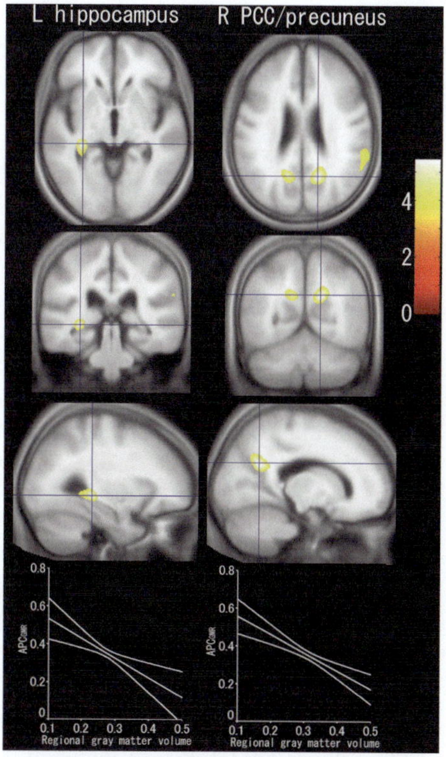

Fig. 12.4 Gray-matter regions showing significant negative correlations with annual percentage change of the gray-matter ratio (APC_{GMR}) adjusted for age, gender, and intracranial volume

size of each subject, we defined the gray-matter ratio (GMR) as the percentage of gray-matter volume divided by the intracranial volume. Next, to reveal the annualized rate of change in GMR with age, we determined the annual percentage change in GMR (APC_{GMR}) for each subject. We determined regional gray-matter volume using VBM. To investigate the correlation between baseline regional gray-matter volume and APC_{GMR}, we performed a multiple regression analysis with age, gender, intracranial volume, and APC_{GMR} as independent valuables and baseline regional gray-matter volume as a dependent valuable. We used the random field theory method to correct for the familywise error rate (FWE); any resulting p value less than 0.05 was considered significant. Next, we tested whether the gray-matter regional volume that showed the significant negative correlation with APC_{GMR} at baseline could predict whether the APC_{GMR} was above or below the mean APC_{GMR} by applying a standard (not stepwise) linear discriminant analysis in SPSS11.5. For the discriminant analysis, we used the mean gray-matter volume over a cluster in each region, and the regional gray-matter volume as defined by multiple regression analysis. We set the significance level at $p < 0.05$.

As a result, the gray-matter regions showing significant negative correlation with APC_{GMR} adjusted for age, gender, and intracranial volume are shown in Fig. 12.4.

Baseline regional gray-matter volumes of the right posterior cingulate cortex (PCC)/precuneus and the left hippocampus showed significant negative correlations with APC_{GMR} after adjusting for age, gender, and intracranial volume (right PCC/precuneus, $t = 5.42, p = 0.020$; left hippocampus, $t = 5.29, p = 0.035$). Therefore, we used the gray-matter regions of the right PCC/precuneus and the left hippocampus in the next discriminant analysis. Baseline regional gray-matter volume of both the right PCC/precuneus and the left hippocampus significantly distinguished, whether APC_{GMR} was above or below the mean APC_{GMR}. The F-value, p value, and discriminant function coefficient were 13.51, 0.001, and 0.833 in the right PCC/precuneus, and 5.71, 0.017, and 0.350 in the left hippocampus respectively. Overall, 58.4% of the APC_{GMR} (55.8% of APC_{GMR} below the mean of APC_{GMR} and 60.9% of APC_{GMR} above the mean of APC_{GMR}) was correctly distinguished using the discriminant function.

To our knowledge, this study provides the first longitudinal findings demonstrating that baseline regional gray-matter volumes in the right PCC/precuneus and the left hippocampus show a significant negative correlation with the rate of global gray-matter volume decline in the following period, as represented by APC_{GMR}, adjusting for age, gender, and intracranial volume. In addition, baseline regional gray-matter volumes of both the right PCC/precuneus and the left hippocampus significantly distinguished whether the APC_{GMR} was above or below the mean APC_{GMR}. These results indicate that subjects who had smaller baseline regional gray-matter volumes in those regions showed a higher rate of global gray-matter volume decline in the following period.

In summary, using a longitudinal design over 6 years in 381 community-dwelling healthy individuals, we examined the correlation between baseline regional gray-matter volume and the rate of global gray-matter volume decline in the following period. We found a significant negative correlation between APC_{GMR} and the baseline regional gray-matter volumes of the right PCC/precunei and the left hippocampus after adjusting for age and gender. In addition, baseline regional gray-matter volume of both the right PCC/precuneus and the left hippocampus significantly distinguished whether the APC_{GMR} was above or below the mean APC_{GMR}. Our results suggest that baseline regional gray-matter volume might predict the rate of global gray-matter volume decline in the following period in healthy subjects. Our study may contribute to distinguishing neurodegenerative diseases from normal aging, and to predicting cognitive decline.

12.3.2 Correlation Between Degree of White-Matter Hyperintensities and Global Gray-Matter Volume Decline Rate

Whether the degree of white-matter hyperintensities (WMHs) shows a significant correlation with the rate of global gray-matter volume decline over a period following initial baseline measurement remains unclear. The purpose of the present study was to reveal the relationship between the degree of WMHs at baseline and the rate of global gray-matter volume decline by applying a longitudinal design (Taki et al. 2011c).

Using a 6-year longitudinal design and MR images of the brains of 160 healthy individuals aged over 50 years and living in the community, we analyzed the correlation between degree of WMHs using Fazekas scaling at baseline and rate of global gray-matter volume decline 6 years later. To obtain the rate of global gray-matter volume decline, we calculated global gray-matter volume and intracranial volume at baseline and at follow-up using a fully automated method.

As a result, the annual percentage change in the GMR (APC_{GMR}), in which GMR represents the percentage of gray-matter volume in the intracranial volume, showed a significant positive correlation with the degree of deep WMHs and periventricular WMHs at baseline, after adjusting for age, gender, present history of hypertension, and diabetes mellitus.

The degree of WMHs, both deep (DWMH) and periventricular (PVWMH), at baseline showed a significant positive correlation with the rate of global gray-matter volume decline, represented by APC_{GMR}, adjusting for age, gender, and present history of hypertension and diabetes mellitus in healthy subjects using longitudinal analysis. To our knowledge, we are the first to show the correlation between the degree of WMHs at baseline and the rate of subsequent global gray-matter volume decline in healthy elderly individuals. Our result is partially consistent with recent studies that showed a significant positive correlation between the degree or load of WMHs and decreases in gray-matter volume in healthy elderly people, although those studies were conducted using a cross-sectional design. However, another recent study using longitudinal analysis has shown that WMHs are not a predictor of brain atrophy rate in elderly subjects. The inconsistency between the findings of the recent study and the present study may have arisen from differences in the volume that was measured. In the present study, we focused on the rate of decline of gray-matter volume, not whole-brain volume, because gray-matter volume correlates significantly with several cognitive functions. Our results suggest that the rate of global gray-matter volume decline could be predicted using the degree of WMHs at baseline, evaluated by simple visual scaling.

In summary, using a longitudinal design over 6 years in 160 community-dwelling healthy individuals, the degree of WMHs was measured at baseline, and the rate of global gray-matter volume decline was obtained. As a result, APC_{GMR} showed a significant positive correlation with the degree of DWMHs and PVWMHs at baseline, adjusting for age, gender, and present history of hypertension and diabetes mellitus. Our results suggest that the degree of WMHs at baseline might predict the rate of subsequent gray-matter volume decline and that simple visual scaling of WMHs could contribute to the prediction of the rate of global gray-matter volume decline.

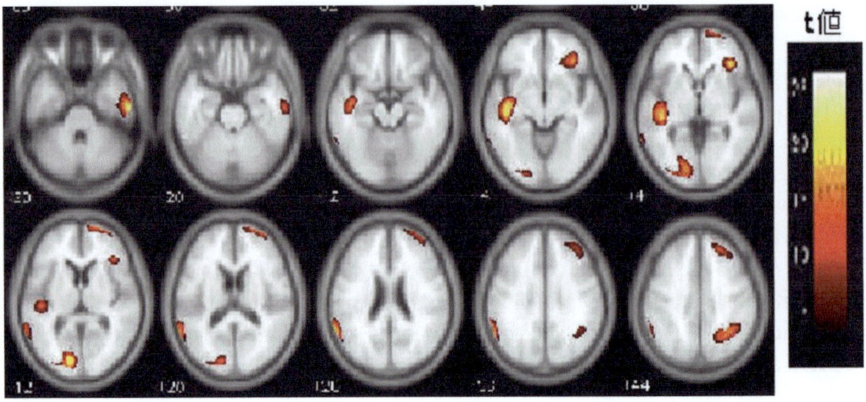

Fig. 12.5 Brain regions that showed a negative correlation between gray-matter volume and systolic blood pressure

12.3.3 Risk Factors for Brain Volume Decrease

12.3.3.1 Hypertension

Using the VBM technique, we analyzed the correlation between the regional gray-matter volume and the subject's characteristics. We found that total gray-matter volume correlated negatively with systolic blood pressure (Taki et al. 2004). Figure 12.5 represents the brain regions that showed a negative correlation between the gray-matter volume and systolic blood pressure. These regions are mainly considered as "watershed regions", which means regions located between the terriories of major arteries.

12.3.3.2 Alcohol Consumption

We also tested the correlation between the GMR and lifetime alcohol intake (Taki et al. 2006). There was a strong negative correlation between the log-transformed lifetime alcohol intake and the GMR. Figure 12.6 shows the gray-matter regions that had a significant negative correlation between the lifetime alcohol intake and the regional gray-matter volume. The gray-matter volume of the bilateral middle frontal gyri showed a significant negative correlation with the log-transformed lifetime alcohol intake.

12.3.3.3 Obesity

We tested the correlation between the GMR and obesity (Taki et al. 2008). As an indicator for obesity, body mass index (BMI) was used. Volumetric analysis revealed that there is a significant negative correlation between BMI and the GMR, which

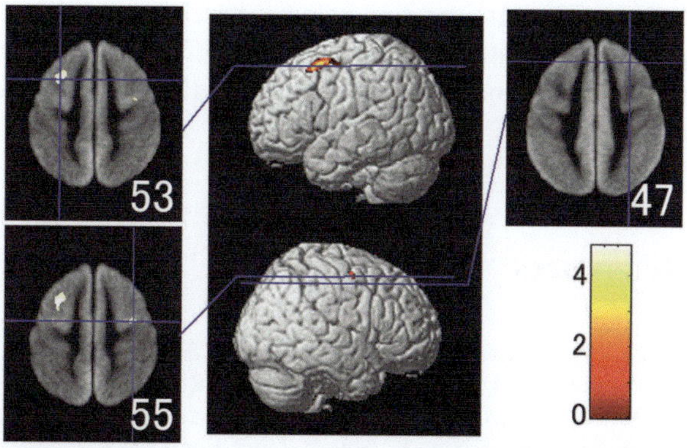

Fig. 12.6 Brain regions that showed a negative correction between gray-matter volume and lifetime alcohol intake

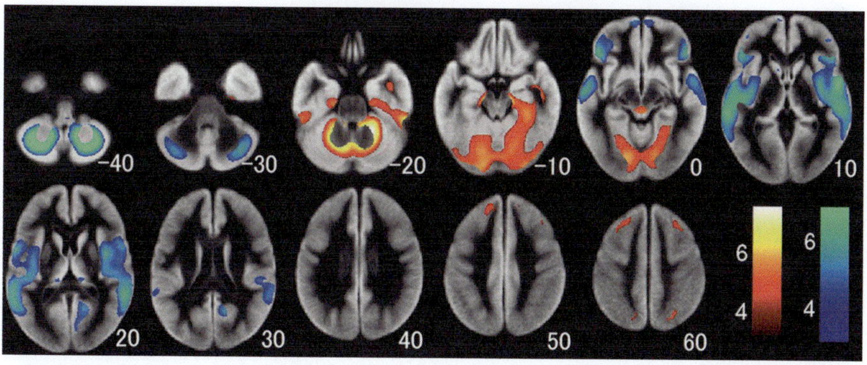

Fig. 12.7 Brain regions that showed correlations between gray-matter volume and body mass index (BMI). Red and blue color indicated negative and positive correlations respectively

represents the percentage of gray-matter volume in the intracranial volume, in men ($p < 0.001$, adjusting for age, systolic blood pressure, and lifetime alcohol intake), but not in women. VBM revealed that regional gray-matter volumes of the bilateral medial temporal lobe, occipital lobe, frontal lobe, and anterior lobe of the cerebellum show a significant negative correlation with BMI, and those of the posterior lobe of the cerebellum, perisylvian regions of the bilateral frontal and temporal lobes, and bilateral orbitofrontal gyri show a significant positive correlation with BMI in men (Fig. 12.7).

12.4 Conclusion

We constructed a large-scale brain MRI database for healthy Japanese subjects and clarified age-related volume changes of the human brain and their risk factors. Several factors such as hypertension, alcohol consumption, and obesity are related to the gray-matter volume reduction of several regions. In addition, we have shown that baseline gray-matter volume structure and white-matter lesions predict the global gray-matter volume decline rate. These results may contribute to the understanding of normal brain aging, and of age-related brain diseases, such as dementia.

References

Taki Y, Goto R, Evans A, Zijdenbos A, Neelin P, Lerch J, Sato K, Ono S, Kinomura S, Nakagawa M, Sugiura M, Watanabe J, Kawashima R, Fukuda H (2004) Voxel-based morphometry of human brain with age and cerebrovascular risk factors. Neurobiol Aging 25(4):455–463

Taki Y, Kinomura S, Sato K, Goto R, Inoue K, Okada K, Ono S, Kawashima R, Fukuda H (2006) Both global gray matter volume and regional gray matter volume negatively correlate with lifetime alcohol intake in non-alcohol-dependent Japanese men: a volumetric analysis and a voxel-based morphometry. Alcohol Clin Exp Res 30(6):1045–1050

Taki Y, Kinomura S, Sato K, Inoue K, Goto R, Okada K, Uchida S, Kawashima R, Fukuda H (2008) Relationship between body mass index and gray matter volume in 1,428 healthy individuals. Obesity (Silver Spring) 16(1):119–124

Taki Y, Hashizume H, Sassa Y, Takeuchi H, Wu K, Asano M, Asano K, Fukuda H, Kawashima R (2011a) Correlation between gray matter density-adjusted brain perfusion and age using brain MR images of 202 healthy children. Hum Brain Mapp 32(11):1973–1985

Taki Y, Kinomura S, Sato K, Goto R, Wu K, Kawashima R, Fukuda H (2011b) Correlation between baseline regional gray matter volume and global gray matter volume decline rate. NeuroImage 54(2):743–749

Taki Y, Kinomura S, Sato K, Goto R, Wu K, Kawashima R, Fukuda H (2011c) Correlation between degree of white matter hyperintensities and global gray matter volume decline rate. Neuroradiology 53(6):397–403

Taki Y, Hashizume H, Thyreau B, Sassa Y, Takeuchi H, Wu K, Kotozaki Y, Nouchi R, Asano M, Asano K, Fukuda H, Kawashima R (2012) Sleep duration during weekdays affects hippocampal gray matter volume in healthy children. NeuroImage 60(1):471–475

Chapter 13
Behavioral and Neural Mechanisms for Memory in Social Contexts

Laura E. Paige and Angela H. Gutchess

Abstract Much of the information we retain and use in our daily life pertains to how we interact with the world around us, including the social environment. Social influences on memory include information that is social in content, such as for impressions or social cues, or for information that is transmitted through social interactions, such as a group working together to remember the plot of a movie. Social neuroscience research has established that distinct neural regions falling along the cortical midline, including the medial prefrontal cortex, respond to social information and predict subsequent memory performance. In this chapter three areas of social influence on memory are reviewed, in addition to the subsequent neural regions involved: the self-reference effect, impression formation, and collaborative memory. Because aging influences the efficiency and functionality of memory systems, these changes can also affect the ways in which social information is prioritized and remembered. This has important implications for memory strategies in older adults and is also addressed in the present chapter.

Keywords Memory • Social • Aging • Self-reference • Medial prefrontal cortex • Impressions • Collaborative memory

13.1 Introduction

Memory is often considered to be a process unique to an individual, shaped by their subjective experiences and perceptions. However, much of the information stored and used in memory relates to how people communicate with others. In the present chapter, memory for information that is social in nature, such as impressions, is reviewed, in addition to memory for information transmitted in social settings, such as through social interactions. There are many ways to consider what is "social" and

L.E. Paige (✉) • A.H. Gutchess
Department of Psychology, Brandeis University, Waltham, MA, USA
e-mail: paigel@brandeis.edu

T. Tsukiura, S. Umeda (eds.), *Memory in a Social Context*,
https://doi.org/10.1007/978-4-431-56591-8_13

this chapter is an introduction to memory for social information and its transmission, and the ways in which memory is influenced by social contexts.

Social neuroscience research using functional magnetic resonance imaging (fMRI) suggests that distinct cortical midline regions, including the medial prefrontal cortex (mPFC) and posterior cingulate cortex (PCC), might be engaged during social tasks. These types of tasks constitute situations such as communicating with or making judgments about the character of another person, unlike nonsocial tasks, which do not incorporate information from or about others. These regions are differentially engaged whether the task involves external or internal focus (Heatherton et al. 2007; Lieberman 2007). For cognitive tasks requiring external focus, such as attention or encoding, the regions are often deactivated, or show a decrease in activation below baseline. For tasks requiring internal focus, such as thinking about others or the self, the regions are engaged, or activated above baseline.

Although it is assumed that these social regions are engaged similarly for everyone, individual differences, such as age, may influence how the regions are involved in tasks. Social networks in the brain are of particular interest to consider with aging, given that they exhibit a different trajectory with age than other networks. Typically, most networks and cognitive processes show a decline in functionality and efficiency with age. However, there is a growing body of literature that investigates whether there is preserved activity within social regions and specifically how *social* memory processes change across the lifespan.

In this chapter the literature on memory in a social context and the effects of aging are discussed. We consider research on three specific types of social processes: memory for the self, memory for impressions, and collaborative memory. We selected these three types because despite the limited research thus far, there is a body of literature on these topics. Moreover, they cover a range of socially relevant topics, including considerations of how thinking about "self" versus "other" has an impact on memory, and how working with another person can influence memory for nonsocial information. The neural correlates involved in the encoding and retrieval of such social information are discussed. In relation to age, these topics are important to consider because of their relevance to everyday problems, ranging from difficulty remembering names all the way to susceptibility to scams, and the likelihood that these processes continue to evolve and change throughout the lifespan.

13.2 Self-Reference Effect

One of the most well-established, socially-relevant strategies for improving memory is by relating information to the self. Self-referencing is an effective encoding strategy for both younger (Rogers et al. 1977; Symons and Johnson 1997) and older adults (Glisky and Marquine 2009; Gutchess et al. 2007a; Mueller et al. 1986). For example, referencing the self enhances both general and specific memory (Hamami et al. 2011; Serbun et al. 2011). Older adults typically show a loss of memory for specific details of information, instead preserving more of the "gist" or general (item-level) theme of the information (Koutstaal and Schacter 1997). However, self-referencing provides a boost for younger and older adults in their ability to remember both general item information and specific details (Hamami et al. 2011).

Behavioral research has manipulated the way in which we test information related to the self versus other. Older adults' performance in later memory tasks is influenced by whether they make judgments only about the self or about both self and other at encoding (Gutchess et al. 2015). On the other hand, younger adults are not influenced by these manipulations (Gutchess et al. 2015). This suggests that younger and older adults think about self and other differently and that these viewpoints differ depending on the situation or context. Older adults do not reap all the mnemonic benefits from self-referencing when they must also reference another person.

The self-reference effect also extends to memory for source details. Source memory goes beyond general memory to incorporate memory for the *who, what, where*, and *when* of information (Johnson et al. 1993). Research into how self-referencing affects source memory has thus shown that the effect extends to more complex information, beyond simply remembering word lists (Leshikar and Duarte 2012, 2014). For example, younger and older adults have better memory for actions (e.g., packing items in a suitcase) performed by self rather than other (Rosa and Gutchess 2011). In some contexts, self-referencing effects carry over to information related to those who are close to us, likely because representation of the self and close other (e.g., mother, sister) overlaps (Aron et al. 1991). Previous work has suggested that referencing a close other can often improve memory to the same extent as referencing the self (Aron et al. 1991; Hamami et al. 2011).

It has been shown that thinking about the self activates several neural regions along the midline of the cortex that are separable from those implicated in thinking about other people (Craik et al. 1999; Heatherton et al. 2006). Research with younger adults has shown that the medial prefrontal cortex is activated more for self-referential processing, whereas the left inferior frontal cortex and anterior cingulate cortex are engaged for other processing (Fig. 13.1a) (Kelley et al. 2002; Macrae et al. 2004). Activity in the medial prefrontal cortex is also engaged for memory-specific processes and shown to be an indicator of successful encoding (Leshikar and Duarte 2012, 2014; Macrae et al. 2004). Engagement of the medial prefrontal cortex predicts both memory performance and judgments of self-relevance (Macrae et al. 2004).

Aging can influence the degree to which the self-reference effect engages neural regions, although the effects seem to depend on the task context. Younger and older adults similarly engage cortical midline neural regions, including the ventromedial prefrontal cortex, when simply making judgments about the self and others (Gutchess et al. 2007b). However, the pattern becomes more complex when considering the successful encoding of information into memory. Similar to behavioral results, there are also discrepancies between younger and older adults in the neural regions implicated in encoding self-referential material, depending on whether people are thinking only about the self or making judgments on both self and other. When

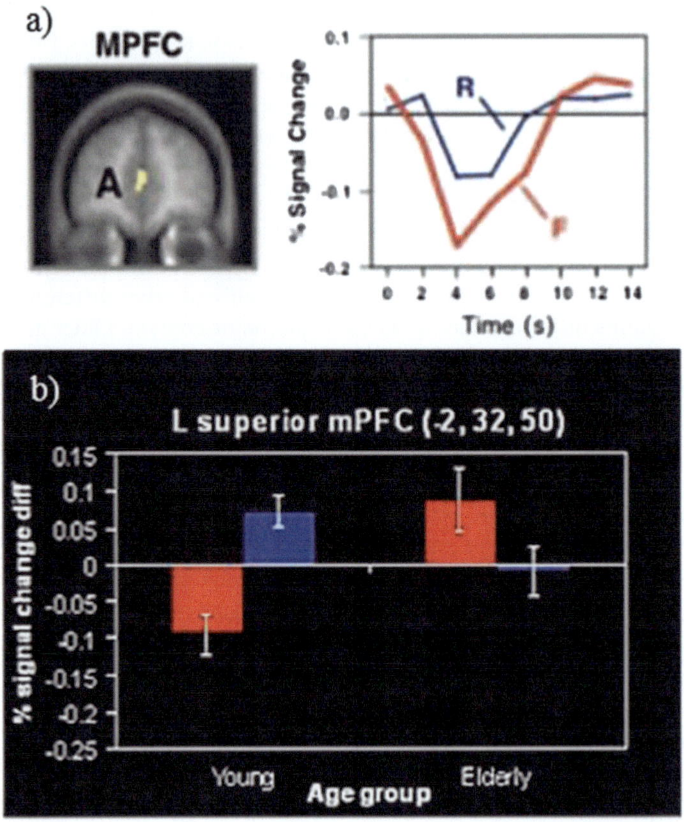

Fig. 13.1 Thinking about the self activates distinct neural regions separable from those activated when thinking about others. (**a**) Participants determined whether personality traits were self-descriptive and brain activation was measured using fMRI. Contrasts were constructed based on whether each word was remembered or forgotten and whether or not the trait was considered to be self-descriptive. There is increased activation observed in the medial prefrontal cortex during self-judgments for words later remembered (*R*) versus forgotten (*F*). (Adapted from Macrae et al. 2004 with permission). (**b**) When making judgments about both self and other simultaneously, activations elicited age-related reversals from previous patterns. Parameter estimates for the left superior medial prefrontal cortex show increased engagement for younger adults for subsequently forgotten self-referenced items. In contrast, older adults show increased engagement for subsequently remembered self-items and forgetting effects for other-referenced items. (Adapted from Gutchess et al. 2010 with permission)

considering the self alone, younger and older adults similarly engage the ventromedial prefrontal cortex for items later successfully remembered (Gutchess et al. 2015). Age-related differences become apparent with more complex judgments, such as when participants must alternate between judgments of self and other. For older adults, regions such as the dorsomedial prefrontal cortex (dmPFC), anterior and posterior cingulate, and left inferior prefrontal cortex, support memory for information related to others (Fig. 13.1b) (Gutchess et al. 2007b, 2010). However, in younger adults, these same regions are involved in forgetting information related to the self (Gutchess et al. 2010). This reversal in pattern of activity for self and other suggests age differences in the engagement of elaborative encoding processes (Gutchess et al. 2010). The authors suggest that older adults might be encoding more ordinary information about the self, whereas younger adults may be encoding more detailed and unique aspects of the self in comparison to others (Gutchess et al. 2010).

The self-reference effect has also been examined using event-related potentials (ERPs), a non-invasive method of measuring electrical brain response from the scalp. Unlike previous fMRI research, which largely focused on the encoding of information into memory, the ERP literature has typically looked at neural activity at memory retrieval. Research has identified "old–new effects" that parse apart recognized studied items (i.e., hits) from correctly rejected new items, where hits are associated with more positive-going deflections in the waveforms (Dulas et al. 2011). Retrieval of self-relevant information reveals an earlier neural signature for the old–new effect than other conditions for both younger and older adults, which suggests that self-referencing might be an effective encoding strategy that leads to more efficient retrieval (Dulas et al. 2011). Topographic plots, which illustrate where effects occur on the scalp, suggest that regions engaged during the retrieval of self-relevant information are similar to those at encoding.

Benefits of the self-reference effect have considerable implications for those with memory impairment, as self-referencing has the potential to be easily implemented into everyday life. However, research has shown that the effect can be specific to certain circumstances or aspects of memory. People with mild cognitive impairment due to Alzheimer's Disease (MCI-AD) can use self-referencing as an effective strategy to improve item memory, improving memory for whether or not an item was studied previously. The MCI-AD group does not benefit memory for the source of information (e.g., who performed an action) for self-related items to the same extent as controls. However, when committing errors, patients are less likely to misattribute new items to the self and are more likely to misattribute new information to others, rather than to the self compared with controls (Rosa et al. 2014). This suggests that there might be limitations to the benefits in source memory with self-referencing, although MCI-AD participants still benefit by decreasing self-related errors. Other work converges in revealing that self-referencing can offer limited benefits to memory for patients with memory impairments related to Alzheimer's disease. Among patients with amnestic mild cognitive impairment (aMCI), self-referencing only boosts memory when the information is positively valenced (Leblond et al. 2016). For negative information, the self-reference effect is dependent upon the valence of participants' self-representations. Negative self-representations correlate with poor memory for negative self-referential adjectives (Leblond et al. 2016). Although the degree of strength of the self-reference effect varies by both content and task, overall research supports the idea that it is a useful strategy for both younger and older adults, and potentially for memory-impaired patients, to improve memory performance. Further research continues to parse apart the conditions in which the strategy is most effective, and the ways in which aging may alter how one thinks about the self and others.

13.3 Impression Formation

A growing line of research investigates memory for information that is social in nature, specifically memory for impressions of others, and how this changes throughout the lifespan. People's impressions of others based on facial appearances converge greatly, but a large question remains whether impression formation is preserved with age, especially because age-related deficits in other cognitive processes could lead to problems with more complex social judgments. It is possible that older adults less accurately rate faces that require emotion recognition accuracy, a process that typically declines with age. Work has shown that older adults' ability to recognize emotion is skewed, as they tend to rate faces as more positive overall, even negatively-valenced faces (Zebrowitz et al. 2013). Other work has shown, however, that emotion recognition accuracy is not related to first impression accuracy (Krendl et al. 2014). An important consideration is that perhaps impression formation does not necessarily rely on facial properties alone and may involve information implied through a person's behavior. In this case and in terms of aging, memory for impressions reveals that younger and older adults form comparable impressions when provided with both behavioral information and facial features (Cassidy and Gutchess 2012a).

Research investigating the effects of aging on memory for impressions of others is still a relatively new area of study. This is important to investigate because in everyday life we must act on impressions stored in the memory (e.g., knowing which person to avoid, ask for help, etc.). Although aging brings declines in memory performance for certain tasks, older adults perform well at remembering impressions of others (Todorov and Olson 2008). When older adults are told to pay attention to the personal meaningfulness of others' behaviors, they can remember impressions as well as younger adults (Cassidy and Gutchess 2012a). This suggests that age-related deficits in other cognitive processes might not be related to the ability to form and remember impressions.

Age differences in memory for impressions could be influenced by the relevance of the presented information. How similar another person is to oneself has an impact on memory for impressions. When another person exhibits a positive trait that is also true of the self, participants are better at remembering impressions (Leshikar and Gutchess 2015). In contrast, negative traits that are also true of the self are more poorly remembered (Leshikar and Gutchess 2015). For behaviors that vary in consistency with attributed personality traits (e.g., someone who is friendly is more likely to invite neighbors over for dinner than someone who is unfriendly), younger adults better remember trait-inconsistent behaviors, whereas older adults show no influence of consistency on memory (Hess and Tate 1991). The authors suggest that younger adults spontaneously produce explanations for inconsistent behaviors,

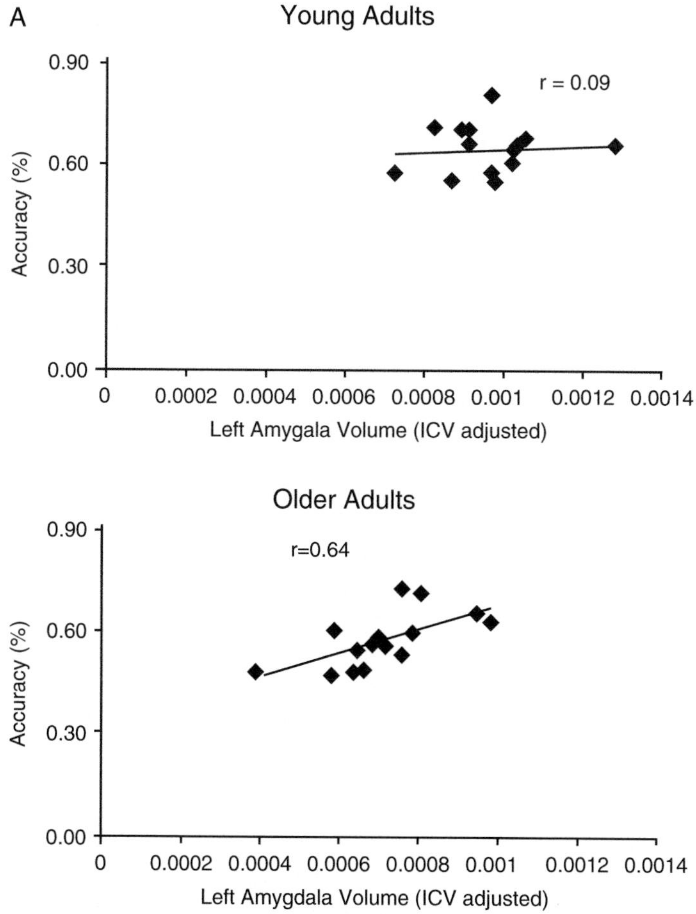

Fig. 13.2 Previous work has shown that structural integrity of the regions implicated in impression formation relates to later memory performance for social information. Anatomical MRI scans were acquired for younger and older adults who completed a behavior-based impressions task and later memory test. Greater left amygdala volume predicts enhanced overall memory for impressions for older adults only. (Adapted from Cassidy et al. (2012b) with permission)

which facilitates access to information in memory at a later time (Hess and Tate 1991). When explanation-based processing is controlled, age differences in consistency effects are eliminated, supporting the authors' explanation (Hess and Tate 1991).

Previous work on neural differences in memory for impression formation across aging has examined structural measures of the brain to compare how the integrity or structural endurance of these neural regions relates to memory performance. This reveals that greater left amygdala volume predicts better overall memory for impressions in older, but not younger adults (Fig. 13.2) (Cassidy and Gutchess 2012b). In older adults compared with younger adults, the thickness of the ventral medial prefrontal cortex (vmPFC) is also related to enhanced memory for impressions of others when they had been learned from socially meaningful judgments (Cassidy and Gutchess 2012b).

Research investigating the neural correlates of impression formation in older adults has largely focused on the same regions implicated in studies of younger adults. Broadly, brain regions involved in impression formation include the amygdala, which is sensitive to facial cues such as trustworthiness, and believed to respond to the valence of the cues (Engell et al. 2007; Todorov and Engell 2008). The amygdala and the dorsomedial prefrontal cortex are also both engaged when faces are paired with sentences indicating behaviors (Baron et al. 2011). When forming impressions of positive and negative face-behavior pairs, younger and older adults exhibit reversals in the patterns of dorsal and ventral medial prefrontal cortex activity corresponding to each valence (Cassidy and Gutchess 2012b; Cassidy et al. 2013). Younger adults show increased activation in these regions when forming negative impressions, whereas older adults show engagement when forming positive impressions (Cassidy et al. 2013). This suggests a shift in the prioritization of positive versus negative social information with age. The dorsomedial prefrontal cortex may also contribute to encoding of information about impressions into memory, as activity in the region has been related to successful memory formation (Gilron and Gutchess 2012; Mitchell et al. 2004).

Results from patients implicate additional regions. Lesion studies show that damage to the amygdala or temporal poles impairs memory for impressions (Todorov and Olson 2008). Interestingly, damage to regions such as the hippocampus and other traditional memory structures does not impair memory for impressions (Todorov and Olson 2008). This suggests that memory for impression formation might rely on distinct memory systems outside of those typically underlying explicit memory. Given that memory systems show a large decline with age, the dissociation of the regions implicated in memory for impressions could be consistent with the relative preservation of social processes across the lifespan.

13.4 Collaborative Memory

Memory can be vastly shaped by information we receive from others. In addition to studying memory for an event that occurred within a group (i.e., collective memory), research has begun to investigate how memories for an episodic event are remembered within a group. In this case, the influence of collaboration is measured by comparing individual memories reported before group conversation to individual memories reported after (Hirst and Echterhoff 2012; Rajaram and Pereira-Pasarin 2010). Discrepancies between original memories and post-collaboration memories are likely owing to internal confusion, where new information from other group members is incorporated into original events (Roediger et al. 2001).

Moving outside of the individual to consider the transactive processes that occur in group settings, there is the potential for both facilitative and disruptive effects of group interaction on memory. When working within a group, collaboration may allow the group to perform better than each individual could on his or her own because the task can be divided, allowing members to fill in the gaps in each other's memory (Hirst and Echterhoff 2012). In contrast, collaboration could also disrupt the use of retrieval strategies beneficial to individual group members, which could result in a decrease in group performance (Hirst and Echterhoff 2012).

The effect of age on collaborative memory tasks has revealed mixed findings. Some research has suggested that both younger and older adults exhibit collaborative inhibition, or reduced recall in pairs of people relative to individual performance. Older adults also show inconsistencies in their susceptibility to remember things that did not happen as a consequence of false information from others. In some instances, older adults show an increase in false memory rates on subsequent recognition tests post-collaboration, but only after a free recall test (Meade and Roediger 2009). In contrast, other work has shown that collaborating in groups leads older adults to have fewer false memories on follow-up recall tests (Ross et al. 2008).

There are many additional factors that can influence whether collaborative memory is beneficial or costly. Two of these factors include whether shared encoding took place and whether there was a pre-existing relationship amongst group members. Groups that encode information together (e.g., the same information at the same time as one another) exhibit diminished collaborative inhibition (Harris et al. 2013). Interestingly, any prior relationship between group members does not influence collaborative inhibition by itself (Harris et al. 2013). However, overall collaborative group recall leads to fewer memory errors than nominal group recall and these benefits carry over into later individual recall (Harris et al. 2013).

Currently, there are no studies investigating the neural underpinnings of collaborative memory. Because social interactions are incredibly complex, advanced technology is required to simultaneously monitor behavior and brain across individuals in a dynamic setting. The ability to measure neural substrates of social interactions has only been developed within the last few years. Coined "Hyperscan," this hardware and software program provides the necessary tools to link scanners through the Internet, allowing for participants to interact with one another (via computer) while brain activity is collected (Montague et al. 2002). Although collaborative memory per se has not been investigated, previous research has examined how memory and its neural correlates are influenced by the recollection of others. Results reveal that exposure to erroneous recollections of group members leads to more memory errors in participants (Edelson et al. 2011). Increased memory errors occur alongside engagement of the amygdala bilaterally and increased amygdalar–hippocampal connectivity (Edelson et al. 2011). This suggests that the amygdala might play a role not only in memory, but also in social cognition. Future research will continue to probe the role of regions such as the amygdala and hippocampus in realistic collaborative memory tasks.

13.5 Conclusion

The study of social neuroscience is an ever-growing field and this chapter touches upon only a few topics of interest related to memory. Social information affects memory both through the sociality of its content and through interactions with others. By applying the use of cognitive neuroscience methods to the investigation of social processes, we gain a better understanding of the behavioral and neural mechanisms underlying memory in social contexts. Further, given the significant age-related changes seen in cognitive processes such as memory, paired with a potential preservation or continued emphasis on social processes, it is reasonable to suggest that there might be deviations in how social information is processed with age. Because social interactions and the necessity of remembering social information occur in everyday life, it is important to comprehend how this changes throughout the lifespan.

References

Aron A, Aron EN, Tudor M, Nelson G (1991) Close relationships as including other in the self. J Pers Soc Psychol 60(2):241–253

Baron SG, Gobbini M, Engell AD, Todorov A (2011) Amygdala and dorsomedial prefrontal cortex responses to appearance-based and behavior-based person impressions. Soc Cogn Affect Neurosci 6(5):572–581

Cassidy BS, Gutchess AH (2012a) Social relevance enhances memory for impressions in older adults. Memory 20(4):332–345

Cassidy BS, Gutchess AH (2012b) Structural variation within the amygdala and ventromedial prefrontal cortex predicts memory for impressions in older adults. Front Psychol 3:319

Cassidy BS, Leshikar ED, Shih JY, Aizenman A, Gutchess AH (2013) Valence-based age differences in medial prefrontal activity during impression formation. Soc Neurosci 8(5):462–473

Craik FI, Moroz TM, Moscovitch M, Stuss DT, Winocur G, Tulving E, Kapur S (1999) In search of the self: a positron emission tomography study. Psychol Sci 10(1):26–34

Dulas MR, Newsome RN, Duarte A (2011) The effects of aging on ERP correlates of source memory retrieval for self-referential information. Brain Res 1377:84–100

Edelson M, Sharot T, Dolan RJ, Dudai Y (2011) Following the crowd: brain substrates of long-term memory conformity. Science 333(6038):108–111

Engell AD, Haxby JV, Todorov A (2007) Implicit trustworthiness decisions: automatic coding of face properties in the human amygdala. J Cogn Neurosci 19(9):1508–1519

Gilron R, Gutchess AH (2012) Remembering first impressions: effects of intentionality and diagnosticity on subsequent memory. Cogn Affect Behav Neurosci 12(1):85–98

Glisky EL, Marquine MJ (2009) Semantic and self-referential processing of positive and negative trait adjectives in older adults. Memory 17(2):144–157

Gutchess AH, Kensinger EA, Yoon C, Schacter DL (2007a) Ageing and the self-reference effect in memory. Memory 15(8):822–837

Gutchess AH, Kensinger EA, Schacter DL (2007b) Aging, self-referencing, and medial prefrontal cortex. Soc Neurosci 2(2):117–133

Gutchess AH, Kensinger EA, Schacter DL (2010) Functional neuroimaging of self-referential encoding with age. Neuropsychologia 48(1):211–219

Gutchess AH, Sokal R, Coleman JA, Gotthilf G, Grewal L, Rosa N (2015) Age differences in self-referencing: evidence for common and distinct encoding strategies. Brain Res 1612:118–127
Hamami A, Serbun SJ, Gutchess AH (2011) Self-referencing enhances memory specificity with age. Psychol Aging 26(3):636
Harris CB, Barnier AJ, Sutton J (2013) Shared encoding and the costs and benefits of collaborative recall. J Exp Psychol Learn Mem Cogn 39(1):183–195
Heatherton TF, Wyland CL, Macrae CN, Demos KE, Denny BT, Kelley WM (2006) Medial prefrontal activity differentiates self from close others. Soc Cogn Affect Neurosci 1(1):18–25
Heatherton TF, Krendl AC, Macrae CN, Kelley WM (2007) A social brain sciences approach to understanding self. In: Sedikides C, Spencer SJ (eds) The self. Psychology Press, New York, pp 3–19
Hess TM, Tate CS (1991) Adult age differences in explanations and memory for behavioral information. Psychol Aging 6(1):86–92
Hirst W, Echterhoff G (2012) Remembering in conversations: the social sharing and reshaping of memories. Annu Rev Psychol 63(1):55–79
Johnson MK, Hashtroudi S, Lindsay DS (1993) Source monitoring. Psychol Bull 114(1):3–28
Kelley WM, Macrae CN, Wyland CL, Caglar S, Inati S, Heatherton TF (2002) Finding the self? An event-related fMRI study. J Cogn Neurosci 14(5):785–794
Koutstaal W, Schacter DL (1997) Gist-based false recognition of pictures in older and younger adults. J Mem Lang 37(4):555–583
Krendl AC, Rule NO, Ambady N (2014) Does aging impair first impression accuracy? Differentiating emotion recognition from complex social inferences. Psychol Aging 29(3):482–490
Leblond M, Laisney M, Lamidey V, Egret S, de La Sayette V, Chételat G, Piolino P, Rauchs G, Desgranges B, Eustache F (2016) Self-reference effect on memory in healthy aging, mild cognitive impairment and Alzheimer's disease: influence of identity valence. Cortex 74:177–190
Leshikar ED, Duarte A (2012) Medial prefrontal cortex supports source memory accuracy for self-referenced items. Soc Neurosci 7(2):126–145
Leshikar ED, Duarte A (2014) Medial prefrontal cortex supports source memory for self-referenced materials in young and older adults. Cogn Affect Behav Neurosci 14(1):236–252
Leshikar ED, Gutchess AH (2015) Similarity to the self affects memory for impressions of others. J Appl Res Mem Cogn 4(1):20–28
Lieberman MD (2007) Social cognitive neuroscience: a review of core processes. Annu Rev Psychol 58:259–289
Macrae CN, Moran JM, Heatherton TF, Banfield JF, Kelley WM (2004) Medial prefrontal activity predicts memory for self. Cereb Cortex 14(6):647–654
Meade ML, Roediger HL (2009) Age differences in collaborative memory: the role of retrieval manipulations. Mem Cogn 37(7):962–975
Mitchell JP, Macrae CN, Banaji MR (2004) Encoding-specific effects of social cognition on the neural correlates of subsequent memory. J Neurosci 24(21):4912–4917
Montague PR, Berns GS, Cohen JD, McClure SM, Pagnoni G, Dhamala M, Apple N (2002) Hyperscanning: simultaneous fMRI during linked social interactions. NeuroImage 16(4):1159–1164
Mueller JH, Wonderlich S, Dugan K (1986) Self-referent processing of age-specific material. Psychol Aging 1(4):293–299
Rajaram S, Pereira-Pasarin LP (2010) Collaborative memory: cognitive research and theory. Perspect Psychol Sci 5(6):649–663
Roediger HL, Meade ML, Bergman ET (2001) Social contagion of memory. Psychon Bull Rev 8(2):365–371
Rogers TB, Kuiper NA, Kirker WS (1977) Self-reference and the encoding of personal information. J Pers Soc Psychol 35(9):677–688
Rosa NM, Gutchess AH (2011) Source memory for action in young and older adults: self vs. close or unknown others. Psychol Aging 26(3):625–630

Rosa NM, Deason RG, Budson AE, Gutchess AH (2014) Source memory for self and other in patients with mild cognitive impairment due to Alzheimer's disease. J Gerontol Ser B Psychol Sci Soc Sci 71(1):59–65

Ross M, Spencer SJ, Blatz CW, Restorick E (2008) Collaboration reduces the frequency of false memories in older and younger adults. Psychol Aging 23(1):85–92

Serbun SJ, Shih JY, Gutchess AH (2011) Memory for details with self-referencing. Memory 19(8):1004–1014

Symons CS, Johnson BT (1997) The self-reference effect in memory: a meta-analysis. Psychol Bull 121(3):371–394

Todorov A, Engell AD (2008) The role of the amygdala in implicit evaluation of emotionally neutral faces. Soc Cogn Affect Neurosci 3(4):303–312

Todorov A, Olson IR (2008) Robust learning of affective trait associations with faces when the hippocampus is damaged, but not when the amygdala and temporal pole are damaged. Soc Cogn Affect Neurosci 3(3):195–203

Zebrowitz LA, Franklin RG Jr, Hillman S, Boc H (2013) Older and younger adults' first impressions from faces: similar in agreement but different in positivity. Psychol Aging 28(1):202–212

Chapter 14
Structural and Functional Changes of Brain Due to Childhood Maltreatment and Adversity

Akemi Tomoda, Shinichiro Takiguchi, Koji Shimada, and Takashi X. Fujisawa

Abstract From birth, parent–infant interaction modulates fundamental brain processes, whereas more complex psychological development occurs when socializing with peers. Childhood maltreatment markedly increases risks for psychopathology and negative outcomes such as depression, post-traumatic stress disorder (PTSD), and reduced cognitive abilities. Our earlier studies elucidated potential discernible effects on the brain morphology of exposure to sexual abuse, parental verbal abuse, interparental violence or harsh corporal punishment during childhood on brain morphology, including gray matter volume (GMV) or cortical thickness in addition to the brain developmental alteration.

Further, our recent studies revealed reduced activity in the bilateral striatum and nucleus accumbens during the high monetary reward condition and reduced GMV in the left primary visual cortex in the reactive attachment disorder (RAD) group compared with the typically developed group. These results suggest that dopaminergic dysfunction occurs in the striatum of children and adolescents with RAD, and the visual cortex GMV abnormalities may also be associated with visual

A. Tomoda (✉)
Division of Developmental Higher Brain Functions, United Graduate School of Child Development, University of Fukui, Fukui, Japan

Research Center for Child Mental Development, University of Fukui, Fukui, Japan

Department of Child and Adolescent Psychological Medicine, University of Fukui Hospital, Fukui, Japan
e-mail: atomoda@u-fukui.ac.jp

S. Takiguchi
Research Center for Child Mental Development, University of Fukui, Fukui, Japan

Department of Child and Adolescent Psychological Medicine, University of Fukui Hospital, Fukui, Japan

K. Shimada • T.X. Fujisawa
Research Center for Child Mental Development, University of Fukui, Fukui, Japan

Division of Developmental Higher Brain Functions, United Graduate School of Child Development, University of Fukui, Fukui, Japan

T. Tsukiura, S. Umeda (eds.), *Memory in a Social Context*,
https://doi.org/10.1007/978-4-431-56591-8_14

stimulus-induced emotion regulation impairments of RAD, leading toward potential future risks for psychopathology.

Brain regions that process and convey the adverse sensory input of the maltreatment may be modified specifically by such experiences. These findings are consistent with preclinical studies showing that sensory cortices are highly plastic structures. However, better understanding of the neural basis of posttraumatic growth may help to find out means of enhancing the resiliency to recover or prevent from traumatic progresses. Further studies are necessary to confirm these findings and to clarify the causality and consequences of such neurodevelopmental mechanisms.

Keywords Child maltreatment • Reactive attachment disorder • Structural magnetic resonance imaging • Functional magnetic resonance imaging • Sensitive period • Posttraumatic growth

14.1 Introduction

Childhood and adolescence are periods of enormous physical and social changes, with considerable implications for both personality and brain development. Considerable progress has been made in the last few decades in elucidating the neurobiological aspects of brain development in human beings.

Exposure to traumatic events such as childhood abuse and neglect have been associated with alterations in the size or functional activity of a variety of brain regions (e.g., Bremner et al. 1997; Teicher et al. 1997; De Bellis et al. 1999, 2002; De Bellis and Kuchibhatla 2006; Richert et al. 2006; Teicher et al. 2004; Andersen et al. 2008; Tomoda et al. 2009a). We have previously conducted a voxel-based morphometry (VBM) study in young adults with histories of exposure to repeated episodes of childhood sexual abuse (CSA) and found that the most significant differences were bilateral reductions in gray matter volume (GMV) in the visual cortex (Tomoda et al. 2009a). Similarly, we conducted an analysis of fiber tract integrity in young adults exposed to parental verbal aggression using diffusion tensor imaging (DTI) and tract-based spatial statistics. We observed a reduction in fractional anisotropy in three fiber tracts, including the arcuate fasciculus that interconnects Wernicke's and the frontal regions (Choi et al. 2009). Further, we have observed regional reductions in GMV and thickness in both susceptible and resilient subjects witnessing domestic violence. Results in subjects witnessing domestic violence were similar to previously reported results in subjects with CSA, as the primary region affected was the visual cortex.

These findings fit with an emerging hypothesis that exposure to early adversity might be associated with alterations in sensory systems that process and convey the adverse sensory experience (Teicher et al. 2006). Brain regions that process and convey the adverse sensory input of abuse may be specifically modified by this experience. However, the impact depends on the type of exposure and developmental stage. Child maltreatment is also a major risk factor for psychopathology, including reactive attachment disorder (RAD). In this chapter, we present the detrimental effects on trajectories of brain development when exposing individuals to harsh corporal punishment (HCP) during childhood and exciting new findings that illustrate just how important these factors are. We also show that there are narrow sensitive periods of markedly increased risk in maltreated children with RAD.

14.2 Structural and Functional Changes of Brain and Psychiatric Problems

14.2.1 Reduced Prefrontal Cortical GMV in Young Adults Exposed to HCP

Harsh corporal punishment during childhood is a chronic developmental stressor associated with depression, aggression, and addictive behaviors. Exposure to traumatic stressors, such as sexual abuse, is associated with alteration in brain structure, but nothing is known about the potential neurobiological consequences of HCP. The aim of this study (Tomoda et al. 2009b) was to investigate whether HCP was associated with discernible alterations in GMV using VBM. A total of 1455 young adults (18–25 years) were screened to identify 23 with exposure to HCP (minimum 3 years' duration, 12 episodes per year, frequently involving objects) and 22 healthy controls. High-resolution T1-weighted MRI datasets were obtained using a Siemens 3-T Trio scanner.

As a result, GMV was reduced by 19.1% in the right medial frontal gyrus (medial prefrontal cortex [MPFC], BA10; $p = 0.037$, corrected cluster level), by 14.5% in the left medial frontal gyrus (dorsolateral prefrontal cortex [DLPFC], BA9; $p = 0.015$, uncorrected cluster level) and by 16.9% in the right anterior cingulate gyrus (BA24; $p < 0.001$, uncorrected cluster level) of HCP subjects (Fig. 14.1). There were significant correlations between GMV in these identified regions and performance IQ on the Wechsler Adult Intelligence Scale III.

Harsh corporal punishment includes the occasional use of objects to induce pain, and is considered an unacceptable form of punishment by the American Academy of Pediatrics (AAP). Results from this study apply to HCP; they do not apply to exposure to ordinary forms of corporal punishment that the AAP considers acceptable (but less effective than alternative forms of discipline). Chronic exposure to HCP was associated with a marked reduction in GMV in the right medial frontal gyrus (MPFC, BA10). There were also possible associations between HCP and reduced GMV in the left medial frontal gyrus (DLPFC, BA9) and the right anterior cingulate gyrus (BA24). Other imaging studies have found those regions to be involved in aspects of addiction (Drexler et al. 2000; Crockford et al. 2005), suicidal behavior and/or depression (Liotti and Mayberg 2001; Bar et al. 2007; Raust et al. 2007), post-traumatic stress disorder (PTSD) (Fennema-Notestine et al. 2002;

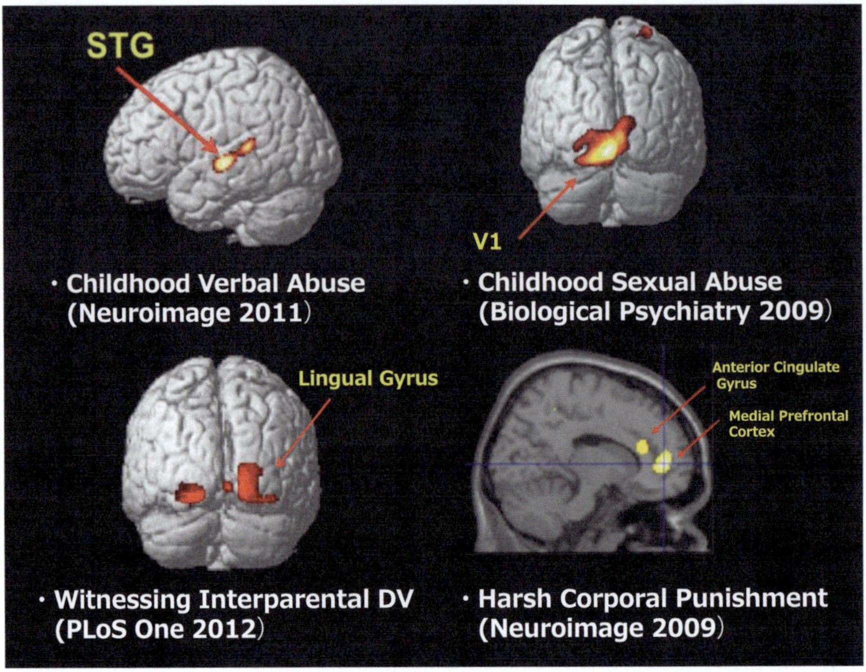

Fig. 14.1 Significant differences in cerebral cortices between maltreated subjects and controls using voxel-based morphometry (Tomoda et al. 2009a, b, 2011, 2012). Significantly higher gray-matter (GM) densities in parental verbal abuse subjects were measured in the superior temporal gyrus (STG; left upper figure), significantly lower GM densities in childhood sexual abuse subjects were measured in the right and left primary visual (V1) and visual association cortices (right upper figure), significantly lower GM densities in subjects witnessing interparental domestic violence were measured in the right lingual gyrus (left lower figure), and significantly lower GM densities in harsh corporal punishment subjects were measured in the right medial prefrontal cortex and the right anterior cingulate gyrus (right lower figure)

Bremner 2003; Liberzon et al. 2003; Geuze et al. 2007; Hou et al. 2007), dissociative disorders (Veltman et al. 2005), and depression (Fitzgerald et al. 2008). HCP may be an aversive and stressful event for human beings that potentially alters the developmental trajectory of some brain regions in which abnormalities have been associated with major forms of psychopathology.

The regions identified with reduced GMV are part of the medial rostral prefrontal cortex (MRPFC). Recent studies have pointed to the MRPFC as a region of the human brain that plays a crucial role in social cognition and functional organization (Amodio and Frith 2006; Gilbert et al. 2007). In particular, medial BA10 and BA32 appear to be involved with *self-knowledge, person perception, and mentalizing* (Amodio and Frith 2006). At its most basic level, *self-knowledge* involves the ability to differentiate the self from other objects and to recognize attributes and preferences related to oneself. The ability to represent another person's psychological perspective is referred to as *mentalizing*, and this capacity allows us to predict the behavior of others. *Person perception* involves judgments about the attributes and behaviors of others. BA24, in contrast appears to be involved in the internal monitoring of our actions to ensure that they are consistent with intentions and the current situational context (Amodio and Frith 2006). The more posterior portion of the MRPFC (primarily BA8 and 9) is activated by cognitive tasks, such as those designed to engage action monitoring and attention. For example, left BA9 is activated in young adults during working memory retrieval tasks (Sun et al. 2005).

14.2.2 Reduced Visual Cortex GMV in Children and Adolescents with Reactive Attachment Disorder

A psychiatric disorder associated with early life abuse and neglect is RAD, where the child displays wary, watchful, and emotionally withdrawn behaviors, according to the fifth edition of the Diagnostic and Statistical Manual of Mental Disorders (DSM-5) (American Psychiatric Association 2013). Because of emotional dampening, RAD closely resembles internalizing disorders with depressive and anxiety symptoms. Despite its clinical importance, there have been very few investigations into the possible neurobiological consequences of RAD.

Previous neuroimaging studies using structural magnetic resonance imaging (sMRI) techniques have revealed that exposure to early adversity is strongly associated with alterations in brain structure, such as gray matter (GM) and white matter (WM) (for reviews, see Hart and Rubia 2012; Lim et al. 2014; McCrory et al. 2011; Teicher and Samson 2013). According to Teicher and Samson (2013), many of the identified neuroanatomical abnormalities are interconnected and are components of a neurocircuit regulating stress response to emotional stimuli, including the thalamus (Liao et al. 2013), visual or auditory sensory cortex (Tomoda et al. 2009a, 2011, 2012), MPFC (Carrion et al. 2009; De Brito et al. 2013; Tomoda et al. 2009b; van Harmelen et al. 2010), hippocampus (Dannlowski et al. 2012; Hanson et al. 2015; Teicher et al. 2012), or amygdala (Edmiston et al. 2011; Hanson et al. 2015; Mehta et al. 2009; Tottenham et al. 2010). Such neuroimaging techniques are helpful in understanding the unknown neurobiological abnormalities in RAD associated with child maltreatment.

We conducted a VBM study in maltreated children with RAD to further assess whether alterations in regional GMV correlated with psychiatric symptom measures (Shimada et al. 2015). Twenty-one right-handed, medication-naïve, 10–17-year-old Japanese children (mean age = 12.76 years) with a clinical diagnosis of RAD were recruited from the Department of Child and Adolescent Psychological Medicine at the University of Fukui Hospital. The diagnosis of RAD was assessed by licensed child and adolescent psychiatrists according to DSM-5 criteria (American Psychiatric Association 2013). All the children had experienced physical, emotional abuse, and/or neglect early in life before coming into care. The children were living within a stable placement (in a child welfare facility), even though they were not living with biological parents (for information about child welfare services in Japan, see Suzuki and Tomoda 2015). Control subjects, 22 typically developed (TD)

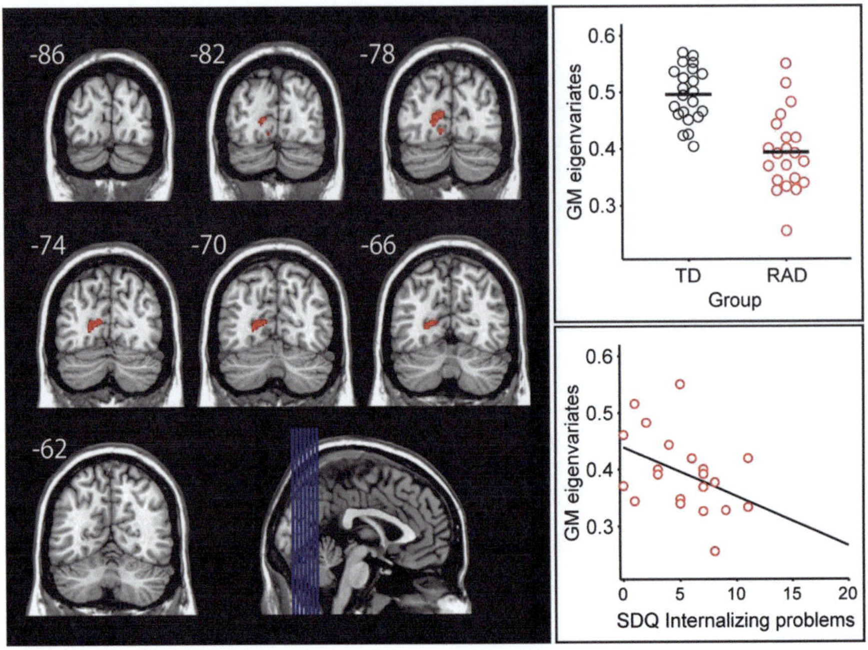

Fig. 14.2 Structural differences in regional GMV between the typically developed (TD) and reactive attachment disorder (RAD) groups (Shimada et al. 2015). The RAD group showed significantly reduced GMV in the left primary visual cortex (BA17) compared with the TD group (p = 0.038, FWE-corrected cluster level). Color scales represent t values. The psychiatric symptom measures for SDQ internalizing problems ($\beta = -0.96$, $t = -3.86$, $p < 0.05$) were a significant predictor for the left visual cortex GMV estimates in the RAD group, which explained approximately 55% of the GMV variance (adjusted $R^2 = 0.55$, $F[10, 8] = 3.16$, $p < 0.10$)

Japanese children (mean age = 12.95 years) with no history of maltreatment, were recruited from local schools, matched on age, gender, and handedness.

As a result, GMV was significantly reduced by 20.6% in the left primary visual cortex (Brodmann area 17) of the RAD group compared with the TD group (p = .038, family-wise error-corrected cluster level) (Fig. 14.2). This GMV reduction was related to a measure of internalizing problems in the Strength and Difficulties Questionnaire (SDQ). The visual cortex has been viewed as a part of the neurocircuit regulating stress response to emotional vision. Combined with the previous studies of adults with childhood maltreatment, early adverse experience (e.g., sensory deprivation) may affect the development of the primary visual system, reflecting the size of the visual cortex in children and adolescents with RAD. This visual cortex GMV abnormality may also be associated with visual emotion regulation impairments of RAD, leading to an increase in risk for later psychopathology.

The notion that the GMV abnormality might play a critical role in RAD symptoms is also supported by the results of multiple regression analysis, showing an association of the visual cortex GMV of the RAD subjects and their internalizing problems (SDQ internalizing) when incorporating other relevant factors (e.g., SDQ externalizing) into the multiple regression model. RAD is considered to resemble internalizing disorders with depressive and anxiety symptoms according to DSM-5. Because of emotional dampening in RAD, it is also reasonable to consider a potential convergence between RAD symptoms and internalizing problems (Zeanah and Gleason 2015). Although the SDQ internalizing problems are likely to reflect inhibited RAD symptoms, they seem to be associated with other questionnaire scores for RAD symptoms. Subscales of the SDQ internalizing (emotional and peer problems) have been associated with the inhibited behavior subscale of the Relation Problems Questionnaire as a useful but limited instrument for screening inhibited and disinhibited RAD symptoms (Vervoort et al. 2013).

Why the left-sided hemisphere of the visual cortex was affected is an interesting question. Given that RAD patients show reduced or absent expression of positive emotions during routine interactions with caregivers (American Psychiatric Association 2013; Zeanah and Gleason 2015), we propose that the reduction of GMV in the left visual cortex identified here might be associated with a malfunction in the neurocircuit regulating positive emotional vision (e.g., a smiling face). In line with our proposal, it has been reported that children exposed to childhood adversity have difficulty recognizing positive emotional expressions (Koizumi and Takagishi 2014). Indeed, the processing of positive emotion has been proposed to be associated with the left hemisphere, although hemispheric lateralization in emotion processing remains controversial between the valence hypothesis and the right hemisphere hypothesis (e.g., Prete et al. 2015). According to the valence hypothesis, a currently leading hypothesis in emotion processing (Ahern and Schwartz 1985; Davidson et al. 1990; Silberman and Weingartner 1986), the left hemisphere is specialized for positive emotions and the right hemisphere is dominant for negative emotions. More recently, behavioral studies using a divided visual field paradigm have found support for this hypothesis, showing that positive emotions were better recognized when presented in the right visual field and negative emotions were better identified when presented in the left visual field (e.g., Jansari et al. 2011). A neurophysiological study using event-related potentials has also found that processing of the mouths of happy faces (i.e., analytical or part-based processing of positive emotional expressions) enhanced left-sided occipito-temporal activity during the early visual processing phase (Calvo and Beltrán 2014). Combined with a previous DTI study showing the left-sided ILF WM abnormalities associated with early adversity (Choi et al. 2012), we suggest that left-sided visual cortex GM abnormalities might lead to a difficulty with regulating positive emotions in RAD.

14.2.3 Ventral Striatum Dysfunction in Children and Adolescents with Reactive Attachment Disorder

Analyses of clinical behavioral and emotional problems have suggested that patients with RAD might have intrinsic reward/pleasure disturbances. However, changes in brain function among individuals with RAD have not yet been investigated. Thus, we examined whether neural activity during reward processing was altered, or not, in children and adolescents with RAD using tasks with high and low monetary rewards. Sixteen children and adolescents with RAD (10–15 years old) and 20 age- and gender-matched individuals with TD performed tasks with high and low monetary rewards while undergoing functional MRI.

Significantly reduced activity in the caudate and nucleus accumbens was observed during a high monetary reward condition in the RAD group compared with the TD group ($p = 0.015$, family-wise error-corrected cluster level). The striatal neural reward activity in the RAD group was also markedly decreased (Fig. 14.3a). Significant negative correlations between bilateral striatal activity and avoidant attachment were observed in the RAD and TD groups.

A growing body of evidence suggests an association between reward circuits and reinforcing processes in social approach and bonding (Aron et al. 2005). In securely attached individuals, social interactions seem to be generally associated with more positive emotion experiences and stronger signals of reward. Conversely, these positive responses are much weaker or even absent in individuals with an avoidant attachment style (Vrticka et al. 2011). Thus, reduced striatal responses to rewards in RAD could be explained by a lack of sensitivity toward social reward (e.g., bonding to a caregiver), leading to deficits in forming secure attachments with significant others.

We also investigated the existence of potential "sensitive periods," during which exposure to childhood maltreatment might be more strongly associated with alterations in striatal responses to high monetary reward (HMR) (Takiguchi et al. 2015). We observed that a significant temporal predictor of bilateral striatal activation to HMR was whether children and adolescents with RAD were exposed to maltreatment at 0–1 year of age, using random forest regressions with conditional trees (Fig. 14.3b). Identifying sensitive periods when adversity has a particularly harmful impact on striatal activity is pivotal for linking childhood experiences to later psychopathology. Interestingly, we have previously discussed possible evidence for sensitive periods during which childhood maltreatment influences regional brain development, indicating a potential causal mechanism (Andersen et al. 2008; Pechtel et al. 2014). Timing of exposure may be extremely important, and overall exposure levels across childhood may be misleading if the maltreatment occurred before or after a sensitive period. Thus, the present findings support the assertion that brain regions may have unique windows of vulnerability toward traumatic stress. In fact, the present results (vulnerability from 0 to 1 year of age) also indicate that very early maternal care and post-birth childcare system interventions are extremely important for preventing RAD and minimizing long-term consequences of abuse.

Thus, striatal neural reward activity in the RAD group was markedly decreased. The results suggest that dopaminergic dysfunction might have occurred in the striatum in children and adolescents with RAD, potentially leading to a future risk for psychiatric disorders such as dependence.

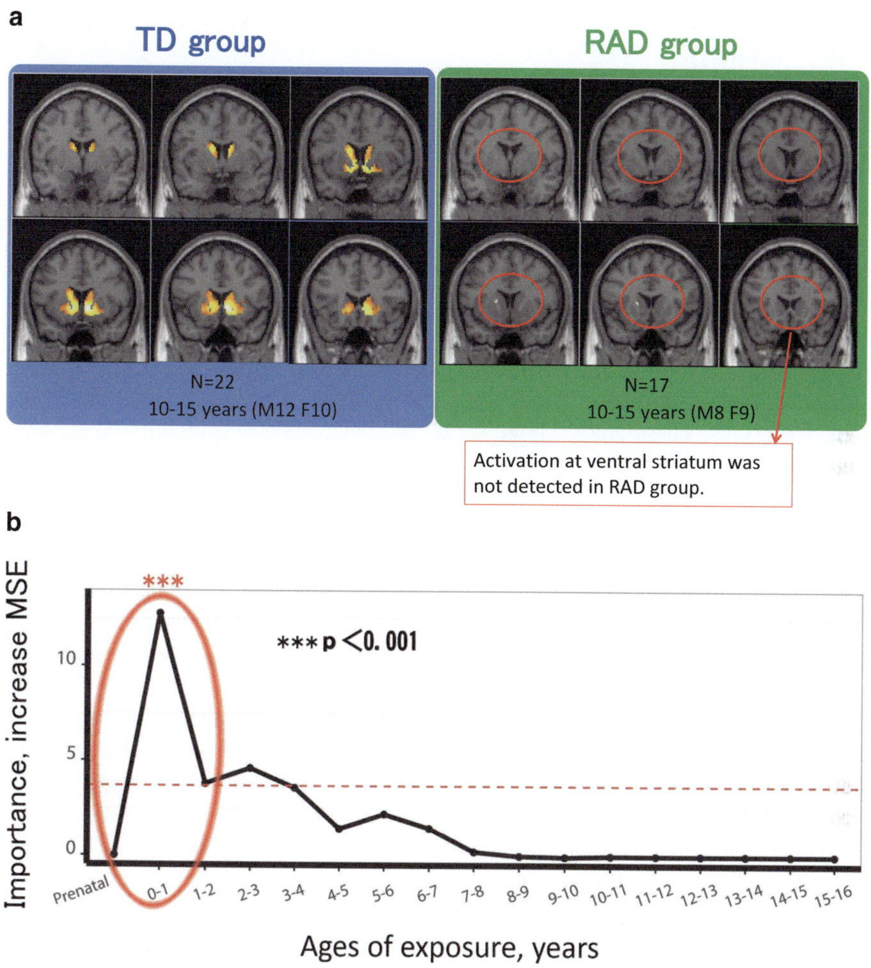

Fig. 14.3 Statistical parametric maps of high monetary reward (HMR minus no monetary reward [NMR]) in TD individuals and patients with RAD (Takiguchi et al. 2015). (**a**) Statistical parametric maps of facing a challenge of HMR in TD, RAD, and TD minus RAD (Takiguchi et al. 2015). The right (R) and left (L) sides and *y*-axis (MNI coordinate) are indicated. The threshold for the analysis was set at $p < 0.001$ at the voxel level and $p < 0.05$ with an FWE correction for multiple comparisons at the cluster level for the entire brain. Each of six representative pieces of data are shown from the TD group ($n = 22$, 12 male and 10 female, 10–15 years old) and the RAD group ($n = 17$, 8 male and 9 female, 10–15 years old). (**b**) Maximal sensitivity by age of exposure (maximal importance of age of exposure, regardless of type) in RAD (Takiguchi et al. 2015). Results of a random forest regression with conditional trees indicated the importance of exposure to early maltreatment from birth to 15 years of age with regard to contrast estimates of high monetary reward (HMR minus NMR) for (**a**) the right and (**b**) the left striatum. Importance is indicated by degradation in fit, as indicated by the increase in mean square error, following effective elimination of each age from the model by permutation. The dashed horizontal line indicates the significance level for the variable importance values

14.3 Functional Changes in the Brain and Post-Traumatic Growth

14.3.1 Neural Basis of Psychological Growth Following Adverse Experiences

Over the past decade, research into the aftereffects of stressful or traumatic events has emphasized the negative outcomes from these experiences. However, the positive outcomes derived from adversity are increasingly being examined, and such positive changes are described as post-traumatic growth (PTG), a term coined by Tedeschi and Calhoun (1996). Evidence gathered during the past decade has increasingly suggested that positive outcomes can be derived from adversity and other negative events. These studies have included a systematic examination of the psychological domains for the positive outcomes and their associations with other social or psychological factors (Shakespeare-Finch and Armstrong 2010). Although the psychological mechanism underlying the growth derived from distress or adversity has been discussed, currently no consensus has been reached regarding the relationship between adversity and positive outcomes (Shakespeare-Finch and Lurie-Beck 2014). Furthermore, studies on PTG tend to focus on the psychological phenomena rather than on the neurological mechanisms; thus, the neural mechanisms underlying PTG remain unclear.

To investigate the relationship between basal whole-brain functional connectivity and PTG, we employed resting-state functional MRI and analyzed the neural networks using independent component analysis in a sample of 33 right-handed, healthy volunteers (21 female, 12 male; mean age = 21.9 ± 5.7 years; age range, 18–48 years) (Fujisawa et al. 2015). To quantify participants' PTG, we used the Japanese version of the Post-Traumatic Growth Inventory (PTGI) (Taku et al. 2007), an instrument used to assess the positive outcomes of people who have experienced traumatic or stressful events. The original PTGI is a 21-item scale that measures the degree of positive change experienced in the aftermath of an identified traumatic event (Tedeschi and Calhoun 1996). The Japanese version has been confirmed to have acceptable validity and reliability of internal consistency (Cronbach's alpha = 0.90) with Japanese samples (Taku et al. 2007). The PTGI consists of four subscales: relating to others; new possibilities; personal strength; and spiritual change and appreciation of life. Each item was rated on a six-point Likert scale, ranging from 0 (not at all) to 5 (a very great degree). Higher scores imply higher levels of PTG. Correlations were calculated between the network connectivity strength and the PTGI score.

14.3.2 Functional Changes in the Central Executive Network with Psychological Growth

As a result, there were positive associations between the PTGI scores and brain activation in the rostral prefrontal cortex and superior parietal lobule (SPL) within the left central executive network (CEN) ($r = 0.41$, $p < 0.001$; $r = 0.49$, $p < 0.001$ respectively) (Fig. 14.4). Individuals with higher psychological growth following adverse experiences had stronger activation of prospective or working memory areas within the executive function network than did individuals with lower psychological growth ($r = 0.40$, $p < 0.001$). Moreover, we found that individuals with higher PTG demonstrated stronger connectivity between the SPL and the supramarginal gyrus (SMG).

Moreover, we found that individuals with higher PTG demonstrated stronger connectivity between the SPL and supramarginal gyrus (SMG). The SMG is one of the brain regions associated with the ability to reason about the mental states of others, otherwise known as mentalizing. These findings suggest that individuals with higher psychological growth might have stronger functional connectivity between memory functions within the CEN and social functioning in the SMG, and that their better sociality might result from using more memory for mentalizing during their daily social interactions.

For a few decades, it has been suggested that the SPL plays significant roles in various executive functions including continuous updating, order memory, and information manipulation (Wager and Smith 2003), and that it contributes to episodic memory (Cabeza et al. 2008). A more recent study also suggested that the SPL is critically important for manipulating information in working memory (Koenigs et al. 2009). Although there is no direct evidence that PTG has beneficial effects on a person's working memory, it has been shown that patients with PTSD demonstrate parietal event-related potentials at 400–800 ms that are smaller in amplitude, which are related to working memory updating (Weber et al. 2005). Therefore, one possible explanation for the activation we observed in the SPL may be that it was induced by the episodic memory of traumatic or stressful events. Meanwhile, it has been suggested that ruminative thoughts in the aftermath of traumatic/stressful events play an important role in PTG (Calhoun et al. 2000). In particular, it is worth noting that one study found that intrusive rumination soon after an event is positively related to PTG, but that "recent" deliberate rumination most strongly predicted participants' current levels of PTG (Taku et al. 2009). Thus, as discussed above, there is no direct evidence that current rumination with PTG affects a person's working memory; however, ruminative thought may potentially explain the brain activation we observed in the SPL within the CEN.

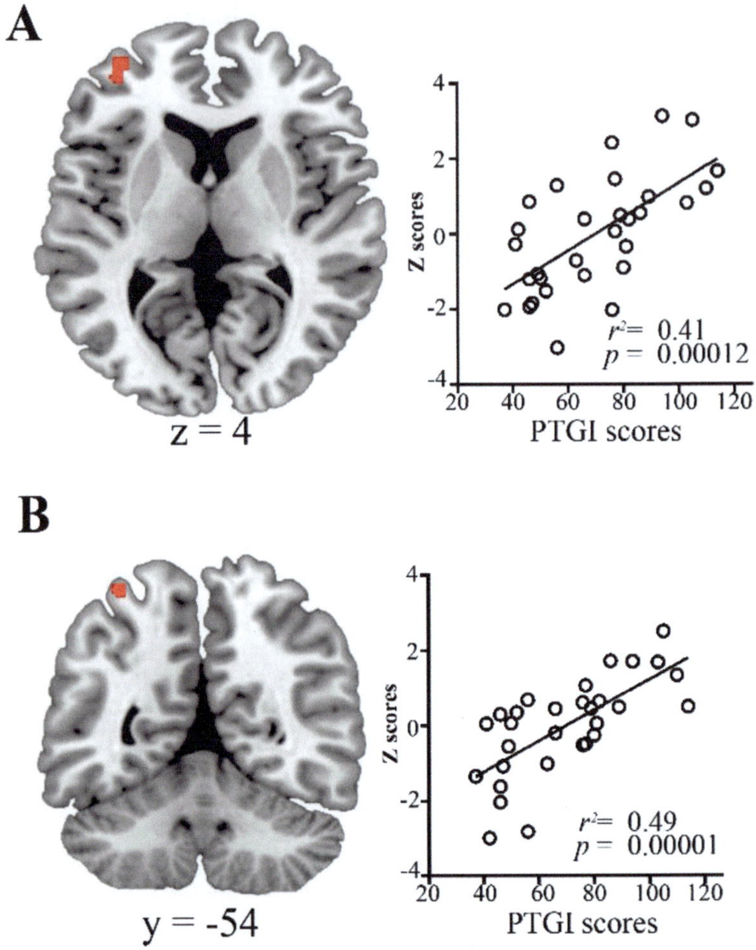

Fig. 14.4 Results of the multiple regression analysis between the central executive network and the scores on the Post-Traumatic Growth Inventory (PTGI) (Fujisawa et al. 2015). Brain regions showing positive correlations between the PTGI scores and the strength of the central executive network activity as determined by multiple regression analysis. The scatter plots show the association between the PTGI scores and the strength of the central executive network activity. (**a**) Rostral prefrontal cortex; (**b**) superior parietal lobule. The statistical threshold for the contrasts was voxel level $p < 0.001$ uncorrected for height and cluster level $p < 0.05$ corrected for multiple comparisons

14.4 Conclusion

A better understanding of the neural basis of PTG may help to establish methods for enhancing an individual's recovery resiliency or prevent them from developing PTSD. Further studies are necessary to confirm these findings and to clarify the causality and consequences of such neurofunctional mechanisms.

Acknowledgements This work was supported, in part, by research grants from the Ministry of Education, Culture, Sports, Science and Technology (MEXT) of Japan (#15 K12720 to A.T. and #15H03106 to A.T.) and by the Japan Science and Technology Corporation (JST)/Research Institute of Science and Technology for Society (RISTEX). This work was also supported by a grant-in-aid for Scientific Research to A.T. from Japan-U.S. Brain Research Cooperation Program. We are grateful to all the children and parents who participated in this study.

References

Ahern GL, Schwartz GE (1985) Differential lateralization for positive and negative emotion in the human brain: EEG spectral analysis. Neuropsychologia 23(6):745–755

American Psychiatric Association (2013) Diagnostic and statistical manual of mental disorders, fifth edition (DSM-5). American Psychiatric Association, Washington, DC

Amodio DM, Frith CD (2006) Meeting of minds: the medial frontal cortex and social cognition. Nat Rev Neurosci 7(4):268–277

Andersen SL, Tomoda A, Vincow ES, Valente E, Polcari A, Teicher MH (2008) Preliminary evidence for sensitive periods in the effect of childhood sexual abuse on regional brain development. J Neuropsychiatry Clin Neurosci 20(3):292–301

Aron A, Fisher H, Mashek DJ, Strong G, Li H, Brown LL (2005) Reward, motivation, and emotion systems associated with early-stage intense romantic love. J Neurophysiol 94(1):327–337

Bar KJ, Wagner G, Koschke M, Boettger S, Boettger MK, Schlosser R, Sauer H (2007) Increased prefrontal activation during pain perception in major depression. Biol Psychiatry 62(11):1281–1287

Bremner JD (2003) Long-term effects of childhood abuse on brain and neurobiology. Child Adolesc Psychiatr Clin N Am 12(2):271–292

Bremner JD, Randall P, Vermetten E, Staib L, Bronen RA, Mazure C, Capelli S, McCarthy G, Innis RB, Charney DS (1997) Magnetic resonance imaging-based measurement of hippocampal volume in posttraumatic stress disorder related to childhood physical and sexual abuse – a preliminary report. Biol Psychiatry 41(1):23–32

Cabeza R, Ciaramelli E, Olson IR, Moscovitch M (2008) The parietal cortex and episodic memory: an attentional account. Nat Rev Neurosci 9(8):613–625

Calhoun LG, Cann A, Tedeschi RG, McMillan J (2000) A correlational test of the relationship between posttraumatic growth, religion, and cognitive processing. J Trauma Stress 13(3):521–527

Calvo MG, Beltrán D (2014) Brain lateralization of holistic versus analytic processing of emotional facial expressions. NeuroImage 92:237–247

Carrion VG, Weems CF, Watson C, Eliez S, Menon V, Reiss AL (2009) Converging evidence for abnormalities of the prefrontal cortex and evaluation of midsagittal structures in pediatric posttraumatic stress disorder: an MRI study. Psychiatry Res Neuroimaging 172(3):226–234

Choi J, Jeong B, Rohan ML, Polcari AM, Teicher MH (2009) Preliminary evidence for white matter tract abnormalities in young adults exposed to parental verbal abuse. Biol Psychiatry 65(3):227–234

Choi J, Jeong B, Polcari A, Rohan ML, Teicher MH (2012) Reduced fractional anisotropy in the visual limbic pathway of young adults witnessing domestic violence in childhood. NeuroImage 59(2):1071–1079

Crockford DN, Goodyear B, Edwards J, Quickfall J, el-Guebaly N (2005) Cue-induced brain activity in pathological gamblers. Biol Psychiatry 58(10):787–795

Dannlowski U, Stuhrmann A, Beutelmann V, Zwanzger P, Lenzen T, Grotegerd D, Domschke K, Hohoff C, Ohrmann P, Bauer J, Lindner C, Postert C, Konrad C, Arolt V, Heindel W, Suslow T,

Kugel H (2012) Limbic scars: long-term consequences of childhood maltreatment revealed by functional and structural magnetic resonance imaging. Biol Psychiatry 71(4):286–293

Davidson RJ, Ekman P, Saron CD, Senulis JA, Friesen WV (1990) Approach-withdrawal and cerebral asymmetry: emotional expression and brain physiology I. J Pers Soc Psychol 58(2):330–341

De Bellis MD, Kuchibhatla M (2006) Cerebellar volumes in pediatric maltreatment-related posttraumatic stress disorder. Biol Psychiatry 60(7):697–703

De Bellis MD, Keshavan MS, Clark DB, Casey BJ, Giedd JN, Boring AM, Frustaci K, Ryan ND (1999) A.E. Bennett research award. Developmental traumatology. Part II: brain development. Biol Psychiatry 45(10):1271–1284

De Bellis MD, Keshavan MS, Shifflett H, Iyengar S, Beers SR, Hall J, Moritz G (2002) Brain structures in pediatric maltreatment-related posttraumatic stress disorder: a sociodemographically matched study. Biol Psychiatry 52(11):1066–1078

De Brito SA, Viding E, Sebastian CL, Kelly PA, Mechelli A, Maris H, McCrory EJ (2013) Reduced orbitofrontal and temporal grey matter in a community sample of maltreated children. J Child Psychol Psychiatry 54(1):105–112

Drexler K, Schweitzer JB, Quinn CK, Gross R, Ely TD, Muhammad F, Kilts CD (2000) Neural activity related to anger in cocaine-dependent men: a possible link to violence and relapse. Am J Addict 9(4):331–339

Edmiston EE, Wang F, Mazure CM, Guiney J, Sinha R, Mayes LC, Blumberg HP (2011) Corticostriatal-limbic gray matter morphology in adolescents with self-reported exposure to childhood maltreatment. Arch Pediatr Adolesc Med 165(12):1069–1077

Fennema-Notestine C, Stein MB, Kennedy CM, Archibald SL, Jernigan TL (2002) Brain morphometry in female victims of intimate partner violence with and without posttraumatic stress disorder. Biol Psychiatry 52(11):1089–1101

Fitzgerald PB, Laird AR, Maller J, Daskalakis ZJ (2008) A meta-analytic study of changes in brain activation in depression. Hum Brain Mapp 29(6):683–695

Fujisawa TX, Jung M, Kojima M, Saito DN, Kosaka H, Tomoda A (2015) Neural basis of psychological growth following adverse experiences: a resting-state functional MRI study. PLoS One 10(8):e0136427

Geuze E, Westenberg HG, Jochims A, de Kloet CS, Bohus M, Vermetten E, Schmahl C (2007) Altered pain processing in veterans with posttraumatic stress disorder. Arch Gen Psychiatry 64(1):76–85

Gilbert SJ, Williamson ID, Dumontheil I, Simons JS, Frith CD, Burgess PW (2007) Distinct regions of medial rostral prefrontal cortex supporting social and nonsocial functions. Soc Cogn Affect Neurosci 2(3):217–226

Hanson JL, Nacewicz BM, Sutterer MJ, Cayo AA, Schaefer SM, Rudolph KD, Shirtcliff EA, Pollak SD, Davidson RJ (2015) Behavioral problems after early life stress: contributions of the hippocampus and amygdala. Biol Psychiatry 77(4):314–323

Hart H, Rubia K (2012) Neuroimaging of child abuse: a critical review. Front Hum Neurosci 6:52

Hou C, Liu J, Wang K, Li L, Liang M, He Z, Liu Y, Zhang Y, Li W, Jiang T (2007) Brain responses to symptom provocation and trauma-related short-term memory recall in coal mining accident survivors with acute severe PTSD. Brain Res 1144:165–174

Jansari A, Rodway P, Goncalves S (2011) Identifying facial emotions: valence specific effects and an exploration of the effects of viewer gender. Brain Cogn 76(3):415–423

Koenigs M, Barbey AK, Postle BR, Grafman J (2009) Superior parietal cortex is critical for the manipulation of information in working memory. J Neurosci 29(47):14980–14986

Koizumi M, Takagishi H (2014) The relationship between child maltreatment and emotion recognition. PLoS One 9(1):e86093

Liao M, Yang F, Zhang Y, He Z, Song M, Jiang T, Li Z, Lu S, Wu W, Su L, Li L (2013) Childhood maltreatment is associated with larger left thalamic gray matter volume in adolescents with generalized anxiety disorder. PLoS One 8(8):e71898

Liberzon I, Britton JC, Phan KL (2003) Neural correlates of traumatic recall in posttraumatic stress disorder. Stress 6(3):151–156

Lim L, Radua J, Rubia K (2014) Gray matter abnormalities in childhood maltreatment: a voxel-wise metaanalysis. Am J Psychiatry 171(8):854–863

Liotti M, Mayberg HS (2001) The role of functional neuroimaging in the neuropsychology of depression. J Clin Exp Neuropsychol 23(1):121–136

McCrory E, De Brito SA, Viding E (2011) The impact of childhood maltreatment: a review of neurobiological and genetic factors. Front Psychiatry 2:48

Mehta MA, Golembo NI, Nosarti C, Colvert E, Mota A, Williams SCR, Rutter M, Sonuga-Barke EJS (2009) Amygdala, hippocampal and corpus callosum size following severe early institutional deprivation: the English and Romanian adoptees study pilot. J Child Psychol Psychiatry 50(8):943–951

Pechtel P, Lyons-Ruth K, Anderson CM, Teicher MH (2014) Sensitive periods of amygdala development: the role of maltreatment in preadolescence. NeuroImage 97:236–244

Prete G, Laeng B, Fabri M, Foschi N, Tommasi L (2015) Right hemisphere or valence hypothesis, or both? The processing of hybrid faces in the intact and callosotomized brain. Neuropsychologia 68:94–106

Raust A, Slama F, Mathieu F, Roy I, Chenu A, Koncke D, Fouques D, Jollant F, Jouvent E, Courtet P, Leboyer M, Bellivier F (2007) Prefrontal cortex dysfunction in patients with suicidal behavior. Psychol Med 37(3):411–419

Richert KA, Carrion VG, Karchemskiy A, Reiss AL (2006) Regional differences of the prefrontal cortex in pediatric PTSD: an MRI study. Depress Anxiety 23(1):17–25

Shakespeare-Finch J, Armstrong D (2010) Trauma type and posttrauma outcomes: differences between survivors of motor vehicle accidents, sexual assault, and bereavement. J Loss Trauma 15(2):69–82

Shakespeare-Finch J, Lurie-Beck J (2014) A meta-analytic clarification of the relationship between posttraumatic growth and symptoms of posttraumatic distress disorder. J Anxiety Disord 28(2):223–229

Shimada K, Takiguchi S, Mizushima S, Fujisawa TX, Saito DN, Kosaka H, Okazawa H, Tomoda A (2015) Reduced visual cortex grey matter volume in children and adolescents with reactive attachment disorder. NeuroImage Clin 9:13–19

Silberman EK, Weingartner H (1986) Hemispheric lateralization of functions related to emotion. Brain Cogn 5(3):322–353

Sun X, Zhang X, Chen X, Zhang P, Bao M, Zhang D, Chen J, He S, Hu X (2005) Age-dependent brain activation during forward and backward digit recall revealed by fMRI. NeuroImage 26(1):36–47

Suzuki H, Tomoda A (2015) Roles of attachment and self-esteem: impact of early life stress on depressive symptoms among Japanese institutionalized children. BMC Psychiatry 15:8

Takiguchi S, Fujisawa TX, Mizushima S, Saito DN, Okamoto Y, Shimada K, Koizumi M, Kumazaki H, Jung M, Kosaka H, Hiratani M, Ohshima Y, Teicher MH, Tomoda A (2015) Ventral striatum dysfunction in children and adolescents with reactive attachment disorder: a functional MRI study. Br J Psychiatry Open 1(2):121–128

Taku K, Calhoun LG, Tedeschi RG, Gil-Rivas V, Kilmer RP, Cann A (2007) Examining posttraumatic growth among Japanese university students. Anxiety Stress Coping 20(4):353–367

Taku K, Cann A, Tedeschi RG, Calhoun LG (2009) Intrusive versus deliberate rumination in posttraumatic growth across US and Japanese samples. Anxiety Stress Coping 22(2):129–136

Tedeschi RG, Calhoun LG (1996) The posttraumatic growth inventory: measuring the positive legacy of trauma. J Trauma Stress 9(3):455–471

Teicher MH, Ito Y, Glod CA, Andersen SL, Dumont N, Ackerman E (1997) Preliminary evidence for abnormal cortical development in physically and sexually abused children using EEG coherence and MRI. Ann N Y Acad Sci 821:160–175

Teicher MH, Dumont NL, Ito Y, Vaituzis C, Giedd JN, Andersen SL (2004) Childhood neglect is associated with reduced corpus callosum area. Biol Psychiatry 56(2):80–85

Teicher MH, Tomoda A, Andersen SL (2006) Neurobiological consequences of early stress and childhood maltreatment: are results from human and animal studies comparable? Ann N Y Acad Sci 1071:313–323

Teicher MH, Anderson CM, Polcari A (2012) Childhood maltreatment is associated with reduced volume in the hippocampal subfields CA3, dentate gyrus, and subiculum. Proc Natl Acad Sci 109(9):E563–E572

Teicher MH, Samson JA (2013) Childhood maltreatment and psychopathology: a case for eco-phenotypic variants as clinically and neurobiologically distinct subtypes. Am J Psychiatry 170(10):1114–1133

Tomoda A, Navalta CP, Polcari A, Sadato N, Teicher MH (2009a) Childhood sexual abuse is associated with reduced gray matter volume in visual cortex of young women. Biol Psychiatry 66(7):642–648

Tomoda A, Suzuki H, Rabi K, Sheu YS, Polcari A, Teicher MH (2009b) Reduced prefrontal cortical gray matter volume in young adults exposed to harsh corporal punishment. NeuroImage 47:T66–T71

Tomoda A, Sheu YS, Rabi K, Suzuki H, Navalta CP, Polcari A, Teicher MH (2011) Exposure to parental verbal abuse is associated with increased gray matter volume in superior temporal gyrus. NeuroImage 54:S280–S286

Tomoda A, Polcari A, Anderson CM, Teicher MH (2012) Reduced visual cortex gray matter volume and thickness in young adults who witnessed domestic violence during childhood. PloS One 7(12):e52528

Tottenham N, Hare TA, Quinn BT, McCarry TW, Nurse M, Gilhooly T, Millner A, Galvan A, Davidson MC, Eigsti I-M, Thomas KM, Freed PJ, Booma ES, Gunnar MR, Altemus M, Aronson J, Casey BJ (2010) Prolonged institutional rearing is associated with atypically large amygdala volume and difficulties in emotion regulation. Dev Sci 13(1):46–61

van Harmelen A-L, van Tol M-J, van der Wee NJA, Veltman DJ, Aleman A, Spinhoven P, van Buchem MA, Zitman FG, Penninx BWJH, Elzinga BM (2010) Reduced medial prefrontal cortex volume in adults reporting childhood emotional maltreatment. Biol Psychiatry 68(9):832–838

Veltman DJ, de Ruiter MB, Rombouts SA, Lazeron RH, Barkhof F, Van Dyck R, Dolan RJ, Phaf RH (2005) Neurophysiological correlates of increased verbal working memory in high-dissociative participants: a functional MRI study. Psychol Med 35(2):175–185

Vervoort E, De Schipper JC, Bosmans G, Verschueren K (2013) Screening symptoms of reactive attachment disorder: evidence for measurement invariance and convergent validity. Int J Methods Psychiatr Res 22(3):256–265

Vrticka P, Bondolfi G, Sander D, Vuilleumier P (2011) The neural substrates of social emotion perception and regulation are modulated by adult attachment style. Soc Neurosci 7(5):473–493

Wager TD, Smith EE (2003) Neuroimaging studies of working memory: a meta-analysis. Cogn Affect Behav Neurosci 3(4):255–274

Weber DL, Clark CR, McFarlane AC, Moores KA, Morris P, Egan GF (2005) Abnormal frontal and parietal activity during working memory updating in post-traumatic stress disorder. Psychiat Res Neuroimaging 140(1):27–44

Zeanah CH, Gleason MM (2015) Attachment disorders in early childhood – clinical presentation, causes, correlates, and treatment. J Child Psychol Psychiatry 56(3):207–222

Part IV
Toward the Application of Memory Research into Our Society

Chapter 15
Benefits of "Smart Ageing" Interventions Using Cognitive Training, Brain Training Games, Exercise, and Nutrition Intake for Aged Memory Functions in Healthy Elderly People

Rui Nouchi and Ryuta Kawashima

Abstract Memory functions decline with age. Many people are interested in methods of improving memory functions. To meet that need, we developed the concept of "smart ageing": a positive acceptance of the later stages in life. It characterizes ageing as a series of developmental stages of intellectual maturity. Based on the smart ageing concept, we introduce the beneficial effects of intervention programs incorporating cognitive training (working memory, episodic memory, and brain training games), exercise training (combination exercise), and nutrition (orange juice consumption) intervention in healthy elderly people. This chapter presents the following evidence. Cognitive training using working memory training and using "Brain Age" can improve working memory and short-term memory performance. Cognitive training using episodic memory training, exercise training using combination exercise, and nutrition intervention using flavonoid consumption can improve episodic memory performance. Finally, we discuss the future directions of intervention programs that can be used to improve memory functions.

Keywords Cognitive improvement • Transfer effect • Cognitive training • Exercise • Nutrition • Brain training games

---

R. Nouchi (✉)
Smart-Aging Research Center (S.A.R.C), Tohoku University, Aobaku, Sendai, Japan

Advanced Brain Science, Institute of Development, Aging and Cancer (I.D.A.C), Tohoku University, Sendai, Japan
e-mail: rui.nouchi.a4@tohoku.ac.jp

R. Kawashima
Advanced Brain Science, Institute of Development, Aging and Cancer (I.D.A.C), Tohoku University, Sendai, Japan

T. Tsukiura, S. Umeda (eds.), *Memory in a Social Context*,
https://doi.org/10.1007/978-4-431-56591-8_15

15.1 Interventions for Aged Memory Functions

Human memory is a key function in basic and social behaviors. For example, human memories are divisible into three types: working memory (WM), semantic memory, and episodic memory. WM is a system for temporarily storing and managing the information needed to perform complex cognitive tasks such as learning, reasoning, and comprehension. Semantic memory is a more structured record of facts, meanings, concepts, and knowledge about the external world. Episodic memory is a personal memory of experiences and specific events.

It is particularly interesting that semantic memory does not affect aging. Semantic memory develops to the age of 60 or 70 (Park et al. 2002). However, elderly people may experience a decline in the number of cognitive functions, including WM and episodic memory (Nilsson 2003; Salthouse 2003). Decline in the cognitive abilities of older people engenders difficulty in performing basic activities of daily living (Barberger-Gateau and Fabrigoule 1997; Cahn-Weiner et al. 2000; Carlson et al. 1999; Grigsby et al. 1998; Lee et al. 2005; Owsley and McGwin 2004). Consequently, many people are actively investigating methods of improving cognitive functions such as WM and episodic memory.

To address the expectations of society and scientific fields related to the improvement of cognitive functions, we developed the new concept of "smart ageing" (Nouchi and Kawashima 2014), which advocates the positive acceptance of the later stages in life and of an outlook related to aging as a series of "developmental stages toward intellectually maturity." Smart ageing is a paradigm shift away from negative concepts such as anti-aging, which simply represents an unwillingness to accept, confront, or adapt to the later stages of life. One purpose of smart ageing research is to investigate strategies for retaining and improving the cognitive functions of healthy elderly people, because cognitive functions play a crucially important role in everyday life. Smart ageing emphasizes cognitive training, nutrition, and exercise because they play key roles in daily life. As with the smart ageing concept, previous studies have demonstrated that daily intervention programs can improve cognitive functions in elderly people. In this chapter, we present summary of the benefits of cognitive training, nutrition, and exercise interventions for cognitive function, especially memory function, in older adults.

15.2 Cognitive Interventions and Transfer Effects

Cognitive intervention and training are defined as interventions that provide structured practice of tasks that are relevant to aspects of cognitive functioning, such as memory, attention, language, or executive function. The many types of cognitive training include WM training (Klingberg 2010; Richmond et al. 2011), processing speed training (Edwards et al. 2005; Takeuchi et al. 2011), memory strategic training (Bjorklund et al. 1997; Carretti et al. 2007; Mahncke et al. 2006; Verhaeghen et al. 1992), and brain training games (Miller and Robertson 2010; Nouchi et al. 2012, 2013). For this discussion, some cognitive intervention studies for memory functions are selected.

Transfer effects are important phenomena for cognitive intervention studies. A transfer effect is defined as "the ability to extend what has been learned in one context to new contexts" (Bransford et al. 2000). Additionally, the transfer effect is classifiable in terms of a near transfer effect and a far transfer effect (Barnett and Ceci 2002; Edwards et al. 2002; Karbach and Kray 2009; Zelinski 2009). A near transfer effect is an improvement in cognitive domains that is closely related to trained cognitive processes. In contrast, a far transfer effect is an improvement in cognitive domains that is not closely related to the trained cognitive processes. For example, WM training using an n-back task is conducted to improve WM performance, and the n-back task, operation span task, and processing speed are improved after the intervention. In this case, improvement of the n-back task is not a transfer effect because the n-back task is the training task. However, improvements in the operation span task are related to the near transfer effect because the operation span task is a WM task. Improvements in processing speed are related to a far transfer effect because processing speed is not trained during the WM training using the n-back task.

15.2.1 Effects of Working Memory Training on Memory Functions

The most famous cognitive training is WM training. For example, Borella et al. investigated the benefits of WM training on memory functions in the healthy older adults (Borella et al. 2010). They recruited 40 healthy elderly people and assigned the participants randomly into WM training (n = 20, mean age 69 years old) and control groups (n = 20, mean age 69.15 years old). Participants of the training and control groups attended five sessions. The first and last sessions were pretest and post-test sessions. WM training groups participated in three WM training sessions over 2 weeks. In the WM training sessions, participants performed the categorization working memory span (CWMS) task (Borella et al. 2008), which is similar to a classic WM task such as the listening span test. The basic instructions were to recall the target words and to tap the hand on the table whenever an animal noun was mentioned. However, the control group did remember common events related to their children, adulthood, and recent events. They used several cognitive functional measures: CWMS and dot matrix are measures for WM tasks; forward span and backward span are measures for short-term memory tasks; the cattle fair test is a fluid intelligence test; the Stroop color task tests inhibition-related processes in executive functions; and the processing speed is measured by a pattern comparison test. Compared with the control group, WM training improved all cognitive functions: working memory span (WMS and dot matrix), short-term memory (forward span and backward span), inhibition process (Stroop color task), processing speed (pattern matrix), and fluid intelligence (cattle fair test). These results demonstrate that WM training had near and far transfer effects.

15.2.2 Effects of Episodic Memory Training on Memory Functions

The benefits of episodic memory training are well known (Hampstead et al. 2014). For example, Cavalini et al. conducted memory strategic training for healthy elderly people (Cavallini et al. 2010). They recruited 62 elderly people and randomly assigned participants into memory strategic training ($n = 34$, mean age 67.94 years old) and control groups ($n = 28$, mean age 68.64 years old). The memory strategic training group participated in four group training sessions (each session was 2h, participant received total 8h training session). Memory strategic training was conducted once weekly for 4 weeks. During the training sessions, two memory mnemonics (sentence generation and interactive imagery) were trained using paired word association and single-word tasks. Sentence generation consisted of the creation of a semantic link among words by making up a sentence, whereas interactive imagery consisted of the mental creation of an active picture of the interacting words. This study used six memory tasks, such as associative learning, list learning, text learning, name–face learning, grocery list learning, and place learning. In these tasks, associative learning and list learning were similar to training tasks. Compared with the control group, memory strategic training improved associative learning, list learning, name–face learning, and grocery list learning. These results demonstrate that strategic memory training showed near transfer effects.

15.2.3 Effects of Brain Training Games on Memory Functions

Video game training is cognitive training (Lustig et al. 2009). Some previous studies have shown that playing video games can improve some cognitive functions of healthy elderly people (Basak et al. 2008). Video game training has attracted much attention because some video game training has engendered improvement of other untrained cognitive functions, which are commonly termed transfer effects (Boot et al. 2008, 2011; Green and Bavelier 2003; Lovden et al. 2010; Takeuchi et al. 2010). In line with these results, commercial brain-training games of many types (e.g., "Brain Age," Big Brain Academy, and Brain Challenge) have been released. Such commercial brain-training games have become popular around the world.

Nouchi et al. investigated the benefits of brain-training games such as Brain Age on the cognitive functions of healthy elderly people (Nouchi et al. 2012). They recruited 32 elderly people and randomly assigned participants to a brain-training group using Brain Age ($n = 16$, mean age 68.86 years old) or a puzzle game training group using Tetris ($n = 16$, mean age 69.31 years old). The study was conducted with 4-week intervention periods (15 min per day, 5 days per week) and the popular brain-training games (Brain Age) as a target intervention.

Brain Age, a popular brain-training game, was developed by Ryuta Kawashima based on earlier neuroscience evidence. Most games in Brain Age include elements of reading aloud and simple arithmetic calculations (Table 15.1). For example, in the calculation 20 game, participants must answer 20 simple arithmetic calculations as quickly as possible. The questions include problems of mathematical addition, subtraction, and multiplication. For reading aloud, participants were asked to read excerpts from Japanese classical literature aloud. For the syllable counting task, some sentences written in *kanji* and *kana* were presented. Participants had to count the kana letters after translating the *kanji* to *kana*. These studies used a popular puzzle game (Tetris) as an active control group. In Tetris, players rotate and move blocks that are descending from the top of the screen so that these blocks form lines at the bottom of the screen. After a complete line with no gaps is formed, the line disappears and points are awarded. If no line is formed, then the blocks pile higher and higher until the block pile reaches the top of the screen, at which point the game ends and the player loses. The goal is to continue the game as long as possible by

Table 15.1 Summary of Brain Age brain-training tasks

Task	Details
Calculation × 20	Participants must answer 20 simple arithmetic calculations as quickly as possible
Calculation × 100	Participants must answer 100 questions as quickly as possible
Reading aloud	Participants must read excerpts from Japanese classical literature aloud.
Low to high	Numbers in boxes are presented for a few seconds. Then, participants must select boxes from the lowest number to the highest number.
Syllable count	Some sentences written in a combination of *kanji* and *kana* are presented. Participants must count the kana letters after translating *kanji* to *kana*
Head count	Participants watch scenes in which some people enter or leave a house. Then participants must report the number of people in the house at the end
Triangle math	Three numbers are presented on the top line (e.g. 5, 7, 2), two mathematical operations on a second line (e.g. +, +), and one mathematical operation (e.g. +) on the bottom line. Participants must solve the first formula (5 + 7) using the first two numbers (5, 7) in the first line and the first mathematical operation (+) in the second line. Next, they must solve the second formula (7 + 2) using the last two numbers (7, 2) in the first line and the last mathematical operation (+) in the second line. The participants must then solve the last formula using the answer of the first formula (12), the answer of the last formula (9), and the mathematical operation (+) in the last line. Last, the participants give the final answer (21)
Time lapse	Two analog clocks are presented. Participants must then calculate the difference in time between those shown on the two clocks
Voice calculation	Participants must answer orally simple arithmetic calculations as quickly as possible

forming complete lines using the descending blocks. As the game progresses, the blocks descend faster, giving players less time to choose where to place the respective blocks.

For elderly people (Nouchi et al. 2012), the brain-training game improved executive functions measured by the frontal assessment battery, a trail-making test, and processing speed as measured by symbol coding and symbol search compared with the puzzle game. However, short term memory performance measured by the digit span did not change after intervention. It is noteworthy that the Nouchi study did not measure episodic memory functions. In addition, other research using 6 weeks of Brain Age intervention for older adults demonstrated significant improvement in WM performance (McDougall and House 2012). Four-week Brain Age intervention for young adults also showed improvements in WM performance (Nouchi et al. 2013). Results show that brain-training games have beneficial effects on WM ability.

15.3 Exercise Training

Exercise training can be any one of three types: aerobic exercise, strength, and combination exercise training (van Uffelen et al. 2008). Aerobic exercise training is a structured exercise program involving the use of large muscle groups for extended periods of time in activities that are rhythmic, such as walking, stepping, running, swimming, cycling, and rowing (Smith et al. 2010). Strength exercise training uses resistance against the force of muscular contraction to build strength, anaerobic endurance, and skeletal muscle mass. Strength exercise training often uses gravity to oppose muscle contraction (Cassilhas et al. 2007; Chang et al. 2012; Peig-Chiello et al. 1998). Combination exercise training is a combination between aerobic and strength exercise training (Colcombe and Kramer 2003). Previous studies using RCTs have revealed that aerobic exercise training alone, strength exercise training alone, and combination exercise training improved cognitive functions (Davis et al. 2011; Peig-Chiello et al. 1998; Smith et al. 2010). An earlier meta-analytical study demonstrated that combination exercise training has a greater effect than aerobic and strength exercise training alone (Colcombe and Kramer 2003). Therefore, we describe the benefits of combination exercise training on memory functions.

Nouchi et al. investigated the benefits of combination exercise training on cognitive functions in healthy elderly people (Nouchi et al. 2014). The combination exercise training included strength, aerobic, and stretching exercises. They recruited 64 elderly people and randomly assigned participants into a combination exercise training group ($n = 32$, mean age 66.75 years old) or a control group ($n = 32$, mean age 67.06 years old). In the combined exercise training, participants performed exercise training 3 days per week for 4 weeks (12 workouts in total). For the strength training parts, participants were informed of the proper use of all equipment and were instructed to complete as many repetitions as possible during a 30-s period. Using 12 machines, they exercised all body parts (chest press/seated row, squat, shoulder press/lat pull, leg extension/leg curl, abdominal crunch/back extension, lateral lift, elbow flexion/extension, horizontal leg press, pectoral deck, oblique, hip abductor/adductor, gluteus). In a continuous interval fashion, participants performed floor-based aerobic training (e.g., running/skipping in place, arm circles) on recovery pads for a 30 s period after each resistance exercise to maintain a consistent exercise heart rate corresponding to 60–80% of their heart maximum heart rate. Finally, participants carried out standardized whole-body stretching training (6 min). Whole-body stretching training consists of 12 stretching exercises designed to stretch 12 body areas: Achilles' tendon, sole of the foot, thigh, armpit, shoulder, shoulder/upper arm, chest/arm, shoulder/chest/arm, waist, back of knee, base of the thigh, and back.

Results of combined exercise training for the older people demonstrated that the combined exercise training improved executive functions measured using a verbal fluency task and the Stroop test, episodic memory measured by logical memory, and processing speed measured by symbol coding and symbol search compared with a waiting list control group (Nouchi et al. 2014). These results show that the combination exercise training has beneficial effects for memory functions such as episodic memory.

15.4 Nutrition Consumption Intervention

Lifestyle patterns such as diet and eating habits affect cognitive functions. For example, epidemiological data suggest that breakfast consumption might be associated with cognitive function (Pollitt and Mathews 1998; Smith 1998). Recently, interest has arisen in the relation between fruit consumption and cognitive function (Lamport et al. 2014). Fruits and juices are well known to provide a rich, easily available source of flavonoids, such as hesperidin and narirutin. Flavonoids found in various fruits, vegetables, and beverages have been recognized as promising plant-based bioactive materials capable of influencing different aspects of synaptic plasticity, thereby improving memory and learning in humans (Rendeiro et al. 2015). For example, fruit juices such as orange juice are rich sources of flavonoids. Some reviews have shown that increased flavonoid consumption over a person's lifespan may attenuate age-associated cognitive decline and the onset of neurodegenerative disease (Macready et al. 2009). Next, we introduce intervention studies using flavonoid intake.

Kean et al. used orange juice to investigate the beneficial effects of flavonoid consumption on cognitive function in healthy older adults (Kean et al. 2015). They used a crossover design RCT to examine 37 healthy elderly people (mean age 66.70 years old). They used two drink conditions: high flavanone (HF) and low flavanone (LF). The HF drink contained 549 mg hesperidin/L and 60 mg narirutin/L (producing a daily 500-mL serving of 305 mg flavanones). The LF drink contained 64 mg hesperidin/L and 10 mg narirutin/L (producing a daily 500-mL serving of 37 mg. Participants were instructed to drink 250 mL twice daily for 8 weeks.

Because of the crossover design, participants received both HF and LF conditions. Participants were tested at weeks 0 (visit 1), 8 (visit 2), 12 (visit 3), and 20 (visit 4). At the end of visits 1 and 3, participants were provided with an 8-week supply of either the LF or the HF drink in accordance with counterbalancing. Washout was set during visit 2 (at 8 weeks) and visit 3 (at 12 weeks). They measured several cognitive functions: go–no-go task, Consortium to Establish a Registry for Alzheimer's disease (CERAD: immediate verbal recall), letter memory, verbal paired associates (VPA: immediate), CERAD (delayed verbal recall), serial sevens, spatial working memory (SWM), digit symbol substitution test (DSST), letter fluency, DSST and letter fluency (both tests performed simultaneously), and VPA (delayed). These cognitive tests were divided into two cognitive measures: executive functions (DSST, go–no-go, serial sevens, letter fluency, letter memory, dual DSST, and dual letter fluency) and episodic memory (immediate CERAD, delayed CREAD, immediate VPA, delayed VPA, and SWM). Consequently, a mean *z* score in executive functions and episodic memory was calculated by averaging all *z* scores obtained for the respective cognitive measures. The global cognitive function was calculated using all cognitive functional measures.

Results showed that the global cognitive function and the mean *z* score in executive functions were better after the HF drink than after the LF drink. For the mean *z* score in episodic memory, no significant difference was found between the HL and the LF drinks. However, the immediate CERAD score was better after HL than it was after LF. These results demonstrate that HF consumption using orange juice has beneficial effects on episodic memory.

15.5 Summary and Future Directions

In this chapter we introduced the concept of "smart ageing" and presented an explanation of the effects of cognitive and brain-training games, exercise, and nutritional intervention on cognitive functions in healthy elderly people (Table 15.2). The summary table shows clearly that the cognitive and brain training, exercise, and

Table 15.2 Summary of the beneficial effects of intervention programs on cognitive functions

Intervention	WM/ST	EM	PS	EF
Working memory training	↑	?	↑	↑
Episodic memory training	?	↑	?	?
Brain-training game (Brain Age)	↑	?	↑	↑
Combination exercise training	=	↑	↑	↑
Flavonoid consumption (Orange juice)	?	↑	?	↑

WM working memory, *ST* short-term memory, *EM* episodic memory, *PS* processing speed, *EF* executive functions, ↑ previous study showed improvements after intervention, = previous study showed no change after intervention; *?* previous study did not measure the cognitive domain nutritional intervention programs can improve memory functions, such as WM and episodic memory, in elderly people. We presented evidence for the benefits of these programs on aged memory functions.

In the final part, we summarized future directions for improvements in memory functions using interventions.

First, assessing the beneficial effects of intervention programs on a wide range of cognitive functions is important. Numerous earlier studies have used only episodic memory measures or WM measures. Therefore, it remains unknown whether the intervention program can improve both episodic and WM functions. For example, brain-training games use WM measures, but not episodic memory measures (Nouchi et al. 2012). Consequently, it remains unclear whether brain-training games can improve episodic memory functions or not. In future studies several cognitive domains should be measured. The results can facilitate a comparison of the benefits of intervention programs on cognitive functions.

Second, it is necessary to investigate how long the beneficial effects will last after the intervention. As presented in Table 15.2, the cognitive and brain-training, exercise, and nutritional intervention can improve memory functions. However, the long-lasting effects of the intervention programs are not well known. Future studies should measure cognitive functions after the intervention period. Third, it is necessary to investigate whether intervention programs can improve performance related to daily behaviors such as activities of daily living (ADL) and instrumental activities of daily living (IADL). One cognitive intervention report described that cognitive training led to less functional decline in IADL than in a control group (Willis et al. 2006). However, it remains unclear whether other intervention programs can improve ADL and IADL in elderly people. Future studies must measure ADL and IADL.

Finally, our review specifically examined the beneficial effects of intervention on cognitive function in healthy elderly people. It remains unclear whether intervention programs can improve cognitive function in nonhealthy populations such as older people with mild cognitive impairments (MCIs) or dementia.

To generalize the results of various studies, it will be necessary to expand and diversify the range of subjects investigated using intervention programs.

References

Barberger-Gateau P, Fabrigoule C (1997) Disability and cognitive impairment in the elderly. Disabil Rehabil 19(5):175–193

Barnett SM, Ceci SJ (2002) When and where do we apply what we learn? A taxonomy for far transfer. Psychol Bull 128(4):612

Basak C, Boot WR, Voss MW, Kramer AF (2008) Can training in a real-time strategy video game attenuate cognitive decline in older adults? Psychol Aging 23(4):765

Bjorklund DF, Miller PH, Coyle TR, Slawinski JL (1997) Instructing children to use memory strategies: evidence of utilization deficiencies in memory training studies. Dev Rev 17(4):411–441

Boot WR, Kramer AF, Simons DJ, Fabiani M, Gratton G (2008) The effects of video game playing on attention, memory, and executive control. Acta Psychol 129(3):387–398

Boot WR, Blakely DP, Simons DJ (2011) Do action video games improve perception and cognition? Front Psychol 2:226

Borella E, Carretti B, De Beni R (2008) Working memory and inhibition across the adult life-span. Acta Psychol 128(1):33–44

Borella E, Carretti B, Riboldi F, De Beni R (2010) Working memory training in older adults: evidence of transfer and maintenance effects. Psychol Aging 25(4):767

Bransford JD, Brown AL, Cocking RR (2000) How people learn. National Academy Press, Washington, DC

Cahn-Weiner DA, Malloy PF, Boyle PA, Marran M, Salloway S (2000) Prediction of functional status from neuropsychological tests in community-dwelling elderly individuals. Clin Neuropsychol 14(2):187–195

Carlson MC, Fried LP, Xue QL, Bandeen-Roche K, Zeger SL, Brandt J (1999) Association between executive attention and physical functional performance in community-dwelling older women. J Gerontol B Psychol Sci Soc Sci 54(5):S262–S270

Carretti B, Borella E, De Beni R (2007) Does strategic memory training improve the working memory performance of younger and older adults? Exp Psychol 54(4):311–320

Cassilhas RC, Viana VA, Grassmann V, Santos RT, Santos RF, Tufik S, Mello MT (2007) The impact of resistance exercise on the cognitive function of the elderly. Med Sci Sports Exerc 39(8):1401–1407

Cavallini E, Dunlosky J, Bottiroli S, Hertzog C, Vecchi T (2010) Promoting transfer in memory training for older adults. Aging Clin Exp Res 22(4):314–323

Chang Y-K, Pan C-Y, Chen F-T, Tsai C-L, Huang C-C (2012) Effect of resistance-exercise training on cognitive function in healthy older adults: a review. J Aging Phys Act 20(4):497–517

Colcombe S, Kramer AF (2003) Fitness effects on the cognitive function of older adults a meta-analytic study. Psychol Sci 14(2):125–130

Davis CL, Tomporowski PD, McDowell JE, Austin BP, Miller PH, Yanasak NE, Allison JD, Naglieri JA (2011) Exercise improves executive function and achievement and alters brain activation in overweight children: a randomized, controlled trial. Health Psychol 30(1):91–98

Edwards JD, Wadley VG, Myers RS, Roenker DL, Cissell GM, Ball KK (2002) Transfer of a speed of processing intervention to near and far cognitive functions. Gerontology 48(5):329–340

Edwards JD, Wadley V, Vance D, Wood K, Roenker D, Ball K (2005) The impact of speed of processing training on cognitive and everyday performance. Aging Ment Health 9(3):262–271

Green CS, Bavelier D (2003) Action video game modifies visual selective attention. Nature 423(6939):534–537

Grigsby J, Kaye K, Baxter J, Shetterly SM, Hamman RF (1998) Executive cognitive abilities and functional status among community-dwelling older persons in the San Luis Valley health and aging study. J Am Geriatr Soc 46(5):590–596

Hampstead BM, Gillis MM, Stringer AY (2014) Cognitive rehabilitation of memory for mild cognitive impairment: a methodological review and model for future research. J Int Neuropsychol Soc 20(2):135–151

Karbach J, Kray J (2009) How useful is executive control training? Age differences in near and far transfer of task-switching training. Dev Sci 12(6):978–990

Kean RJ, Lamport DJ, Dodd GF, Freeman JE, Williams CM, Ellis JA, Butler LT, Spencer JP (2015) Chronic consumption of flavanone-rich orange juice is associated with cognitive benefits: an 8-week, randomized, double-blind, placebo-controlled trial in healthy older adults. Am J Clin Nutr 101(3):506–514

Klingberg T (2010) Training and plasticity of working memory. Trends Cogn Sci 14(7):317–324

Lamport DJ, Saunders C, Butler LT, Spencer JP (2014) Fruits, vegetables, 100% juices, and cognitive function. Nutr Rev 72(12):774–789

Lee Y, Kim JH, Lee KJ, Han G, Kim JL (2005) Association of cognitive status with functional limitation and disability in older adults. Aging Clin Exp Res 17(1):20–28

Lovden M, Backman L, Lindenberger U, Schaefer S, Schmiedek F (2010) A theoretical framework for the study of adult cognitive plasticity. Psychol Bull 136(4):659–676
Lustig C, Shah P, Seidler R, Reuter-Lorenz PA (2009) Aging, training, and the brain: a review and future directions. Neuropsychol Rev 19(4):504–522
Macready AL, Kennedy OB, Ellis JA, Williams CM, Spencer JP, Butler LT (2009) Flavonoids and cognitive function: a review of human randomized controlled trial studies and recommendations for future studies. Genes Nutr 4(4):227–242
Mahncke HW, Connor BB, Appelman J, Ahsanuddin ON, Hardy JL, Wood RA, Joyce NM, Boniske T, Atkins SM, Merzenich MM (2006) Memory enhancement in healthy older adults using a brain plasticity-based training program: a randomized, controlled study. Proc Natl Acad Sci U S A 103(33):12523–12528
McDougall S, House B (2012) Brain training in older adults: evidence of transfer to memory span performance and pseudo-Matthew effects. Aging Neuropsychol Cognit 19(1–2):195–221
Miller DJ, Robertson DP (2010) Using a games console in the primary classroom: effects of 'brain Training' programme on computation and self-esteem. Br J Educ Technol 41(2):242–255
Nilsson LG (2003) Memory function in normal aging. Acta Neurol Scand Suppl 179:7–13
Nouchi R, Kawashima R (2014) Improving cognitive function from children to old age: a systematic review of recent smart ageing intervention studies. Adv Neurosci 2014:Article ID 235479
Nouchi R, Taki Y, Takeuchi H, Hashizume H, Akitsuki Y, Shigemune Y, Sekiguchi A, Kotozaki Y, Tsukiura T, Yomogida Y (2012) Brain training game improves executive functions and processing speed in the elderly: a randomized controlled trial. PLoS One 7(1):e29676
Nouchi R, Taki Y, Takeuchi H, Hashizume H, Nozawa T, Kambara T, Sekiguchi A, Miyauchi CM, Kotozaki Y, Nouchi H, Kawashima R (2013) Brain training game boosts executive functions, working memory and processing speed in the young adults: a randomized controlled trial. PLoS One 8(2):e55518
Nouchi R, Taki Y, Takeuchi H, Sekiguchi A, Hashizume H, Nozawa T, Nouchi H, Kawashima R (2014) Four weeks of combination exercise training improved executive functions, episodic memory, and processing speed in healthy elderly people: evidence from a randomized controlled trial. Age 36(2):787–799
Owsley C, McGwin G Jr (2004) Association between visual attention and mobility in older adults. J Am Geriatr Soc 52(11):1901–1906
Park DC, Lautenschlager G, Hedden T, Davidson NS, Smith AD, Smith PK (2002) Models of visuospatial and verbal memory across the adult life span. Psychol Aging 17(2):299–320
Peig-Chiello P, Perrig WJ, Ehrsam R, Staehelin HB, Krings F (1998) The effects of resistance training on well-being and memory in elderly volunteers. Age Ageing 27(4):469–475
Pollitt E, Mathews R (1998) Breakfast and cognition: an integrative summary. Am J Clin Nutr 67(4):804S–813S
Rendeiro C, Rhodes JS, Spencer JP (2015) The mechanisms of action of flavonoids in the brain: direct versus indirect effects. Neurochem Int 89:126–139
Richmond LL, Morrison AB, Chein JM, Olson IR (2011) Working memory training and transfer in older adults. Psychol Aging 26(4):813
Salthouse TA (2003) Memory aging from 18 to 80. Alzheimer Dis Assoc Disord 17(3):162–167
Smith A (1998) Breakfast consumption and intelligence in elderly persons. Psychol Rep 82(2):424–426
Smith PJ, Blumenthal JA, Hoffman BM, Cooper H, Strauman TA, Welsh-Bohmer K, Browndyke JN, Sherwood A (2010) Aerobic exercise and neurocognitive performance: a meta-analytic review of randomized controlled trials. Psychosom Med 72(3):239–252
Takeuchi H, Taki Y, Kawashima R (2010) Effects of working memory training on cognitive functions and neural systems. Rev Neurosci 21(6):427–449
Takeuchi H, Taki Y, Hashizume H, Sassa Y, Nagase T, Nouchi R, Kawashima R (2011) Effects of training of processing speed on neural systems. J Neurosci 31(34):12139–12148

Van Uffelen JG, Paw MJCA, Hopman-Rock M, van Mechelen W (2008) The effects of exercise on cognition in older adults with and without cognitive decline: a systematic review. Clin J Sport Med 18(6):486–500

Verhaeghen P, Marcoen A, Goossens L (1992) Improving memory performance in the aged through mnemonic training: a meta-analytic study. Psychol Aging 7(2):242

Willis SL, Tennstedt SL, Marsiske M, Ball K, Elias J, Koepke KM, Morris JN, Rebok GW, Unverzagt FW, Stoddard AM (2006) Long-term effects of cognitive training on everyday functional outcomes in older adults. JAMA 296(23):2805–2814

Zelinski EM (2009) Far transfer in cognitive training of older adults. Restor Neurol Neurosci 27(5):455–471

Chapter 16
Cultural Psychology as a Form of Memory Research

Masataka Nakayama, Yoshiyuki Ueda, Pamela Marie Taylor, Hitoshi Tominaga, and Yukiko Uchida

Abstract Cultural psychology investigates the mutual construction of culture and the mind, in which culture creates the mind and the mind creates culture. With the goal of opening a dialogue between memory researchers and cultural psychologists, this chapter argues that cultural psychology is a type of memory research that investigates how culturally acquired knowledge and cognitive strategies (such as independent/interdependent self-construal; holistic/analytical thinking style in cognition) is reiterated and maintained by internalization as automatic long-term memory processes affecting human behavior and psychological processing. We provide an overview of the main findings and theories of cultural psychology and illustrate how cultural psychologists have used theoretical frameworks and methodological paradigms from the field of memory research to explain these findings. Finally, we discuss areas for future research, such as the role of memory systems in the emergence, maintenance, and dynamic change of cultural systems.

Keywords Memory • Culture • Cognition • Attention • Self-construal • Cognitive style

16.1 Introduction

Memory is not a record, but a re-constructive process of meaning-making. This was proposed Frederic Bartlett in his classic book *Remembering* (Bartlett 1932, 1995), in which he laid out a constructivist theory of human memory that emphasized the dynamic role of socio-cultural factors on memory. His work has inspired various independent lines of research in the field of memory, not only regarding its cognitive aspects (Botvinick 2005), but also its relevance to socio-cultural phenomena

M. Nakayama • Y. Ueda • Y. Uchida (✉)
Kokoro Research Center, Kyoto University, Sakyo-ku, Kyoto, Japan
e-mail: uchida.yukiko.6m@kyoto-u.ac.jp

P.M. Taylor • H. Tominaga
Graduate School of Human and Environmental Studies, Kyoto University, Kyoto, Japan

T. Tsukiura, S. Umeda (eds.), *Memory in a Social Context*,
https://doi.org/10.1007/978-4-431-56591-8_16

(Kashima 2000), such as the influence of socio-cultural knowledge on the mind and behavior. In this chapter we aim to show how cultural psychology is also a form of memory research. In doing so, we hope to facilitate a dialogue that will open up new areas of investigation for both fields.

16.1.1 The Constructivist View of Memory

Similar to cultural psychology, memory research has asserted that human cognitive processes are dynamic and constructive (Baddeley et al. 1998; McClelland et al. 2010; Patterson et al. 2007). Cognition and behavior are affected not only by immediate input, but also by past experiences accumulated in long-term memory (LTM). Identical stimuli do not produce identical responses, but are influenced by LTM, which affects how they are processed. The results of such stimulus processing accumulate in LTM, which then influences the processing of subsequent stimuli in a dynamic but reiterative cycle.

For example, after hearing an English sentence, English speakers can easily reproduce it but non-English speakers cannot. Despite identical input, the memory (i.e., sentence retention) functions differently depending on LTM. English speakers automatically intuit each word's meaning (i.e., lexical/semantic knowledge), their proper order (i.e., syntactic knowledge), and how the mouth should move to pronounce the heard sequence of sounds (i.e., procedural memory). Repeated exposure to English thus contributes to the processing of the language via LTM. Research has indicated that phonological knowledge supports verbal working memory functionality (Nakayama et al. 2015), such that learned patterns utilized in a working memory task produce errors consistent with those patterns (Botvinick and Bylsma 2005). Similar processes can be observed in the nonverbal domain, such as tool use (Bozeat et al. 2002).

16.1.2 The Constructivist View of Memory in the Context of Cultural Psychology

Similarly, in focusing on the influence that socio-cultural factors have on the mind and behavior, in cultural psychology a constructivist stance on human mind and behavior is taken. Culture can be defined as a meaningful system of symbols, such as customs, institutional systems and norms (Markus and Kitayama 2010). A given community with a history of interaction via such shared symbolic systems can be treated as a unit of "culture." The most common unit of culture analyzed by cultural psychology is the "nation" (e.g., Japan), but smaller, more localized communities (e.g., a district or an organization) and larger composite groups (e.g., "Asia") can also be treated as cultural units.

In cultural psychology, it is asserted that culture and the mind "make each other up" (Shweder 1995) through the construction and transmission of meaning systems

(Bruner 1994) and how this process works is investigated by comparing cultures to detect which cultural functions are associated with certain psychological tendencies. Owing to the inter-relatedness of the various aspects of cultural knowledge, it may be difficult to clearly categorize it via the taxonomy of memory (e.g., semantic memory, procedural memory). However, in many key aspects, cultural psychology is a form of memory research that investigates "cultural LTM" and its influence on the mind and behavior (Wang and Ross 2007). Importantly, not only do cultural environments shape human mind and behavior, but these learned patterns in turn create and maintain the cultural environment in accordance with how they are represented in our LTM.

16.1.3 What Can Memory Researchers Learn from Cultural Psychology?

Traditionally, memory researchers have emphasized universal learning principles and mechanisms over the complex and context-dependent information we learn and utilize in real-life situations (hence their use of controlled experimental materials in the laboratory, such as "nonsense syllables," which separate meaning from the mechanisms of memory; Nakayama et al. 2015). In contrast, in cultural psychology it has been investigated how our mind and behavior vary as a consequence of adaptation to the cultural context, challenging such universality (Kitayama and Cohen 2007) and attempts have been made to find a mechanism of cultural learning (Kitayama et al. 2014). In cultural psychology, patterns in and constructed theories about culture-specific targets of learning (e.g., differential attention to visual narratives; Senzaki et al. 2016), motivational factors of learning (e.g., differential effects of failure/success feedback on motivation; Heine et al. 2001), and the consequences of learning have been found. Thus, theories and findings from cultural psychology provide an opportunity for memory researchers to reevaluate the ecological validity of their research. Considering such culturally mediated memory processes would ultimately help to identify which aspects of memory are universal.

Despite recent attention to "collective memory" (Roediger and Abel 2015; Stone et al. 2015), memory researchers have largely focused on internal processes within the individual and omitted the collective aspect of memory from their research. However, issues such how the socio-cultural environment affects what memories are shared, how they are shared, and why they are shared have been investigated in cultural psychology. The dynamic co-construction of culture and mind is achieved not only through formal and informal education practices (Senzaki et al. 2016), but also via value systems implicitly embedded in, affirmed, shared, and learned through cultural products, such as advertisements and books (Morling and Lamoreaux 2008). Thus, cultural psychology helps memory researchers to ask and answer how and why collective memory (which is central to culture) arises from individual memory processes.

16.1.4 What Can Cultural Psychologists Learn from Memory Research?

The nature of memory systems within the individual is one key factor in the co-constructive process of mind and culture. Through memory, we acquire and internalize cultural systems and normative schemata that help us to engage in behaviors that are efficacious, given the surrounding context (Kitayama and Uskul 2011). Thus, memory systems facilitate individual adaptation to socio-cultural environments. At the individual level, memory systems have limitations that bias learning, in which previous knowledge affects new information to make it consistent with an individual's existing cultural knowledge (e.g., misremembering, such as correspondence bias). Adaptation through memory systems also works at the collective level. The accumulation of such biases at the collective level constructively reinforces beliefs currently shared within the culture (Kashima 2000; Martin et al. 2014). Thus, meaning search conducted by memory system helps to create and shape central aspects of culture, such as collective memory, historical memory, and autobiographical life scripts (for a review, see Boyer and Wertsch 2009).

Owing to the importance of memory in the co-constitution of mind and culture, theoretical and experimental paradigms from memory research are useful to cultural psychologists. Of particular utility is the distinction between two different memory systems: the flexible and controlled system of working memory, and the habitual and automatic system of LTM (Baddeley 2012; Norman and Shallice 1986). The working memory system retains and controls information in accordance with the current context, whereas the LTM maintains knowledge and habits proven to be heuristically adaptive by experience. However, these two distinct systems operate together, allowing us to flexibly accommodate our current knowledge to be optimally effective in the prevailing situation. For example, although we automatically know that it is both incorrect and inefficient to open a door by breaking it down (LTM function), given certain contexts, we can flexibly adapt this knowledge to effectively handle the current situation (working memory function), such that if we are in a burning building, we realize that breaking down the door might be optimal.

This distinction between flexible working memory and automatic LTM is useful for cultural psychology because the locus of cultural difference is likely to be in the habitual LTM system. Wagar and Cohen (2003) distinguished between the online "working self" that flexibly adapts to the immediate socio-cultural environment, and the "long-term self-concept" that emerges from self-understanding in LTM. The working self, which was measured by looking at reaction times for the online judgment of personal traits (e.g., "Does this word describe you?"), showed no cultural differences between European Canadians and Asian Canadians. However, cultural differences were found in the long-term self-concept (which is called a "self-schema") with European Canadians showing more elaborate personal selves and Asian Canadians showing more elaborate collective self-representations, as measured by reaction time in recognition tasks involving incidentally learned traits.

16.2 Theories of Cultural Differences

16.2.1 Independent and Interdependent Self-Construal

To date, most cultural psychological research has made nation-level comparisons, from which several theoretical paradigms have emerged. Markus and Kitayama (1991) proposed the concept of the cultural construal of self, with independent selves emerging from the US cultural context and interdependent selves from the Japanese cultural context (Markus and Kitayama 1991).

In the independent model of the self, which is prevalent in North America, people define themselves as the locus of agency and action is understood as originating in the autonomous force of internal traits, motivations, and beliefs (Kitayama and Uchida 2005). That is, each individual is seen as a disjoint agent (Markus et al. 2006) acting toward his or her own goals. Independent selves tend to encode behavior (both their own and that of others) in terms of the dispositions and emotions of the actor (Miller 1984; Miyamoto and Kitayama 2002).

In contrast, in the interdependent model of the self, which is prevalent in East Asia, the self is understood in terms of an individual's relationships with others, and each individual is seen as a conjoint agent acting in attunement with the goals and desires of those around them. Within this model, others are seen as an important source of the self's action and underlying motivations. For example, Japanese are less likely to explain the other person's behavior in terms of dispositional factors and more likely to explain it in terms of situational factors (Masuda and Kitayama 2004; Miyamoto and Kitayama 2002).

To see how these two cultures view the self in similar circumstances, Markus et al. (2006) analyzed Japanese and US media coverage of the Olympic Games (Markus et al. 2006). They found that US media coverage framed American gold medalists as the result of the individual effort and talent of the athlete. However, Japanese coverage showed medalists as resulting not solely from the athlete's effort or talent, but also from the collaborative effort of them with those who offered practical and social support, such as coaches and family members.

16.2.2 Holistic and Analytical Thinking Styles

Adaptive cognitive processing of the environment is necessary for survival. As such, by interacting with our environment, we develop styles of thought that effectively derive meaning from the sensory and social input, which in turn allows us to usefully navigate it. This leads to the culturally mediated construction of certain styles of cognition. This affects not only what information is more accessible in our memory when we make sense of the world, but also what kinds of higher level inferences and world views we derive from it.

Indeed, evidence suggests that cognitive processing might differ across cultures. For example, people from independent cultures, such as European Americans, are more likely to use a style of cognitive processing that is "analytical" (Masuda and Nisbett 2001; Nisbett 2003; Nisbett et al. 2001). Analytical styles of cognition preferentially attend to the focal objects of a scene over the background, thus detaching salient objects from their surrounding context. Analytical thinking also tends to organize the world via taxonomic and rule-based categorization (Norenzayan et al. 2002). Conversely, people from interdependent cultures, such as East Asians, engage in a more "holistic" style of thought (Masuda and Nisbett 2001; Nisbett 2003; Nisbett et al. 2001). Holistic processing attends to both background and focal information, processing them together to understand the relationships between objects and field in which they are embedded. Consequently, East Asians are more dependent on the field in visual processing than Westerners (Kitayama et al. 2003; Witkin and Berry 1975). Holistic processing also categorizes objects based on their relational qualities, such as family resemblance (Norenzayan et al. 2002).

Although no individual-level correlation between interdependence–independence and holistic–analytical thinking styles has been found (Na et al. 2010), evidence for macro-level (nation-level) correlations does exist (Kitayama et al. 2009; Uskul et al. 2008), indicating that people from East Asian cultures are more likely to be interdependent and exhibit a holistic thinking style, whereas European Americans are more likely to be independent and exhibit an analytical thinking style.

This macro-level correlation may be due to the different social tasks that are vital in each culture. Interdependent cultures call for more group-based decision-making, which fosters holistic thinking by requiring the integration of broad social cues. However, independent cultures require autonomy and individual decision making, thus fostering an analytical thinking style by locating agency in individual thought and reasoning (Kuhnen et al. 2001).

Uskul et al. (2008) suggested that current styles of cognition might have originated in the traditional economic activities of a culture, with traditionally agriculture-based cultures showing a more holistic style and traditionally herding-based cultures, showing an analytical style (Uskul et al. 2008). As farming requires harmonious group collaboration (especially rice farming; Talhelm et al. 2014) and adjustment to the environment (farming binds people to the land, resulting in low levels of social/residential mobility that require adjustment to the social and physical environments), it is adaptive for people in farming cultures to spread their attention across the contextual environment. By contrast, herding activities are largely solitary and require a more migratory lifestyle. Thus, herding does not require cooperation within groups, but does require individual decision making to survive in the unstable conditions of high social/residential mobility. These cognitive styles emerged to adapt to pre-modern ecological pressures, yet they persist in contemporary culture, although few people participate in agricultural or herding activities (Talhelm et al. 2014). Therefore, even in the absence of immediate ecological pressures, the collective memory processes may be maintaining and reproducing the culture.

Another area where cultural differences may arise because of the habitual and automatic processes of LTM is in how we allocate our attention. Interdependent selves attend to social information more holistically, whereas independent selves attend to it more analytically. In a study with eye-tracer, Masuda et al. (2008) showed that when looking at a target person in a group, Japanese looked at the facial emotions of both the target and of the surrounding figures, whereas Americans focused on the emotional expression of the target figure (Masuda et al. 2008). Further, when evaluating the focal target's emotion, Japanese judgments were influenced by background facial expressions, but those of Americans were not (see Sect. 16.4 for detailed explanations on culture and attention).

Holistic and analytical thinking styles are also found in causal inference and reasoning. When explaining facts, East Asians are more likely to assume that the underlying causes are complex and various (e.g., that situational contingencies or background history have effects) than are European Americans (Markus et al. 2006). Experimental research on causal attribution has shown that East Asians consider situational factors to be as important as personal factors (e.g., attitudes, opinions), whereas European Americans consider personal factors to be more important than situational factors (Miller 1984).

East Asian thinking also tends to harmoniously integrate factors that may seem distinct or contradictory to Westerners. For example, Japanese people believe that happiness and unhappiness must coincide (Uchida and Kitayama 2009), and even describe their ideal level of happiness as one that includes a degree of unhappiness (Uchida et al. 2015). Such "yin and yang" concepts arose from the philosophical ideas of Confucianism, Taoism, and Buddhism, which emphasize a holistic or dialectical world order in which everything is assumed to be interconnected (Kitayama and Markus 1999; Peng and Nisbett 1999). Looking at such lay beliefs in dialectical ambivalence, Ji et al. (2001) suggested that relative to Westerners, East Asians assume that change is inevitable, which leads to nonlinear prediction-making (e.g., assumptions that current trends will reverse).

16.3 Cultural Differences in Memory

16.3.1 Memory and Self-Construal

How can theories from cultural psychology help us to expand our understanding of the functions of human memory? As long-term memories of an individual's interactions with the socio-cultural environment are key in shaping self-construals, one area of memory research that has received significant attention from cultural psychologists is autobiographical memory, or memory of the self (for a review, see Ross and Wang 2010).

Indeed, as the contents of autobiographical memory assist in understanding how culture affects lived experience, cultural psychology often uses research methods that probe differences in autobiographical memory, such as situation sampling and diary methods. For example, Japanese consistently report lower levels of subjective well-being compared with North Americans (Diener and Diener 1995; Inglehart et al. 2008). Oishi (2002) examined the origin of this difference by comparing online evaluations (i.e., experience sampling or daily diary) of positive emotion (e.g., life satisfaction) with subsequent retrospective evaluations of these same experiences made 1 week later. Although there were no cultural differences in online reports, differences did emerge in retrospective accounts that relied on the memory of the experiences. Retrospectively, Americans rated their subjective well-being as having been higher than the Japanese subjects did. However, this difference was not due to differences in explicit memories, but rather differences in how much weight positive and negative events were implicitly given in overall evaluations of the past.

Using such memory-based methodology, researchers have identified how self-construal can affect memory. Americans report their first memories as occurring earlier than did the Chinese (Wang 2001). However, this tendency may be mediated by self-construal, as priming of the private (i.e., independent) self lead Asian Americans to report earlier first memories than collective self primes (Wang and Ross 2005). European Americans' memories also include more individual actions and self descriptions based on personal traits than those of Asian participants, which described more collective actions and self-descriptions based on social roles (Wang 2001; Wang and Ross 2005). European American memories also showed greater self-focus (e.g., writing about personal needs and desires) and mentioned the self (versus others) more frequently than did Asians (Wang 2001; Wang and Ross 2005). Similarly, in written reports of "self-defining memories" (i.e., events that shaped who they are) Australians and Asians preferentially elaborated recollections that supported predominance of the self in their culture, with Australians providing more elaborate autonomous personal memories, and Asians more elaborate relational personal memories (Jobson and O'Kearney 2008).

Cultural psychologists have thus used memory paradigms to help separate automatic from controlled processes, revealing that self-construal is internalized as habitual and automatic memory processes. For example, cultural differences in recognition after incidental learning show that reference effects can be mediated or attenuated by culture or cultural priming. In Western (i.e., independent) cultural contexts, items encoded in reference to the self tend to be recognized more quickly and/or more accurately. However, in East Asian (i.e., interdependent) contexts, similar or even stronger reference effects for significant others (e.g., the mother for Chinese) are observed (Sui et al. 2007; Wagar and Cohen 2003). Uchida et al. (2014) used the retrieval-induced forgetting paradigm, where the retrieval of a subset of items leads to the forgetting of learned items that are categorically related to the retrieved items. They demonstrated that not only items encoded in association with the self, but also items encoded in association with a close other (e.g., best friend) were resistant to retrieval-induced forgetting in Japanese participants. This runs in contrast to Western studies reporting that information encoded in reference to close others is vulnerable to retrieval-induced forgetting, but not information encoded as relevant to the self (Macrae and Roseveare 2002).

Cross-cultural differences in self-construal are also reflected in the self-perspective through which events are recalled. East Asian participants are more likely to recall episodes in which they were the focus of attention from the third-person perspective than episodes when they were not the center of attention. However, Western participants showed similar levels of third-person perspective in both types of episodes (Cohen and Gunz 2002; Martin and Jones 2012). Such differences in perspective are also influenced by online cognitive processes. He et al. (2014) indicated a positive correlation between collectivism and how much attention in a visual search was guided by task-irrelevant information on what is in the working memory of *their partner*, indicating that they automatically took the perspective of their partner.

16.3.2 Memory and Cognitive Style

Cultural differences in cognitive style are also reflected in memory. The holistic style of cognition typical of East Asians may make them more likely to segment events more coarsely, leading events to be remembered with less specificity. Wang (2009) demonstrated that in a daily diary, Asians reported fewer events at the end of the day than European Americans, and recalled fewer of these events after both 1 and 2 weeks (Wang 2009). They also remembered fewer episodes in fictional stories than European Americans. This pattern does not reflect general memory ability, but a holistic style of encoding that breaks the story into a smaller number of events (Wang 2009).

Memory paradigms have also allowed cultural psychologists to examine cultural differences in memory errors that reflect automatic processes, thus showing that an individual's cognitive style is internalized. The mnemonic context effect is a phenomenon where people's judgments of stimuli qualia (e.g., size) are influenced by previously presented stimuli (i.e., information accumulated in memory). Using this paradigm, Duffy and Kitayama (2007) showed that this context effect is stronger for Japanese than for Americans, suggesting that Japanese might (automatically) integrate preceding stimuli/events into judgments of the current stimulus. Schwartz et al. (2014) tested errors in the paired association of categorically related or unrelated words, comparing Americans with Turks (whose culture is more similar to Asian culture). Similar to the findings of Duffy and Kitayama (2007), the false memories of Turks consisted of more intrusions from other lists than those of Americans. In contrast, the false memories of Americans consisted of more categorical errors (i.e., incorrect items from the same category as the prompt or the target), reflecting a more analytical style by which judgments are implicitly based on categories.

16.4 Cultural Differences in Attention

Culturally mediated cognitive styles affect not only higher-level cognitive processes, but also lower-level perceptual processes involving visual attention. Studies demonstrating cultural differences in attention assume that these differences are consistent with cultural differences in thinking and reasoning (Chua et al. 2005; Ji et al. 2000; Kitayama et al. 2003; Masuda and Nisbett 2001).

In a narrative recall of a visual animation, Japanese participants recalled more background and relational information than did American participants. In another incidental learning experiment, Japanese recognition of focal objects was more impaired by background changes than it was for Americans, suggesting that Americans used a focused/analytical style of attention that encoded the focal object independent of the background, whereas Japanese used a diffuse/holistic style that integrated them during encoding (Masuda and Nisbett 2001). Using a similar procedure, Chua et al. (2005) showed that this difference was the result of differences in eye movement during encoding, indicating that East Asians and Westerners spontaneously focus on different areas when encoding visual information, although these results should be interpreted with caution because another study failed to replicate this finding (Evans et al. 2009).

Kitayama et al. (2003) showed that there are cultural differences in how attention is engaged during encoding and reproduction, even in the absence of socio-semantic information. In their "framed line test" study, a vertical line within a square frame was presented to participants, who were given a second blank square frame that was larger, smaller, or the same size as the first. Their task was to remember the length of the line in the first frame and reproduce it in the second frame. In the absolute reproduction condition, participants were asked to draw a line that was absolutely the same length as the line in the first frame. In the relative condition, they were asked to draw a line whose length was in equal proportion to the frame as the line in the first square was to its frame. American participants were more accurate in the absolute task, whereas Japanese participants were more accurate in the relative task. This indicates that judgments of length were based on different information, with Americans focusing on salient focal information (the line itself) and Japanese focusing on contextual and relational information (the proportional length of the line to frame).

To prevent thinking, reasoning, and knowledge-based strategizing (i.e., more controlled processes) from interfering in perceptual (i.e., automatic) task performance, Ueda et al. (2017) employed a speeded task of visual attention that largely excluded the effects of conscious reasoning (Rauschenberger and Chu 2006; Saiki et al. 2013; Shen and Reingold 2001). North American and Japanese participants searched for either a long line among short lines or a short line among long lines. Western participants exhibited "search asymmetry," where searches for long lines among short ones were faster than vice versa. However, East Asian participants showed no search asymmetry, suggesting that cultural differences in attention might be found even in perceptual tasks that do not require conscious thinking or reasoning.

Importantly, Ueda et al. (2017) showed that other kinds of stimuli (a circle versus a circle with an intersecting line, and a vertical line versus a tilted line) resulted in a more nuanced pattern of cultural difference. Similar to the line length search task, North Americans showed greater search asymmetry than Japanese in the circle/circle-with-line searches; however, in the vertical/tilted line searches, they showed a smaller search asymmetry than Japanese. As these tasks were identical except for the stimuli shape, this suggests that cultural differences in attention are not due to simple strategic factors such as focused/analytical versus diffuse/holistic processing strategy. Rather, visual attention and stimulus encoding may be affected by cultural visual environments, such as orthographical systems. For example, Japanese writing systems are sensitive to line length (e.g., 土 versus 士), intersection points (e.g., さ versus き), and the presence of small distinguishing elements (e.g., 大 versus 犬 versus 太), whereas the characters of the Western alphabet are sensitive to orientation (e.g., "u" versus "v"). These critical features lead to equivalent stimulus activation and smaller differences in grouping activities (a pooling of stimuli) in one target among distractors and vice versa (Treisman and Gormican 1988), thus resulting in weak search asymmetry.

Research using tasks other than visual searches also supports the argument that environmental factors – such as natural scenes encountered in daily life – can incite culture-specific adaptive tunings of visual attention (Miyamoto et al. 2006; Ueda and Komiya 2012). Some studies indicate that attention processing may slowly adapt after immigration into a new culture or environment (Kitayama et al. 2003). Such proposals are supported by findings from brain imaging and computational modeling studies. Familiarity with the Western alphabet can lead to increased activation in primary visual area V1, which relates to the processing of line orientation, whereas familiarity with Chinese characters leads to increased activation in extrastriate cortical areas such as V3 and V4, which relates to local grouping and the correlation of simple linear filters to perceive intersection points and the co-occurrence of multiple features (Freeman et al. 2013; Rensink and Enns 1995; Szwed et al. 2014). Thus, cultural differences in attention can be observed even after minimizing non-perceptual factors such as thinking, reasoning, and conscious strategy selection based on preexisting knowledge, and may be closely related to the processing of everyday environments in more automatic memory systems.

16.5 Future Directions and Conclusion

We have shown how memory and culture are inextricably intertwined. The human mind and behavior emerge from the interaction of the cultural LTM (in which experiences of the socio-cultural environment are accumulated) and the more controlled memory systems, such as working memory. This review has shown how theories and methodology from memory research help to empirically explain the theories of cultural psychology by illustrating cultural differences in habitual automatic processes (i.e., cultural LTM). At the same time, these findings provide an opportunity for memory researchers to reconsider the universality of observed phenomena and theories based on them (e.g., retrieval-induced forgetting).

This review has also raised the question of the role that memory systems play in the mind's construction of culture. The distinction between controlled versus habitual systems may provide an insight into this process. Hopefully, by integrating an awareness of the influence culture has on memory into their research, memory researchers may help to uncover how memories create and maintain cultural patterns, and conversely, how culture affects the structure and processes that make up memory.

References

Baddeley AD (2012) Working memory: theories, models, and controversies. Annu Rev Psychol 63:1–29

Baddeley AD, Gathercole S, Papagno C (1998) The phonological loop as a language learning device. Psychol Rev 105(1):158–173

Bartlett FC (1932) Remembering: a study in experimental and social psychology, 1st edn. Cambridge University Press, Cambridge

Bartlett FC (1995) Remembering: a study in experimental and social psychology, 2nd edn. Cambridge University Press, Cambridge

Botvinick MM (2005) Effects of domain-specific knowledge on memory for serial order. Cognition 97(2):135–151

Botvinick MM, Bylsma LM (2005) Regularization in short-term memory for serial order. J Exp Psychol Learn Mem Cogn 31(2):351–358

Boyer P, Wertsch JV (2009) Memory in mind and culture. Cambridge University Press, Cambridge

Bozeat S, Lambon Ralph MA, Patterson K, Hodges JR (2002) When objects lose their meaning: what happens to their use? Cogn Affect Behav Neurosci 2(3):236–251

Bruner J (1994) The "remembered" self. In: Neisser U, Fivush R (eds) The remembering self: construction and accuracy in the self-narrative. Cambridge University Press, Cambridge, pp 41–54

Chua HF, Boland JE, Nisbett RE (2005) Cultural variation in eye movements during scene perception. Proc Natl Acad Sci U S A 102(35):12629–12633

Cohen D, Gunz A (2002) As seen by the other…: perspectives on the self in the memories and emotional perceptions of easterners and westerners. Psychol Sci 13(1):55–59

Diener E, Diener C (1995) The wealth of nations revisited: income and quality of life. Soc Indic Res 36(3):275–286

Duffy S, Kitayama S (2007) Mnemonic context effect in two cultures: attention to memory representations? Cogn Sci 31(6):1009–1020

Evans K, Rotello CM, Li XS, Rayner K (2009) Scene perception and memory revealed by eye movements and receiver-operating characteristic analyses: does a cultural difference truly exist? Q J Exp Psychol 62(2):276–285

Freeman J, Ziemba CM, Heeger DJ, Simoncelli EP, Movshon A (2013) A functional and perceptual signature of the second visual area in primates. Nat Neurosci 16(7):974–981

He X, Sebanz N, Sui J, Humphreys GW (2014) Individualism-collectivism and interpersonal memory guidance of attention. J Exp Soc Psychol 54:102–114

Heine SJ, Kitayama S, Lehman DR, Takata T, Ide E, Leung C, Matsumoto H (2001) Divergent consequences of success and failure in Japan and North America: an investigation of self-improving motivations and malleable selves. J Pers Soc Psychol 81(4):599–615

Inglehart R, Foa R, Peterson C, Welzel C (2008) Development, freedom, and rising happiness: a global perspective (1981–2007). Perspect Psychol Sci 3(4):264–285

Ji LJ, Peng KP, Nisbett RE (2000) Culture, control, and perception of relationships in the environment. J Pers Soc Psychol 78(5):943–955

Ji LJ, Nisbett RE, Su YJ (2001) Culture, change, and prediction. Psychol Sci 12(6):450–456

Jobson L, O'Kearney R (2008) Cultural differences in retrieval of self-defining memories. J Cross-Cult Psychol 39(1):75–80

Kashima Y (2000) Maintaining cultural stereotypes in the serial reproduction of narratives. Personal Soc Psychol Bull 26(5):594–604

Kitayama S, Cohen D (2007) Handbook of cultural psychology. Guilford Press, New York

Kitayama S, Markus HR (1999) Yin and yang of the Japanese self. In: Cervone D, Shoda Y (eds) The coherence of personality: social-cognitive bases of consistency, variability, and organization. Guilford Press, New York, pp 242–302

Kitayama S, Uchida Y (2005) Interdependent agency: an alternative system for action. In: Sorrentino RM, Cohen D, Olson JM, Zanna MP (eds) Cultural and social behavior: the Ontario symposium. Lawrence Erlbaum Associates, Mahwah, pp 137–164

Kitayama S, Uskul AK (2011) Culture, mind, and the brain: current evidence and future directions. Annu Rev Psychol 62:419–449

Kitayama S, Duffy S, Kawamura T, Larsen JT (2003) Perceiving an object and its context in different cultures: a cultural look at new look. Psychol Sci 14(3):201–206

Kitayama S, Park H, Sevincer AT, Karasawa M, Uskul AK (2009) A cultural task analysis of implicit independence: comparing north America, western Europe, and east Asia. J Pers Soc Psychol 97(2):236–255

Kitayama S, King A, Yoon C, Tompson S, Huff S, Liberzon I (2014) The dopamine D4 receptor gene (DRD4) moderates cultural difference in independent versus interdependent social orientation. Psychol Sci 25(6):1169–1177

Kuhnen U, Hannover B, Roeder U, Shah AA, Schubert B, Upmeyer A, Zakaria S (2001) Cross-cultural variations in identifying embedded figures: comparisons from the United States, Germany, Russia, and Malaysia. J Cross-Cult Psychol 32(3):365–371

Macrae CN, Roseveare TA (2002) I was always on my mind: the self and temporary forgetting. Psychon Bull Rev 9(3):611–614

Markus HR, Kitayama S (1991) Culture and the self: implications for cognition, emotion, and motivation. Psychol Rev 98(2):224–253

Markus HR, Kitayama S (2010) Cultures and selves: a cycle of mutual constitution. Perspect Psychol Sci 5(4):420–430

Markus HR, Uchida Y, Omoregie H, Townsend SSM, Kitayama S (2006) Going for the gold: models of agency in Japanese and American contexts. Psychol Sci 17(2):103–112

Martin M, Jones GV (2012) Individualism and the field viewpoint: cultural influences on memory perspective. Conscious Cogn 21(3):1498–1503

Martin D, Hutchison J, Slessor G, Urquhart J, Cunningham SJ, Smith K (2014) The spontaneous formation of stereotypes via cumulative cultural evolution. Psychol Sci 25(9):1777–1786

Masuda T, Kitayama S (2004) Perceiver-induced constraint and attitude attribution in Japan and the US: a case for the cultural dependence of the correspondence bias. J Exp Soc Psychol 40(3):409–416

Masuda T, Nisbett RE (2001) Attending holistically versus analytically: comparing the context sensitivity of Japanese and Americans. J Pers Soc Psychol 81(5):922–934

Masuda T, Ellsworth PC, Mesquita B, Leu J, Tanida S, De Veerdonk EV (2008) Placing the face in context: cultural differences in the perception of facial emotion. J Pers Soc Psychol 94(3):365–381

McClelland JL, Botvinick MM, Noelle DC, Plaut DC, Rogers TT, Seidenberg MS, Smith LB (2010) Letting structure emerge: connectionist and dynamical systems approaches to cognition. Trends Cogn Sci 14(8):348–356

Miller JG (1984) Culture and the development of everyday social explanation. J Pers Soc Psychol 46(5):961–978

Miyamoto Y, Kitayama S (2002) Cultural variation in correspondence bias: the critical role of attitude diagnosticity of socially constrained behavior. J Pers Soc Psychol 83(5):1239–1248
Miyamoto Y, Nisbett RE, Masuda T (2006) Culture and the physical environment – holistic versus analytic perceptual affordances. Psychol Sci 17(2):113–119
Morling B, Lamoreaux M (2008) Measuring culture outside the head: a meta-analysis of individualism-collectivism in cultural products. Personal Soc Psychol Rev 12(3):199–221
Na J, Grossmann I, Varnum MEW, Kitayama S, Gonzalez R, Nisbett RE (2010) Cultural differences are not always reducible to individual differences. Proc Natl Acad Sci U S A 107(14):6192–6197
Nakayama M, Tanida Y, Saito S (2015) Long-term phonological knowledge supports serial ordering in working memory. J Exp Psychol Learn Mem Cogn 41(5):1570–1578
Nisbett RE (2003) The geography of thought: how Asians and westerners think differently… And why. Free Press, New York
Nisbett RE, Peng KP, Choi I, Norenzayan A (2001) Culture and systems of thought: holistic versus analytic cognition. Psychol Rev 108(2):291–310
Norenzayan A, Smith EE, Kim BJ, Nisbett RE (2002) Cultural preferences for formal versus intuitive reasoning. Cogn Sci 26(5):653–684
Norman D, Shallice T (1986) Attention to action. In: Davidson RJ, Schwartz GE, Shapiro D (eds) Consciousness and self regulation. Springer US, Boston, pp 1–18
Oishi S (2002) The experiencing and remembering of well-being: a cross-cultural analysis. Personal Soc Psychol Bull 28(10):1398–1406
Patterson K, Nestor PJ, Rogers TT (2007) Where do you know what you know? The representation of semantic knowledge in the human brain. Nat Rev Neurosci 8(12):976–987
Peng KP, Nisbett RE (1999) Culture, dialectics, and reasoning about contradiction. Am Psychol 54(9):741–754
Rauschenberger R, Chu HQ (2006) The effects of stimulus rotation and familiarity in visual search. Percept Psychophys 68(5):770–775
Rensink RA, Enns JT (1995) Preemption effects in visual-search: evidence for low-level grouping. Psychol Rev 102(1):101–130
Roediger III HL, Abel M (2015) Collective memory: a new arena of cognitive study. Trends Cogn Sci 19(7):359–361
Ross M, Wang Q (2010) Why we remember and what we remember: culture and autobiographical memory. Perspect Psychol Sci 5(4):401–409
Saiki J, Ueda Y, Chen R, Kitayama S (2013) Cultural differences in visual search with culturally neutral items. J Vis 13(9):1254–1254
Schwartz AJ, Boduroglu A, Gutchess AH (2014) Cross-cultural differences in categorical memory errors. Cogn Sci 38(5):997–1007
Senzaki S, Masuda T, Takada A, Okada H (2016) The communication of culturally dominant modes of attention from parents to children: a comparison of Canadian and Japanese parent-child conversations during a joint scene description task. PLoS One 11(1):e0147199
Shen J, Reingold EM (2001) Visual search asymmetry: the influence of stimulus familiarity and low-level features. Percept Psychophys 63(3):464–475
Shweder RA (1995) The confessions of a methodological individualist. Cult Psychol 1(1):115–122
Stone CB, Luminet O, Takahashi M (2015) Remembering public, political events: a cross-cultural and – sectional examination of Australian and Japanese public memories. Appl Cogn Psychol 29(2):280–290
Sui J, Zhu Y, Chiu CY (2007) Bicultural mind, self-construal, and self- and mother-reference effects: consequences of cultural priming on recognition memory. J Exp Soc Psychol 43(5):818–824
Szwed M, Qiao E, Jobert A, Dehaene S, Cohen L (2014) Effects of literacy in early visual and occipitotemporal areas of Chinese and French readers. J Cogn Neurosci 26(3):459–475
Talhelm T, Zhang X, Oishi S, Shimin C, Duan D, Lan X, Kitayama S (2014) Large-scale psychological differences within China explained by rice versus wheat agriculture. Science 344(6184):603–608

Treisman A, Gormican S (1988) Feature analysis in early vision: evidence from search asymmetries. Psychol Rev 95(1):15–48

Uchida Y, Kitayama S (2009) Happiness and unhappiness in east and west: themes and variations. Emotion 9(4):441–456

Uchida Y, Ueno T, Miyamoto Y (2014) You were always on my mind: the importance of "significant others" in the attenuation of retrieval-induced forgetting in Japan. Jpn Psychol Res 56(3):263–274

Uchida Y, Ogihara Y, Fukushima S (2015) Cultural construal of wellbeing: theories and empirical evidence. In: Glatzer W, Camfield L, Moller V, Rojas M (eds) Global handbook of quality of life: exploration of well-being of nations and continents. Springer Netherlands, Heidelberg, pp 823–837

Ueda Y, Komiya A (2012) Cultural adaptation of visual attention: calibration of the oculomotor control system in accordance with cultural scenes. PLoS One 7(11):e50282

Ueda Y, Chen L, Kopecky J, Cramer ES, Rensink RA, Meyer DE, Kitayama S, Saiki J (2017) Cultural differences in visual search for geometric figures. Cogn Sci, Advance online publication.

Uskul AK, Kitayama S, Nisbett RE (2008) Ecocultural basis of cognition: farmers and fishermen are more holistic than herders. Proc Natl Acad Sci U S A 105(25):8552–8556

Wagar BM, Cohen D (2003) Culture, memory, and the self: an analysis of the personal and collective self in long-term memory. J Exp Soc Psychol 39(5):468–475

Wang Q (2001) Culture effects on adults' earliest childhood recollection and self-description: implications for the relation between memory and the self. J Pers Soc Psychol 81(2):220–233

Wang Q (2009) Are Asians forgetful? perception, retention, and recall in episodic remembering. Cognition 111(1):123–131

Wang Q, Ross M (2005) What we remember and what we tell: the effects of culture and self-priming on memory representations and narratives. Memory 13(6):594–606

Wang Q, Ross M (2007) Culture and memory. In: Kitayama S, Cohen D (eds) The handbook of cultural Psychology. Guilford Press, New York, pp 645–667

Witkin HA, Berry JW (1975) Psychological differentiation in cross-cultural perspective. J Cross-Cult Psychol 6(1):4–87

Chapter 17
Memory Practice in Society: Eyewitness Memory in Children and Forensic Interviews

Makiko Naka

Abstract Child witness memory is the focus of this chapter. We first described the dispositions of child memory for experienced events, and then introduced forensic/investigative on which method was designed to elicit as accurate information as possible on trying to reduce the child's stress at being interviewed. The topic includes the effects of training in forensic interviews on the quantity and quality of information obtained from interviewees; effects of training on professionals' perception of information to collect and convey in the forensic interviews; and evaluation of child witness testimony obtained through closed questions and open-ended questions. Finally, we discussed the need for a multidisciplinary team approach to make the most of forensic interviews.

Keywords Child witness memory • Forensic/investigative interviews • Training • Effects and evaluation of forensic/investigative interviews

17.1 Child Witness Memory

In this chapter on memory practice in society, the focus is on child witness memory. It has been shown repeatedly that children have poor memories. Younger children learn fewer things than older children, and also remember less than older children. They are more likely to be misled than older children, and when being misled, they are less likely to be aware of being misled (Lamb et al. 2015; Naka 2016a for review). Such constraints may be linked to brain development. Shing et al. (2010) suggested that the strategic components of memory functions, namely, the cognitive

Makiko Naka: We are grateful for the support from Grant in Aid for Scientific Research in Innovative Areas "Law and Human Sciences", JST/RISTEX "Creating a Safe and Secure Living Environment in the Changing Public and Private Spheres", and the author's previous affiliation, Hokkaido University, Graduate School of Letters, where many of the studies were conducted.

M. Naka (✉)
Ritsumeikan University, Ibaraki, Osaka, Japan
e-mail: m-naka@fc.ritsumei.ac.jp

T. Tsukiura, S. Umeda (eds.), *Memory in a Social Context*,
https://doi.org/10.1007/978-4-431-56591-8_17

control operations to encode, maintain, and retrieve memories related to the frontal cortex grow slower than the automatic components that are mediated by the medial temporal region. Schacter et al. (1995) argued that because of the underdevelopment of the pre-frontal lobe, children may present symptoms that are similar to those who suffer brain impairment such as distortion of memory, distortion of orders, and suggestibility. However, the latter is not always the case. Although absolute quantity and quality of memory may be inferior in children, their memories can be restored better in some ways than others. For instance, in one experiment we conducted, second and fifth grade children were presented with a video clip and then asked to remember what they had seen in the video in one of four conditions: free recall, free recall with context reinstatement (i.e., closing their eyes for 1 min so that they can remember and then engage in free recall), questioning, or an interview with open-ended questions followed by when, where, who, etc. (WH) questions and closed questions, if necessary. Results showed that both older and younger children produced a greater amount of accurate information in interviews than in the other conditions, even though the older children generally produced more information than the younger ones. The worst condition was the questioning, where both the older and younger children provided the minimum amount of information necessary to answer the questions. No age differences were found in this condition. These results support the general findings that open-ended questions as opposed to closed questions elicited more accurate information (Lamb et al. 2008).

17.2 A Better Way of Eliciting Child Witness Memory

Conducting forensic or investigative interviews is the method employed to elicit as much accurate information as possible from alleged child victims and witnesses, while trying to reduce their psychological burden as much as possible. To minimize the chance of leading or contaminating memory, interviewers were encouraged to use open-ended questions to elicit free narratives from the interviewees (Ministry of Justice 2011; Lamb et al. 2007a, b). The *open-ended questions* that are known to be efficient were as follows (Lamb et al. 2007a, 2008; Naka 2011 for Japanese):

Invitation: Tell me everything that happened.
Time segmentation: Tell me what happened from time A until time B.
Cued recall: Tell me more about [what the interviewee said].
Follow-up: And then? What else?

Interviewers are discouraged to make use of *WH questions* (*directives*) until the last phase of the interview (Lamb et al. 2008), because such questions take the control of the conversation away from the interviewee. In other words, the interviewer takes control of the conversation by asking questions about what he/she wants to hear rather than let the interviewee talk freely. Furthermore, WH questions direct the interviewee's focus to the specific aspects of an event, that is, when, where, who, etc. This may discourage the interviewee from relating other aspects of an event.

Closed questions (*option posing*) or multiple-choice questions that are used to elicit only "yes" or "no," or "A" or "B" are used only when necessary. Furthermore, interviewers are discouraged from using leading or suggestive questions (e.g., You saw it, didn't you?), which may mislead the interviewee.

Forensic interviews are structured in a flexible manner so as to motivate the child to share more accurate information. Typically, an interviewer introduces herself/himself and explains the purpose of an interview. This is followed by ground rules such as "Tell me only the truth"; "If you do not understand my question, tell me you don't understand"; "If you don't know the answer, tell me you don't know"; and "If I made a mistake, tell me it was wrong." Then, to build a rapport, the interviewer asks the child what she/he likes to do; this is followed by episodic memory training where the child is asked to describe what happened from the time she/he woke up until the time she/he came to see the interviewer (e.g., Lamb et al. 2007a, b).

17.2.1 Benefits of Using Forensic Interviews

As described previously, forensic interviewing is a promising way of eliciting more accurate information from child victims and witnesses. However, it is not easy to conduct such an interview. It is important for professionals to learn the skills and become acquainted with them.

Besides conducting research on the basics of memory and communication, we developed a training program for forensic interviews (e.g., Naka 2014a, 2015). In the first training course for professionals in 2008, we invited social workers working at the Child Guidance Center of Hokkaido University. It was a 3-day course, comprising two sessions with an interval of 1 month. Once we received feedback from the trainees, we improved our program. In the meantime, the Japanese police agency and prosecutor's office started to video-record suspects, increasing their attendance at our training. By the end of 2016, we had trained approximately 4800 professionals.

There are several benefits of using forensic interviews. First, the interviewing method increases the quantity and quality of information elicited from interviewees. Second, forensic interviews may ensure that professionals are more focused on what information to collect to establish a case. Third, free narratives to open-ended questions have been evaluated as being more credible than the answers to closed questions. Allow me to describe some studies conducted in my laboratory.

17.2.2 Quantity and Quality of Information

In the study described above (Naka 2011), the participants were 32 social workers and psychologists who worked at the Child Guidance Center in Japan. They took part in a training course conducted by the author based on the National Institute of Child Health and Human Development (NICHD) protocol. Before attending the

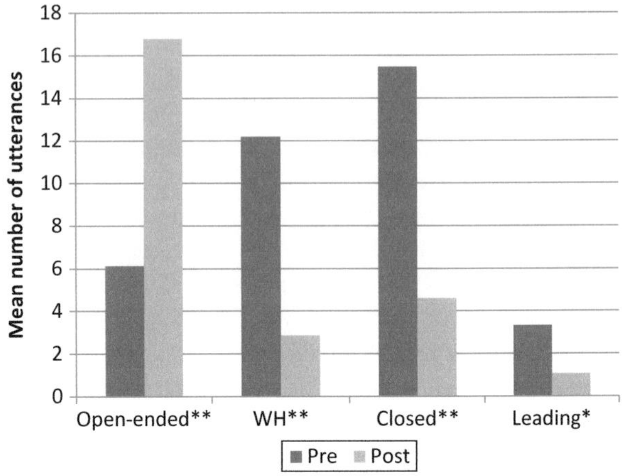

Fig. 17.1 Mean numbers of each question type *(top)* in pre- and post- training. *Indicates significant differences between pre- and post- training ($p < 0.05$); **Indicates significant differences between pre- and post-training ($p < 0.01$).

course, they received one of three videos via mail, each of which lasted approximately 1 min, and using which the participant conducted an interview using the video as a stimulus. That is, a volunteer (interviewee) watched the video without the interviewer (participant) waching it, and then the interviewer interviewed the interviewee on the content of the video. The procedure was also repeated after attending court (post). Different videos were used in the pre- and post-interviews. The video was counterbalanced between participants and pre- and post-interviews. The duration of the pre- and post-interviews were 999 s and 1070 s respectively; there was no significant difference between them.

In Fig. 17.1, the pre- and post-interview results are shown.[1] The number of open-ended questions increased, whereas the number of WH, closed, and suggestive questions decreased from pre- to post-interviews.

In Fig. 17.2, the amount of speech elicited by children in terms of letters[2] is depicted. More information was elicited by means of open-ended questions in the post-interview. Although there was no time difference between pre- and post- interviews as mentioned previously, there was a significant increase in the amount of information elicited the mean number of letters in the pre-interview was 570.53, whereas that of the post-interview was 732.25.

We also counted the accurate details elicited in the pre- and post-interviews. The tentative analysis showed that the amount of accurate information in the post-interview (29.09) was significantly greater than that in the pre-interview (24.06), whereas the amount of inaccurate information in the post-interview (2.09) was

[1] Figures 17.1 and 17.2 show the results of reanalysis of data from Naka (2011).

[2] We counted the number of letters instead of words. For instance, "私は男を見ました" (I saw a man) is counted as eight. The reason we used the number of letters is that it is not easy to count words in Japanese: we need to divide a sentence into words manually or by running a computer program (e.g., 私/は/男/を/見/ま/した). Because the number of letters can be considered to correlate with the number of details, we used the number of letters, which can be counted by Excel functions.

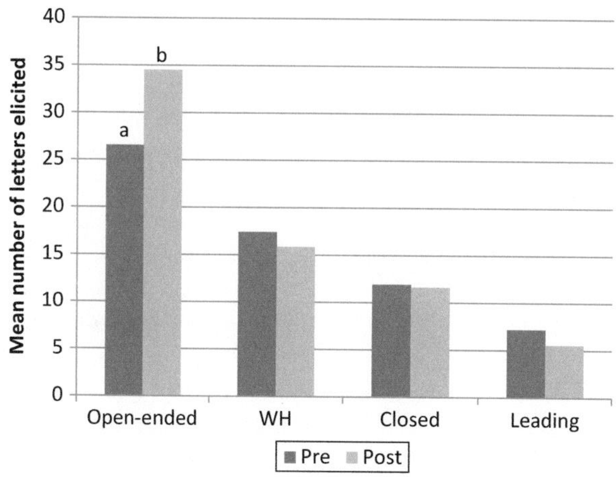

Fig. 17.2 Mean number of letters elicited by each question type (*bottom*) in pre- and post- training. (**a**) Before the training (pre), open-ended and WH questions elicited more information than leading questions ($p < 0.01$). WH questions elicited more information than closed questions ($p < 0.01$). Closed questions elicited more information than leading questions ($p < 0.01$). (**b**) After the training (post), an open-ended question elicited more information than WH- questions ($p < 0.05$), closed, and leading questions ($p < 0.05$). WH questions elicited more information than closed and leading questions ($p < 0.01$). Closed questions elicited more information than leading questions ($p < 0.01$).

significantly less than that in the pre-interview (3.03). Thus, information elicited by the forensic interview increased in quantity and quality.

17.2.3 *Information to Collect*

Second, forensic interviews can ensure that professionals focus on collecting more relevant information to establish the case (Naka 2014b). The purpose of forensic interviews is to collect information on a specific event, that is, it involves an individual's episodic memory. However, we noticed in the training and in the real interviews that the interviewers' attention could shift from episodic memory (i.e., what happened) to more general, script-like memory (i.e., what usually happens). The latter information may be important for social work, but priority should be given to collecting the episodic memory of an event.

Furthermore, we noticed that interviewers sometimes confused forensic interviews with counseling or therapy. In such a case, an interviewer may tell a child, "You are not wrong," and "I am sorry to ask details. You don't have to say anything if you don't want to." However, forensic interviews are not a form of counseling and therapy (Home Office 1992). The aim of forensic interviews is to elicit facts rather than soothe the interviewee's subjective feelings and emotions. If there is anything that an interviewer should convey to the child, it is the ground rules rather than empathy, compassion, comments, and opinions.

We conducted a study through a questionnaire to ascertain whether the training of forensic interviewers ensures that professionals are more focused on the facts rather than on semantic or script-like memory and subjective feelings. The participants were 39 clinical psychologists, 33 social workers, and 39 police officers and prosecutors. They were shown a mock case in which a child said that she was hit by her father. The professionals were asked to rate the importance of 15 facts that had to be collected, in addition to the importance of 12 facts that they had to convey. The 15 facts included information about the father (age, job, characteristics, intention, etc.), the event of hitting (the name of father, time, place, body part being hit, etc.), and routines and circumstances (last instance of hitting, one time or more than one time, routine of hitting, child's feelings, family information, needs of the child, etc.). The 12 facts included the instructions and ground rules ("Tell me the truth"; "If you don't understand, tell me you don't understand."; "If you don't know the answer, tell me you don't know"; "Correct me if I said something wrong"; "Tell me everything.," etc.) and the interviewer's opinions and compassion ("If you don't want to talk, you don't have to talk"; "You are not wrong"; and empathy, comments, decisions, promises, etc.)

Before and after the training, the participants completed the questionnaire. As shown in Fig. 17.3 (top), after the training, the participants became more focused on the information about the event rather than that about the father, his routines, and circumstances. In Fig. 17.3 (bottom), the information that had to be conveyed is shown. After the training, the participants rated the ground rules higher and the interviewer's opinion and compassion lower. Thus, forensic interviews seemed to assist professionals to focus on more relevant information to establish a case.

17.2.4 Evaluation of Testimony

As described previously, open-ended questions are more likely to elicit accurate information than closed questions (Lamb and Fauchier 2001; Lamb et al. 2007a, b; Naka 2012; Orbach and Lamb 2001). To determine whether lay people actually evaluated the information elicited by open-ended questions as being more credible, we conducted an experiment using a mock interview. The participants were 92 undergraduates (Naka 2013, 2017).

There were two versions of the mock interview. Although the information provided in both versions of the interview was the same, in one condition (open-ended question condition [OQ]), a child provided the information to open-ended questions. In the other condition (closed question condition [CQ]), the interviewer provided information by asking closed questions. This is shown in the following excerpts:

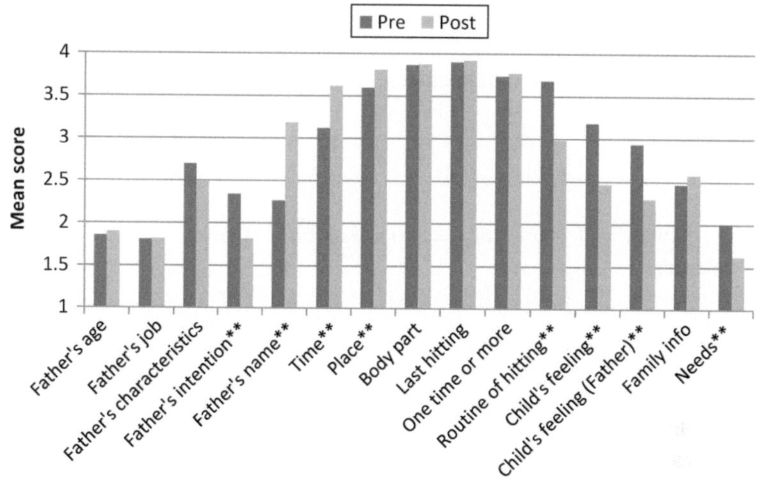

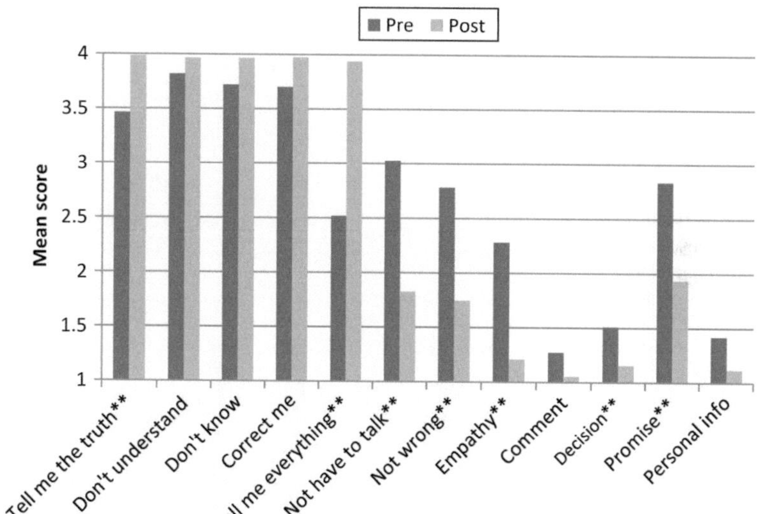

Fig. 17.3 Professionals' perspective on the information to be collected (*top*) and information to be conveyed (*bottom*). *F* denotes father, *C* denotes child. **Indicates the significant differences between pre- and post-training ($p < 0.01$)

OQ condition.

> Interviewer: What did you come here for?
> Child: To talk about Father.
> Interviewer: What about Father?
> Child: Father did Sachan (the name of victim) "Ei!" (kicking action) with foot.
> Interviewer: Where?
> Child: In the tummy.

Interviewer: What happened then?
Child: Sachan cried.
Interviewer: Then what happened?
Ei! Again.
…

CQ condition.

Interviewer: Did you come here to talk about Father?
Child: Yes.
Interviewer: Did Father do 'Ei!' to Sachan with foot?
Child: Yes.
Interviewer: In the tummy?
Child: Yeah.
Interviewer: Then did Sachan cry?
Child: Right.
Interviewer: Did he do 'Ei!' again?
Child: Yes.
…

After being presented with either of the interviews,[3] the participants were asked to decide whether or not the father was guilty or not, and determine the sentence.[4] Then, they were asked to rate the interview on a seven-point Likert scale. The items

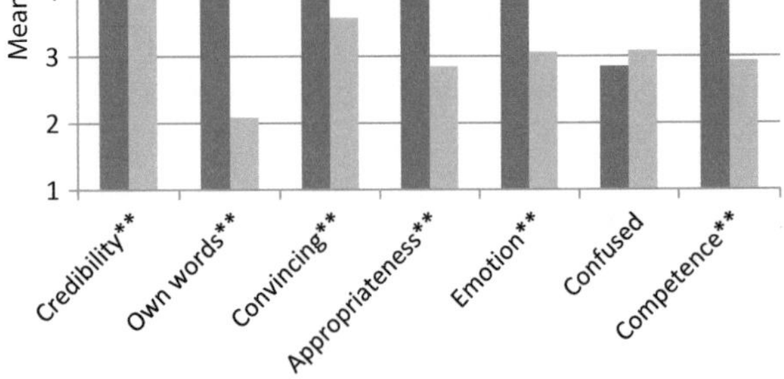

Fig. 17.4 Evaluation of child testimony in the open-ended questions (OQ) and closed questions (CQ) conditions. **Indicates a significant difference between open-ended and closed questions ($p < 0.01$)

[3] There were three conditions in the presentation: a closer perspective, a distant perspective, or no picture. Because the manipulation is beyond the scope of this article, we collapsed the data and described the results for the type of questions.

[4] In Japanese Saiban-in Seido (lay judge system), lay judges deliberate on the case with (a) professional judge(s) and make decisions regarding the verdict of guilty or not guilty, before deciding on the sentence.

included the child's credibility, whether the child used her own words, how convincing the testimony was, the appropriateness of the interview, the participant's emotive reaction on hearing the testimony, if the child's testimony was confusing, and the competence of the child.

The results showed that a guilty judgment was significantly higher for the OQ condition rather than for the CQ condition (0.91 vs 0.66), although there was no difference in the sentence. As shown in Fig. 17.4, the ratings for credibility, using own words, convincingness, appropriateness, emotion, and competence were significantly higher in the OQ condition than in the CQ condition. Lay participants seem to be aware of the difference between the information elicited by open-ended questions and that elicited by closed questions.

17.3 Necessity for Cooperation

It appears that forensic interviews are a promising way of eliciting and presenting memory in the judicial context. However, there remains the problem of repeated interviews. In Japan, child victims are typically interviewed in a medical center or at the Child Guidance Center where they are first brought in, and then by the police, followed by the prosecutor's office, and finally in court. It is said that a child needs to be interviewed 10-20 times in total; this could confuse the child's memory, and may cause psychological sensitization (Fulcher 2004). To prevent such problematic procedures, a multidisciplinary/multi-agency team approach is inevitable.

To grasp the professionals' perception of a multidisciplinary team (MDT), we conducted a survey in 2013 and 2014 in which we asked the professionals to write down the factors that they thought hindered the MDT approach ($n = 83$ and $n = 103$ respectively) (Naka 2016b). Responses were categorized using the KJ (Jiro Kawakita) method. Five major categories emerged: "the lack of system" (lack of law and system, lack of directions of organization, etc.); "differences in agencies" (differences in goals, positions, methods, etc.); "lack of knowledge and understanding" (lack of knowledge, understanding, awareness, skills, etc.); "practical reasons" (lack of time, people, and apparatus, etc.); and "no problems." The percentage of entries in each of the above categories were 0.29, 0.16, 0.19, 0.10, and 0.09 in 2013, and 0.43, 0.12, 0.40, 0.06, and 0.03 in 2014; this suggests that a lack of system is a major barrier to promoting collaboration.

In 2015, there was a breakthrough. On 28 October 2015 the Ministry of Health, Labor and Welfare (MHLW), the National Police Agency, and the Supreme Prosecutors Office issued papers to recommend a *cooperative interview* with a child victim or witness (Ministry of Health, Labor and Welfare 2015; National Police Agency 2015; Supreme Public Prosecutors Office 2015). The paper issued by police states:

> "In cases where interviews are conducted repeatedly, a child might undergo an excessive mental and physical burden. In addition, children's characteristics of being easily affected by leading or suggestions cause suspicion of the credibility of the statement.

> ...In order to consider an interviewing method that contributes to reducing the child's burden and to guarantee credibility, cooperation between the prosecutors and child guidance offices shall be strengthened."

This recommendation may lessen the disadvantage of a "lack of system" and foster collaboration.

17.4 Further Problems

Although three agencies pictured a future system for collaboration, specific methods of collaboration are left for individual prefectures and counties. According to the paper issued by MHLW (2013), the cooperative interview is a forensic interview conducted by an interviewer representing one of the three agencies, with professionals from the other two agencies sitting as monitors to support the interview in the monitor room (MHLW 2015). However, it does not specify who should serve as the interviewer.

Social workers may say that they want to conduct an interview because not all the cases reported to the Child Guidance Center result in criminal procedures. Police officers may want to be interviewers as well, because they are entitled to initiate a criminal investigation (Criminal Procedure Code, 189(2)). However, in Japanese criminal proceedings, prosecutors also interview victims, witnesses, and suspects to decide whether they should go to court. The statements made by prosecutors have more evidential power than those made by the police or other professionals (Criminal Procedure Code, 320(2)). Thus, a prosecutor may declare that he/she should be an interviewer. One solution is to conduct a forensic interview in a team that have been collaborating for a long period, and not just a group of people working together. If they work as a team to plan and conduct an interview, that is, the interviewer elicits information by free narratives and the others support the interview by checking what information was and was not obtained, and by helping the interviewer as it progresses, then who becomes an interviewer may not be a big issue.

17.5 Conclusion

Although we are still at the initial stages, memory research in the field and the laboratory seem to be one of key factors needed to realize a better system for dealing with human memory in the real world.

References

Fulcher G (2004) Litigation-induced trauma sensitisation (LITS): a potential negative outcome of the process of litigation. Psychiatry Psychol Law 11(1):79–86

Home Office (1992) Memorandum of good practice on video recorded interviews with child witnesses for criminal proceedings. The Stationery Office, London

Lamb ME, Fauchier A (2001) The effects of question type on self-contradictions by children in the course of forensic interviews. Appl Cogn Psychol 15(5):483–491

Lamb ME, Orbach Y, Hershkowitz I, Esplin PW, Horowitz D (2007a) A structured forensic interviewprotocol improves the quality and informativeness ofinvestigative interviews with children: a review of research using the NICHD Investigative Interview Protocol. Child Abuse Negl 31(11–12):1201–1231

Lamb ME, Orbach Y, Hershkowitz I, Horowitz D, Abbott CB (2007b) Does the type of prompt affect the accuracy of information provided by alleged victims of abuse in forensic interviews? Appl Cogn Psychol 21(9):1117–1130

Lamb EM, Hershkowitz I, Orbach Y, Esplin PW (2008) Tell me what happened: structured investigative interviews of child victims and witnesses. Wiley, Hoboken

Lamb ME, Malloy LC, Hershkowitz I, La Rooy D (2015) Children and the law. In: Lamb ME, Lerner RM (eds) Handbook of child psychology and developmental science: socioemotional processes. Wiley, Hoboken, pp 464–512

Ministry of Health, Labor, and Welfare(2013) Kodomo gyakutai taiou no tebiki (Heisei 25. 8 kaisei) [On the guideline of how to respond to child abuse. Revised in August, October 8th, 2013]. http://www.mhlw.go.jp/seisakunitsuite/bunya/kodomo/kodomo_kosodate/dv/dl/130823-01c.pdf

Ministry of Health, Labor, and Welfare (2015) Kodomo no shinriteki hutan tou ni hairyosita mensetsu no torikumi ni muketa keisatsu kensatsu tono saranaru renkei kyouka nit suite (Heisei 27, Oct 8th) [Strengthening further collaboration with police agency and prosecutor office to accomplish the interviews with children considering their psychological burden. Oct 8th, 2015] http://www.moj.go.jp/keiji1/keiji10_00006.html

Ministry of Justice, U.K (2011) Achieving best evidence in criminal proceedings: guidance on interviewing victims and witnesses, and guidance on using special measures https://www.cps.gov.uk/publications/docs/best_evidence_in_criminal_proceedings.pdf

Naka M (2011) The effect of forensic interview training based on the NICHD structured protocol. Jpn J Child Abuse Negl 13(3):316–325. (In Japanese)

Naka M (2012) The effect of different ways of interviewing on children's reports and subsequent memories of an eye-witnessed event. Jpn J Psychol 83(4):303–313. (In Japanese)

Naka M (2013) Lay judges' evaluation of a child interview: effects of how a child talks and camera perspectives. Paper presented at International Investigative Interviewing Research Group (iIIRG), Maastricht, the Netherlands, July 3–5, 2013

Naka M (2014a) A training program for investigative interviewing of children. In: Bull R (ed) Investigative interviewing. Springer, New York, pp 103–122

Naka M (2014b) Information to deal with in forensic interviews: practitioner's view before and after the training. Paper presented at international congress of applied psychology, Paris, France, 10 July 2014

Naka M (2015) Interviews with victims and witnesses of crime in Japan: research and practice. In: Walsh D, Oxburgh GE, Redlich AD, Myklebust T (eds) International developments and practices in investigative interviewing and interrogation, volume 1: victims and witnesses. Routledge, London, pp 43–57

Naka M(2016a) Where developmental psychology meets law: forensic interviews with alleged child victims and witnesses. In Japan Society for Developmental Psychology (ed) Frontiers in developmental psychology research: Japanese perspectives. Hitsuzi Shobo, Tokyo, pp 251–264

Naka M (2016b) The development of forensic interviews: towards the multi-disciplinary team approach. Hou to Shinri 16(1):24–30. (in Japanese)

Naka M (2017) Rokuon rokuga mensetsu ni okeru kodomo no kyojutsu: Shitsumon no shikata, camera perspective, senmonka shogen ga shinyosei handan ni oyobosu kouka. In Ueishi K, Otsuka H, Musashi K (eds) Gendai Nihon no Houkatei. Shinzansha, Tokyo, pp 345–368 [Videotaped child testimony: effects of question type, camera perspective, and expert testimony on the credibility judgment. In Ueishi K, Otsuka H, Musashi K (eds) Legal processes in modern Japan. Shinzansha, Tokyo, pp 345–368]

National Police Agency (2015) Jidou wo higaisha tou to suru jian heno taiou ni okeru kensatsu oyobi jidou soudansho tono saranaru renkei kyouka nit suite. (October 28, Heisei 27). [Strengthening further collaboration with Child Guidance Centers and Prosecutors Office in response to the cases of child victims (October 28, 2015)] http://www.moj.go.jp/keiji1/keiji10_00006.html

Orbach Y, Lamb ME (2001) The relationship between within-interview contradictions and eliciting interviewer utterances. Child Abuse Negl 25(3):323–333

Schacter DL, Kagan J, Leichtman MD (1995) True and false memories in children and adults: a cognitive neuroscience perspective. Psychol Public Policy Law 1(2):411–428

Shing YL, Werkle-Bergner M, Brehmer Y, Müller V, Li SC, Lindenberger U (2010) Episodic memory across the lifespan: the contributions of associative and strategic components. Neurosci Behav Rev 34(7):1080–1091

Supreme Public Prosecutors Office (2015) Keisatsu oyobi jidou soudansho tono saranaru renkei kyouka nit suite (Oct 28, Heisei 27) [Strengthening Further collaboration with Police Agency and Child Guidance Centers (October 28, 2015)]

Chapter 18
Social Ageing, Informatization, and Cognitive Tasks Required for Human Memory in Daily Lives: Problems Observed in a Usability Test for Home-Use Medical Equipment

Etsuko T. Harada

Abstract Various phenomena related to older people's lives are now being widely observed in Japan's super-aged society. It is important to know that those phenomena are not only related to cognitive ageing, but also to many circumstantial changes accompanying social ageing and social digital–informatization. After introducing some topics, concepts, or models to increase our understanding of these compound situations, we report an experiment using the usability testing method on home use medical equipment, as an example of the complex combining of those three factors.

Keywords Social ageing • (digital) informatization of society • Cognitive ageing • Usability test • Medical equipment for home use

18.1 Introduction: Human Memory and Society

Many people currently believe that brain activities inside the head carry out cognitive processes and/or create cognitive experiences. This belief makes people sometimes feel that everything is predetermined biologically or genetically; however, (of course) the truth is that brain activities, and brain structures in turn, are greatly affected by circumstances outside the head, which arise from the physical or social world. Some of these changes have their origins in nature, but many originate with human beings and their societies. The adaptations of the brain to the outside world may be realized through interactions of brain cognition and circumstances, and susceptibility to adaptation may be more salient with higher cognitive activities,

E.T. Harada (✉)
Division of Psychology, Faculty of Human Sciences, University of Tsukuba, Tsukuba, Japan
e-mail: etharada@human.tsukuba.ac.jp

T. Tsukiura, S. Umeda (eds.), *Memory in a Social Context*,
https://doi.org/10.1007/978-4-431-56591-8_18

because of their late development, and of their greater dependency on artifacts in the outside world, for example, symbol systems such as languages and tools such as paper and pencils or information gadgets.

As humans living in the twenty-first century, we are now facing two drastic changes in society that have created new circumstances for our brains to interact with: one is social ageing, which appears in the demographics of population change, and the other is social informatization or electrical digitization. How are these social changes affecting our cognitive activities?

18.2 Cognitive Ageing Is not the Only Source of Problems

Research into ageing and human memory has a long history. Healthy older adults have been investigated mainly as control groups in studies of memory and cognitive functioning in patients with various type of dementia or other age-related diseases, to identify pathological characteristics, mechanisms, and medical treatments or interventions for prevention. Research interest has recently shifted to focus on healthy ageing itself, and many useful research results have accumulated (for example, see changes in the 1992, 2000, and 2008 versions of Craik & Salthouse's handbook). Many of those studies, however, have treated targeted phenomena one-by-one in a microscopic way, and although they have produced convincing results in laboratory settings, they are inapplicable for consideration of human memory and social ageing as a whole.

In reality, social ageing is becoming more and more salient. Japan, for example, is now a super-aged society, in which more than 25% of the population are aged 65 + years. In such a situation, not only researchers in psychology laboratories, but also ordinary people in the community, are now realizing that we are facing social change and that social ageing has arrived. Along with this realization, the media is picking up on many social problems related to older adults that are attracting attention, e.g., traffic accidents caused by older adults, cases of fraud targeting the elderly, and even stalking or violent crimes committed by them.

These problems are frequently mentioned in relation to cognitive ageing, especially a decline in prefrontal functions. However, problems arise not only from such direct relationships to individual brain/cognitive ageing, but also from interactions with social problems of ageing. The living circumstances of a large proportion of aged people in society are changing drastically, and their brains/cognitive processes are making (or failing to make) changes in connection with these situational changes. One such change is the number of people living together: many older adults in Japan are now living alone, or only with their spouse, in so-called "old people's" homes. This is a strong cause of the increasing numbers of fraud victims, because phone calls are targeting older people in these homes. Increasing numbers of older adults as criminals in violent or stalking cases may also be reflecting such situations, because the circumstances of living alone can easily cause rumination, which can lead to criminal behavior (Ochi, personal communication).

Another example of the major impact of social changes is the relative capital diminution of the ageing society, because of both the decline in numbers of the working population and the increase in expenses for social services for the care of the elderly. This social impoverishment is one of the reasons why many older adults dwelling in the community should continue to drive automobiles, even when they know their driving skills have deteriorated. The decline has become a main cause of traffic accidents among older adults; however, without additional public transportation, older people living in the countryside have to drive themselves to hospitals or grocery stores. In general, social situations force older adults to confront cognitively demanding tasks by themselves with little social assistance to remedy their individual changes.

A similar situation is occurring in the area of health management. The growing ratio of older adults in the population requires not only larger budgets for the health/medical security system, but also greater numbers of beds in hospitals. One possible answer to those "low capital – high needs" problems is to shift from hospital-centered medical treatment to medical treatment at home (Ikai 2010). However, a "home-centered" medical strategy is a drastic change that is producing many serious, compound problems, especially in old people's homes. These problems are also connected to problems with social informatization: in an old people's home, is it possible to use medical-equipment safely and without creating anxiety and undue cognitive load? This problem is addressed in a later part of this chapter.

18.3 How Are Digitized Circumstances Forcing Humans to Adapt?

Before reporting our research into medical equipment for home use, we introduce the problems of informatization and human cognition in general. From an economic viewpoint, social informatization is frequently refered to as the Information Revolution, analogous to the Industrial Revolution. Like the Industrial Revolution, the Information Revolution is now having far-reaching effects on human lives. However, unlike the Industrial Revolution, social informatization has special direct effects inside the head, too, i.e., on our cognitive activities and experiences.

Why does informatization generate such dynamic effects on humans? There are two very closely connected reasons. First, informatization or digitization changes almost all aspects of social and economic activities; in other words, informatization has redefined all kinds of human social activities as "transformations of information." Thus, almost all aspects of human activities are now captured as operations on information and have been changed to a digitized mode. The other reason is that human brain activities are themselves aggregates of information operations or transformations, and requirements from the outside in the form of information processing are directly changing the way we carry out activities in our heads. Therefore, informatization has had such a great effect on human social activities that it is directly related to, and changing, what human brains are doing.

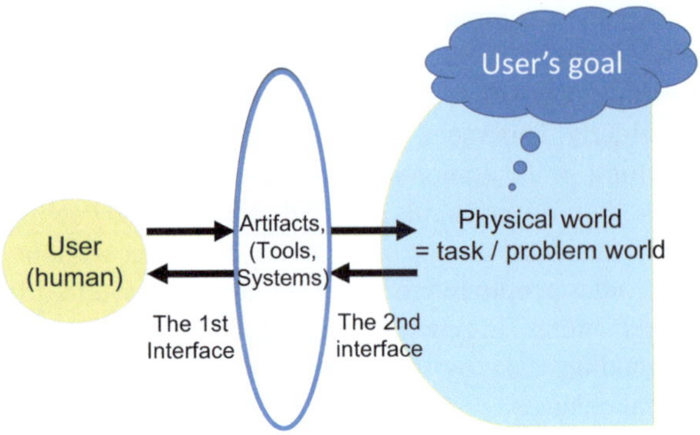

Fig. 18.1 Model of cognitive engineering: dual interface theory (Sayeki 1988)

18.3.1 Informatization and Human Cognition in General: A Model

If so, what is the difference between informatized, digital information processing and the "natural" or original style of human information processing that the brain offers? In other words, how can we describe those effects on human cognition?

A useful framework is provided by Sayeki's model, the dual interface theory of using artifacts (Sayeki 1988). This model is merely metaphorical, but it provides a good starting point to think about what digital informatization is and what effects it has (Fig. 18.1).

At baseline, it is presupposed that the principal aspect of human cognitive activities is problem-solving by interacting with physical worlds/problem spaces. Because electronic informatization transforms all information to electronic signals (on/off, or 1/0), which are invisible to human beings, any activities that result from digitized information require mediated artifacts (tools), which transform human input into electronic signals and transform status information about the world, represented by electronic signals, back into symbols that the human can see or hear and understand. In the model, these activities are represented as the first interface, between the humans (users) and artifacts/systems.

On the other hand, the purposes of human activities, such as the goal of problem-solving, are in the world behind the mediated artifacts/tools; thus, humans must also understand the second interface, the relationships between the artifacts and the physical world, i.e., "what will happen if I say/ask to do that?" and "what is happening if the world situation looks like that?" Compared with natural problem-solving activities, in which humans are trying to use their body and senses directly, in the digitized world with mediated artifacts, effects and results between human activities and physical world responses are indirect and strongly depend upon the relationships of the first and second interfaces. In addition, those relationships can be made quite arbitrary, because any connections at the first and second interfaces can be constructed or designed by someone else, e.g., the system designers. This fact forces human users to learn the relationships between the first and second interfaces and integrate the two, in order to use the artifacts to accomplish goals effectively.

This indirectness (Norman 1991), combined with the fact that goals themselves are sometimes events in invisible electronic fashion and are therefore difficult to understand (Can you explain what you are doing when you connect to public Wi-Fi from your mobile phone?), means that human problem-solving is now highly mediated and complicated, and that there are large amounts of information "to be learned" to accomplish the goal. It is necessary to use every artifact: in short, in this informatization age, human problem-solving requires translation to electronic information every time, and we must learn and understand what to do and how to do this any time we do something. These situations are placing a large burden on human cognition, especially on working memory.

18.3.2 Informatization and Human Cognition: Personal Identification and Memory Externalization

In addition to those general characteristics of activities within the electronic, digitized world, Harada (2008) identified problems of human memory with social informatization or digitization in specific areas: memory loads for the purposes of personal identification, externalization of personal memories, and highly concentrated and accelerated environments.

Personal identification load is a direct change in the digitized information society; people must now verify their identity for every social activity using their memory for passwords.[1] For safety, users are strongly urged to create unique and unguessable passwords and to change every password frequently (e.g., every 3 months); thus, demands for memorizing and using their own passwords are becoming more prominent in people's lives. For many people, those requirements are so difficult to handle that they try to decrease the demands by using the same simple password for every target for long periods, or by writing down the passwords on paper (and putting a yellow post-it on the system itself!), all of which make the world of personal and social information dangerous and fragile.

Recently, to make personal identification systems more useful (or to make them less error-prone), it has become more common to require users to enter personal, autobiographical semantic memory as answers to secret key questions. That is, as it became clear that systems should be more supportive of human memory and at lower cost, the system required more information for the sake of identification. Ironically, however, those secret keys result in additional cognitive loads. For example, I have the experience of being asked to input the name of the teacher of my kindergarten class. Unfortunately, at that time it was very difficult for me to recall her name, even though I myself had set up the secret key question at some earlier

[1] Although in the place of passwords, identification systems based on physical possession of keys or some means of biometric matching (e.g., fingerprints or retinal scans) have been proposed and used, each method has its own disadvantages, and the password method remains the main means of identification.

point, when I thought that recalling her name would be an easy task for me. Another problem, specific to the Japanese language, is that we use several different writing systems (*hiragana*, *katakana*, *kanji*) and alphabets and we must recall in which writing system we input the key answer; this is sometimes a significant memory problem.

In the present day, not only explicit or episodic memory for one's own passwords, but also more autobiographical or personal history data must be stored in massive ways and in a detailed manner, in the outer world of our heads. For information systems to work, they must gather huge amounts of information, bit by bit, every day with every action. Many of these are not erased until someone decides to erase them. In addition, with the recent realization of the value of the analysis of *Big Data*, or massive amounts of information, these data are now stored and reanalyzed independently of the users' intentions (but of course they agree to this on the consent form). I can see what I bought on an online store 3 years ago, and how many times I bought the same book (oops!). The same thing is happening, for example, with many paying systems (e.g., credit cards or e-tickets with IC cards for public transportation), so that facts about where I was and what I did are automatically being stored to a company's database. This trend is not restricted to personal economic activities, but is the same with all activities on the Internet, e.g., writing and reading on social networks or even just searching in an internet community. Data gathered by one company are "under control" of that company. However, data distributed on the Internet are almost uncontrollable for now; they cannot be erased, and we can never know where and by whom that information is being accessed. Are these circumstances good or bad for our cognitive system?

This world situation supports our memory system from outside of our head (I can see where I went last night from my iPhone's GPS information without using my own memory). In addition, we now rely on search engines even while chatting face-to-face: my students google while we are talking, and they can instantly show me related interesting facts or pictures. This is truly the time in which knowledge "outside our head" have come true. It is convenient, and with very low memory cost we can obtain a great deal of knowledge from mobile devices (and the Internet). However, if the Internet connection is poor or absent, or the mobile device has problems, we suddenly realize that we are standing alone and naked in the world, and it is sometimes not easy to remedy those problematic situations. With such a strong dependency, people in our society would now be unable to act efficiently without network systems, and any condition that interferes with the comfortable use of networks and information causes problems. It remains an open question whether or not such dependencies on the outer informational world is a great boon to human beings. However, because the status of information is not under each human's control, it sometimes causes people anxiety about their safety or stability, emotional harms, or impairments to human rights, all of which can become cognitive loads. One of the most extreme examples may be the struggle for the right to be forgotten (European Commission 2012).

18.3.3 Informatization and Human Cognition: Concentration and Acceleration

Because the main motivation for the introduction of informatized or digital systems in social activities was the pursuit of efficiency in business, the highly concentrated and accelerated environment may be thought of as an advantage from an economic viewpoint. However, it can at the same time be a stress and load on human cognition. Although those heavier loads are felt by many people and have been described in various commentaries from many perspectives, it is difficult to demonstrate those changes as psychological data using numbers; thus, I provide an example of a symbolic phenomenon, multitasking.

As mobile information gadgets are introduced into our daily lives, aspirations for higher speeds and more condensed activities are escalating. A new concept in cognitive psychology, the media multitasker index, has therefore been introduced (Ophir et al. 2009). This method measures how a person is using multiple media at the same time. Surprisingly, when looking into individual differences in this index, results showed that higher multiple media users executed less attentional control: in other words, heavy media multitaskers did not exhibit a greater ability for attentional control, but rather poorer ability. It implies that now, humans are required to use information gadgets with high cognitive controllability.

Although the effects of informatization on human memory are wide-spread and multifaceted, many of them converge on a single point: using information technology changes the human processing of information. From my own perspective, the importance of cognitive or attentional control by each human user is growing.

18.4 Characteristics of Older Adults' Behavior While Using ICT Tools: Two Facets of the Phenomenon

What happens when informatization and ageing come together? Many studies have pointed out that the performance of healthy older adults is almost the same as that of younger adults when their tasks and stimuli are simple, on the other hand, when tasks and stimuli become a more complex and require attentional processing, the their performance shows a huge decline compared with younger adults (Verhaeghen et al. 2003). Based on the situation of informatizational effects and ageing, there are many concerns about what happens to older adults who use ICT equipment.

These phenomena are readily observed in our society at present. Harada and Akatsu (2003) summarized seven characteristics of older adults' behaviors while using ICT equipment (see Harada et al. 2012), and then proposed a four-layered model of the factors of older adults required for those behaviors (Fig. 18.2, modified by Harada 2009). In fact, older adults have much more troubles and make more

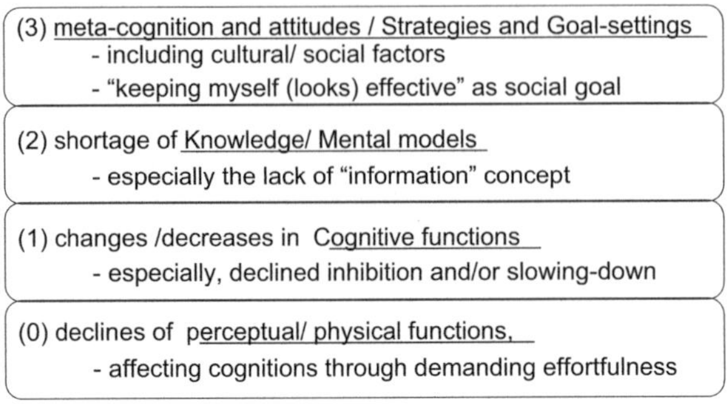

Fig. 18.2 Hypothetical sources of difficulties in using IT artifacts with cognitive ageing: a four-layered model (Harada 2009)

errors when interacting with equipment, and the causes are not simple, but are rather compounds of mechanisms on different layers.

Layer 0 in Fig. 18.2 shows declines in perception and/or physical abilities and reflects the effortfulness hypothesis (McCoy et al. 2005; Wingfield et al. 2005); namely, that older adults faced with such problems require more effort and energy to regulate perception and their physical selves. This results in fewer resources remaining for higher cognitive processes, which in turn causes poor performance on cognitive tasks.

Layer 1 holds problems arising from the cognitive processes associated with ageing. Such ageing changes include a decrease in working memory capacity (Craik 1986), the slowing of cognitive processing speed (Salthouse 1991), and a decline in inhibition (Hasher and Zacks 1988). Effects of these cognitive changes have been directly observed in many interactions when older adults use ICT equipment, for example, failure to notice information changes, or the need for more time to understand displayed messages than younger adults (Hara et al. 2009).

Layer 2 describes difficulties coming from the potential lack of procedural knowledge (how to use those equipment) and/or declarative knowledge (general concepts underlying electrical information technology). Adults who are 70 or older in Japan are especially likely to have relatively poor mental models of information technology, perhaps because they have encountered the computational concept of *information* late in life or have had only limited opportunities to learn the concept, which are crucial for the use of such technology.

Layer 3 represents another causal features of older adults' deficits, which is related to higher cognition. Older adults often have metacognition about their cognitive changes and the areas in which they lack knowledge. Such metacognition, blended with social/ cultural cognition and motivation, results in changes in goal setting, attitudes, and strategies in older adults. For instance, older adults are likely to avoid using computer-like IT equipment or to rely heavily on instructions. These kinds of behaviors illustrate computer-stereotype threat (Harada 2009) as a variant of ageing-related stereotype threat (Chasteen et al. 2005).

This model focused on differences between older and younger adults in their interactions with ICT equipment. However, emphasizing those differences does not show important facts about the relationships between human cognition and the design of artifacts and tools. Actually, as Harada (2009, 2012) pointed out, there are interesting double facets of the problems older adults show in a usability testing laboratory. When older adults made errors, or had trouble at one stage in using a system, the very same problems, but on a much smaller scale, occurred with younger adults (e.g., university students). For example, when older adults made serious errors using one display design of an automatic teller machine (ATM), younger adults showed a micro-slip, moving their finger close to the incorrect button, but just before pushing, turning it and moving to the right one, or showed their hesitation in making responses, requiring a longer time to respond than usual. In short, the *universal design* principle is commonly observed in the usability test laboratory: *bad design is bad for all*. Harada and Hashimoto (2016) found that evaluation of signboard design was common to older and younger adults. Although there were complicated interactions between icon design and arrows in the quality of signboard design, ageing effects did not show any interactions with such factors. Taken together, these results imply that if someone wants to identify design problems for users, it will be much easier to detect with testing older adults compared with younger adults. In other words, older adults are more prone to respond to problematic designs, which in turn means that we can create better designs based on data from older adults (Fig. 18.3).

Why do these interesting combinations occur? That is, why do older adults have much more trouble interacting with information technology-based artifacts, but at the same time, the causes of problems are common to younger adults, who are observed to use these artifacts easily? As a hypothesis, we suggest that only older adults appear to have trouble because of changes in learning efficacy with ageing. As the universal design principle says, design problems are common and affect younger adults too; however, they can solve and overcome those problems by themselves, sometimes using their own knowledge ("those buttons should have different shapes") and sometimes through error experiences. Indeed, when younger adults

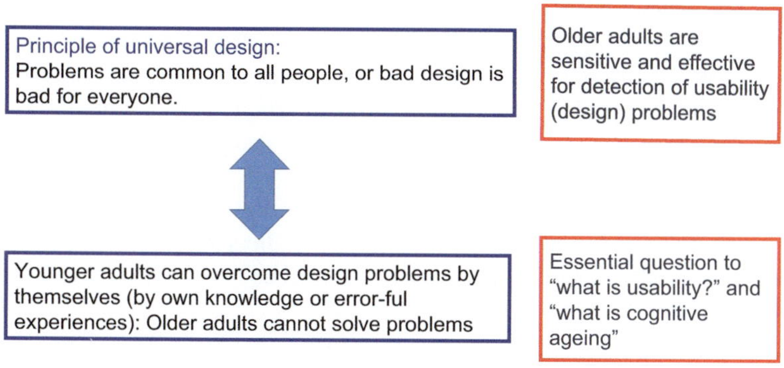

Fig. 18.3 Dual aspects of usability and ageing

make an error and return to the previous stage, they make a decision about the next candidate by saying "maybe this one, because I experienced that that one was incorrect". In contrast, older adults frequently make repetitive error responses, saying that they do not know why this is not the correct choice (Harada et al. 2012). Here, we can see the cognitive ageing phenomenon at layer 1, changes in errorful/errorless learning with ageing (Anderson and Craik 2006). This is just one example, however; these kinds of processes in learning to overcome design problems may exist in ageing differences, and the necessity of learning the relation between the first and second interfaces to use an informatized system may emphasize those ageing differences strongly. In other words, differences between older and younger adults trying to use a new, difficult tool may be useful resources, reflecting the essence of which tools are usable for humans and how ageing affects real-world activities.

18.5 Combined Effects of Social Ageing and Informatized Society: Home-Use Medical Equipment

Here, we report research into the usability testing of home-use medical equipment, as an example of cognitive effects coming from a combination of individual ageing and social changes in the informatized society, and social ageing as a whole.

Usability testing is the most basic method for observing a typical user as a participant performs a typical task in a laboratory setting, which is set up to resemble as closely as possible real circumstances of usage. To trace a participant's problem-solving process while using the target system to discover their goals, the thinking aloud method is commonly used; participants are told to tell what they are thinking in a monologue, without concern about communicating or reporting to experimenters (Ericsson and Simon 1993; Kaiho and Harada 1993). To make the laboratory setting more like the real situation, a cover story and a realistic task sequence are prepared within a scenario.

In Japan, hospitals have been the center of medical treatment for a long time, but with the growing costs of social ageing, the government has recently proposed to shift the principles of home- or community-centered medical treatments. Originally, this was not meant to apply to complex medical treatment, but many forms of medical equipment are now being brought into the home from the hospital. With infusion pumps and oxygen inhalation therapy as typical examples, more complex systems are now moving to the home. We were interested in many aspects of this problem, but we are especially concerned with discovering whether there are specific usability problems with medical equipment and its interactions with older users, because many patients are older adults who are living by themselves or with an elderly spouse. Older adults should be able to use this medical equipment if this new trend in medicine is to work out.

18.5.1 A Cognitive Usability Test of Home-Use Medical Equipment: A NPPV-Type Respirator

To explore these problems, we executed a usability test of a piece of medical equipment for home use. The term "home use" means that normal people, not medical professionals, are expected to use it by their own efforts. Users of home-use medical equipment include the patient or their families or care-takers. In the context of Japan now, most of such users should be older adults. As we described above, older adults are poor at using new kinds of equipment, and at the same time, they frequently show timidity or avoidance in using such equipment by themselves (Tanaka & Harada 2016; Tanaka et al. 2017). However, because it is directly related to medical treatments, it may be very rare for patients or their families to refuse to use medical equipment; it must be almost compulsory to use. Thus, we think that usability testing is especially important. What kind of problems could happen when an older adult uses new medical equipment? To identify the ageing problem, we also recruited younger adult participants to compare two age groups in performance and other characteristics in interaction with the equipment.

As a target, we selected a respirator of the non-invasive positive pressure ventilation (NPPV) type for home use (Fig. 18.4), because this type of equipment is now increasingly used to treat chronic obstructive pulmonary disease (COPD), which is also increasing as the population ages. The mechanism of an NPPV-type respirator is not simple: it is used without any surgery, and the gap between the two levels of air pressure (high and low) to the mask (and in turn, to the lungs) helps patients to ventilate, with low (relatively negative) pressure to the lungs helping to expel CO_2. Although it looks like an oxygen inhaler, the mechanism is quite different, and completeness of sealing of the mask is important for adequate usage.

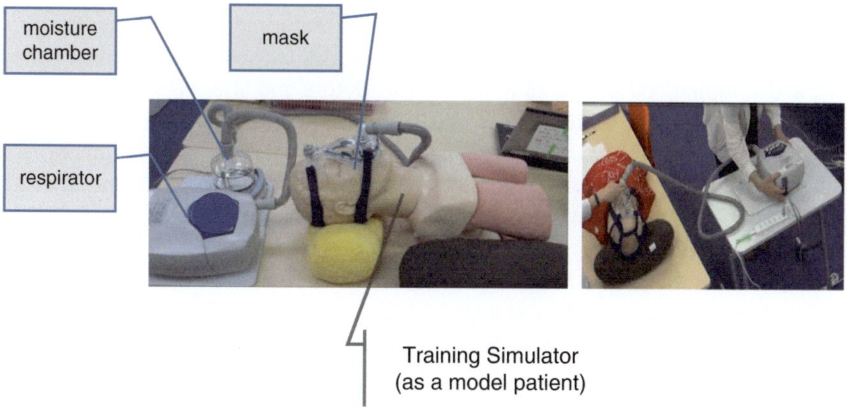

Fig. 18.4 Usablity test scean of a NPPV (non-invasive positive pressure ventilation)-type respirator

To use the system as an effective medical intervention, many important parameters should be set up, including ventilation frequencies and strength of pressures. Those are set up according to a prescription by doctors or medical engineers before the system is brought into the home, and it is often said that the only things patients or care-takers need to do at home are to connect the system to electricity and to turn it on. However, this is not true. Because this equipment should be cleaned as part of daily maintenance, assembling and dismantling of parts are necessary skills for proper use. Handling system alarms is another important operation required of home users. Based on these operations and requirements for handling the system, multiple tasks for the cognitive usability test were generated.

18.5.2 Method Overview

Sixteen older adults (including six female participants: average age 73 ± 2.15 years) were recruited from the database of voluntary participants at the Center of Usability and Aging Research at the University of Tsukuba. In addition, ten undergraduate students (including five female participants, age 21 ± 1.06 years) participated voluntarily in the group of younger adults. All subjects were paid for participation. None had any professional medical knowledge, experience in using or watching the use of respirators, or experience with hospital admission.

The experiment was executed individually. Because participants could not use the respirator by themselves without a prescription, each participant was supposed to be a family member of a patient, whose role was played by a simulator training model for tracheal intubation (a doll with lungs). After the participant signed the consent form, a cover story for the experimental situation was given: "Your uncle (or grand uncle, for younger participants), whose name is *KYOTA*, is suffering from COPD and is now back from the hospital in your home with a new type of respirator, and you are asked to help him use the respirator for one day". There were four main tasks: assembling and starting to use the respirator; adding a moisture chamber to the system; handling an alarm (for an air leak caused by the mask coming off); and stopping and dismantling the device. Before and after each task, levels of anxiety and cognitive loads were evaluated using a visual analog scale (VAS). Questionnaires with semantic differential (SD) evaluations of the equipment (ten pairs of adjectives, rated on six-point scales) were completed before and after all tasks to check subjective impressions. In addition, to capture changes in the concept of breathing through using and understanding the NPPV-type respirator, a request was made twice, before and after all tasks, "please explain what breathing is", with a diagram of the human body and a pen.

Before the task started, the thinking aloud method was instructed and demonstrated, and participants were asked to perform some practice exercises. Throughout the experiment, the participant was given neutral prompting to think aloud.

Behaviors and utterances were logged by two video cameras. The experiment lasted about 1 h for younger adults, and 1.5 h for older adults.

18.5.3 Design Problems Common to Both Age Groups

There were apparent differences between older adults and younger adults in how smoothly they executed the tasks. Older adults needed a longer time to perform the task and showed a variety of errors, which required intervention by the experimenter. Table 18.1 lists important errors, along with the number of participants who needed intervention by the experimenters when the participant's execution of the task was almost at a standstill.

Some of these errors look similar based on suspected error causes or problem-solving. The error "electricity turning ON" means that participants had difficulties finding the power switch, which was on the back, in the vertical direction, and partly covered by a black plastic bar, whose design prevents it from being accidentally switched off. Even though the overall size and shape of the respirator itself was very

Table 18.1 Errors observed in the usability test of NPPV-type respirator with results of statistical tests of number of participates who needed intervention by an experimenter between age groups (***, p <.001; **, p < .01; *, p<.05; ns, non-significant)

Error category	Observed behaviors (error descriptions)	Number of patients with error/*N*
Electricity turn ON	Cannot find the electricity switch on the back of the body: cannot turn on because the switch lay vertical, and was covered with a bar, to prevent accidental turning off	OA 14/16 YA 1/10***
Start	Trying to start the respirator, but cannot push the start button; after pushing the start botton. During the delay of start, pushes other buttons and change some conditions of medical treatment.	OA 8/16 YA 2/10 ns
Alarm on electricity OFF	To stop the respirator, turn off the electricity switch and a system alarm occurs, which is the failsafe system prevent inadequate turning off	OA 7/16 YA 0/10 **
Stop–YES sequence (1)	To stop the respirator, pushes the stop button correctly, then cannot push YES key in response to the confirmation message "Stop respirator: OK?"	OA 9/16 YA 2/10*
Stop–YES sequence (2)	In response to the alarm, tries to STOP the respirator, and Stop–YES sequence error is shown	OA 7/16 YA 1/10*
Construction: adding moisture chamber	Tries to add moisture chamber to the main circuit which are already set up, cannot understand where the chamber should go. (Error in the sequence of respirator→moist chamber→mask→patient)	OA 8/16 YA 1/10**
Setting up of moisture chamber	Cannot understand the way the chamber slides into the base.	OA 9/16 YA 0/10***

similar to a home-use electrical device, for example, a rice cooker, the switch's shape (with marks of input/output along vertical lines) and location were quite different from "typical" home-use electrical devices. No one, including younger adults, directly reached for the switch and turned it on without any hesitation; even younger adults frequently said that "is this really the switch?" The different shape and location is also related to the errors for the alarm handling task problem. Many participants who were aware of the alarm tried to shut down the equipment by turning off the electricity (maybe because the stopping procedures were complicated and needed confirmation). Because this equipment must be connected to electricity to perform its life-sustaining function, turning it off without stopping the system first is a medical equipment error, and this behavior triggers an alarm. These specific characteristics and tacit rules of operation confused both older and younger groups, although only older adults showed "deadlock," with experimenter intervention necessary.

Here, we can see the same double facets of usability and ageing as those shown in Fig. 18.2, but at the same time, the errors and problems observed were specific to the medical equipment.

18.5.4 Common but Specific-To-Medical Equipment Reactions in Both Age Groups

The behavioral results observed during task performances looked similar to the results of the subjective evaluations. Usually in usability testing experiments, users asked to use a quite new device typically show greater anxiety and a heavier load *before* the task; however, just after the first task trial, both younger and older adults often show steep declines in anxiety and cognitive loads. The present results were different. Evaluation of anxiety and cognitive load using the VAS (0–10) was rather high at first and remained high until all the tasks for handling the respirator were completed. There appears to be a small age difference in the means, but this was not statistically significant (Fig. 18.5).

These results imply that using medical equipment causes special anxiety and feeling of load that are different from those caused by the usual gadgets and devices. This may arise from the different tacit rules such as "never unplug it before stopping the machine" or the special purpose of the equipment, which looks like "only an expert can touch it." Using medical equipment at home is not the same as using a home electrical device such as a rice-cooker, and it raises the user's mood or anxiety to a special level or status. These results are very similar to the response in an anecdotal report of a skillful elderly user not being able to respond to an alarm caused by a simple problem of unconnected electricity (Suto in press).

There were no differences in subjective evaluation between the two age groups. However, considering the existence of age differences in emotional and motivational control (e.g., Barber and Mather 2013; Mather and Carstensen 2005), further investigation into the outcomes of these internal states is necessary, in more natural settings, and in situations of longer continuous use (Suto et al. 2014).

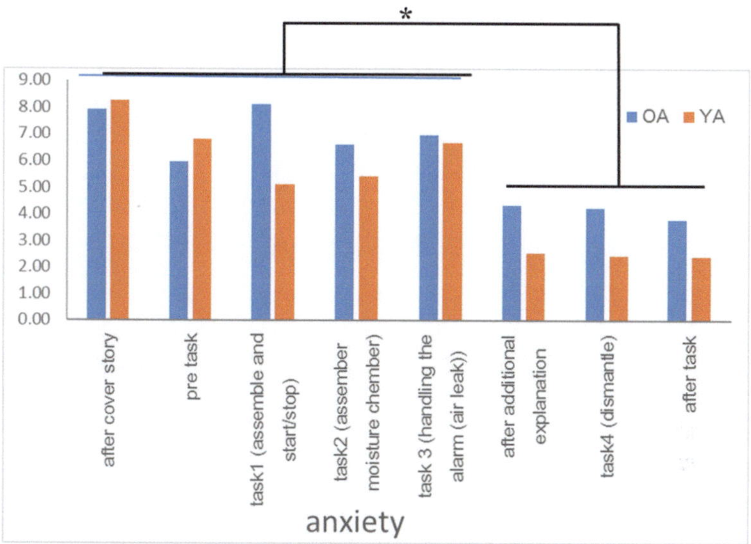

Fig. 18.5 Subjective evaluation of anxiety: no ageing differences, and participants showed declines in anxieties only after all tasks were completed (Harada and Oishi 2014)

18.5.5 Ageing and Crystalized Knowledge

So far, the results of task performance and subjective evaluations showed characteristics specific to using medical equipment, with no special age differences compared with the usual usability testing experiments. However, when we looked at the explanations for breathing, we were surprised at the results. Figure 18.6 shows examples of the results. Most participants drew lines from the nose/mouth, through the throat, to the lungs, and coming back to the nose/mouth (panel B). However, some of the older participants, 6 out of 16, produced a drawing in which lines reached to the stomach/bowel (panel A), and they also explained that the air goes down to the abdomen. Interestingly, they were not self-confident about their explanations: for example, one participant wondered "after the lung, where does it go?", then she said, "yeah, a fart is air from the belly." Another participant said, "the trainer in my gym class always tells us that belly breathing (abdominal respiration) is important. That means air goes this way, doesn't it?"

In the experiment, those six participants showed relative poorer performance in two construction tasks (with and without a moisture chamber), compared with the other ten older participants. However, they all scored over 26 on the MMSE (Folstein et al. 1975), showing no difference from the other 10 participants, and none showed any symptoms of dementia or mild cognitive impairment. In addition, all participants had more than 12 years of school education and thus should have had a science class in which the breathing process was explained.

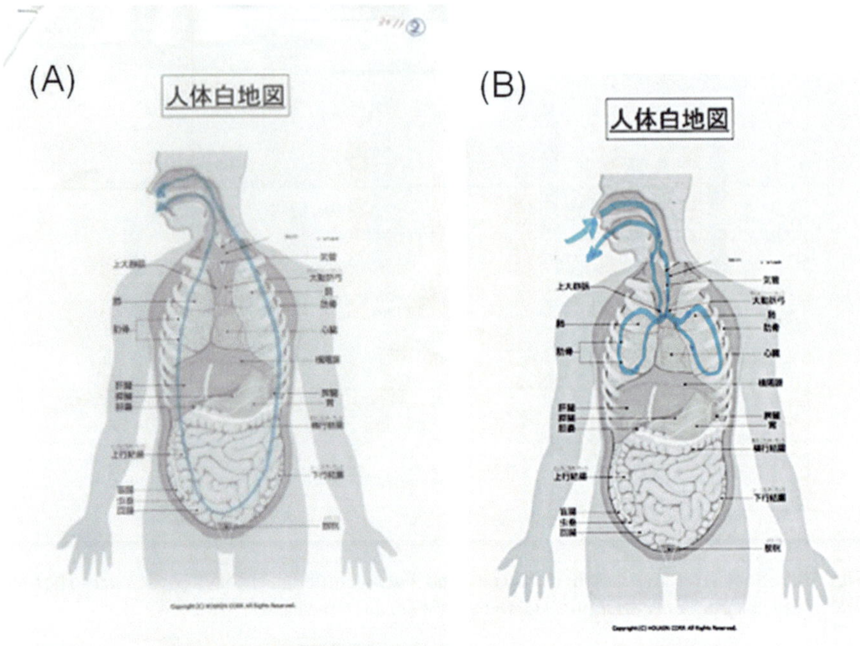

Fig. 18.6 Examples of drawing (blue lines on the diagram) while explaining "breathing". (**a**) Air is going down to the "belly," (**b**) usual pattern of drawing with younger adults

This is an interesting result for cognitive ageing research, because it is widely believed that even though fluid intelligence declines with age, crystallized intelligence does not, remaining intact until the late 80s (Schaie 1994). The fact that one third of our older participants showed unstable knowledge about physiology indicates that the hypothesis of intact crystallized intelligence with ageing may have at least some limitations; perhaps tests of crystallized intelligence have tended to be based on simple questions, for which older adults can easily use their own knowledge elicited using the simple stimulus. In our experiment, the test situation was complex, and a detailed map of the human body may have disturbed those processes when they did not have tacit knowledge. Although the result was strongly affected by the experimental situation and the style of the question, it suggests that it might be sometimes difficult for older adults to use their (not so strong) knowledge when that knowledge is important. Even if "crystallized" knowledge is in an older adult's head, it is useless if it cannot be recalled in a stable fashion, and our data imply that such knowledge use might be dependent upon a cognitive control process (or central executive function in working memory), which can be impaired by ageing.

An important practical implication is that older adults may not be able to understand explanations of medical equipment presented by medical staff in the same way as younger adults, because of their lack of knowledge in using and understanding the function and the meaning of the equipment. This fact is important because it is difficult to know what kinds of problems patients or care-givers have in understanding medical explanations and machinery instructions. Even though nurses and medical engineering staff know that older people have some trouble understanding how to use and handle equipment in daily activities, they usually attributed this trouble to older adults' dislike of machine usage, and this leads them to explain "only operations, no mechanisms." However, our results suggest that there might be a lack of basic knowledge about how the equipment works with the physical body. Because we have no data about how medical staff explain the equipment and what kinds of problems older adults have in understanding those explanations, further investigation will be important and interesting, both from a practical view and from the perspective of cognitive ageing research.

18.6 General Discussion

We have examined social changes and their relationships to and effects on human cognition and memory. With such drastic changes in circumstances, there are many factors to be considered, and it is very difficult to determine which cognitive components or indicators should be measured. Some researchers say that these situations are too difficult for investigation of the *pure* cognitive process or brain functions. Is this true? As components of living beings, both brains and their cognitive functions are governed by circumstances that keep changing, and the characteristics of adaptations and change in accordance with circumstances should be important traits of brain and cognition. Investigating these targets under changing circumstances, or at least always comparing them under different circumstances, may be a good way of pursuing the investigation of the human brain and cognitive functions.

To execute such a research endeavor, a wide and deep observation of human behavior in "natural" settings is important. Japanese society may be a very good showcase to the world, not only socio-economically, but also for cognitive researchers.

References

Anderson ND, Craik FIM (2006) The mnemonic mechanisms of errorless learning. Neuropsychologia 44(14):2806–2813

Barber SJ, Mather M (2013) Stereotype threat can both enhance and impair older adults' memory. Psychol Sci 24(12):2522–2529

Chasteen AL, Bhattacharyya S, Horhota M, Tam R, Hasher L (2005) How feelings of stereotype threat influence older adults' memory performance. Exp Aging Res 31(3):235–260

Craik FIM (1986) A functional account of age differences in memory. In: Klix F, Hagendorf H (eds) Human memory and cognitive capabilities: mechanisms and performances. North-Holland, Amsterdam, pp 409–422

Craik FIM, Salthouse TA (eds) (1992) The handbook of aging and cognition. Lawrence Erlbaum Associates Publishers, Mahwah
Craik FIM, Salthouse TA (eds) (2000) The handbook of aging and cognition, 2nd edn. Lawrence Erlbaum Associates Publishers, Mahwah
Craik FIM, Salthouse TA (eds) (2008) The handbook of aging and cognition, 3rd edn. Psychology Press, New York/Hove
Ericsson K, Simon H (1993) Protocol analysis: verbal reports as data, 2nd edn. MIT Press, Boston. ISBN 0-262-05029-3
European Commission. (2012). Proposal for a regulation of the European parliament and of the council, p 9. http://ec.europa.eu/justice/data-protection/document/review2012/com_2012_11_en.pdf
Folstein MF, Folstein E, McHugh PR (1975) "Mini-mental state". A practical method for grading the cognitive state of patients for the clinician. J Psychiatr Res 12(3):189–198
Harada ET, Akatsu H (2003) Usability for old users. In: Harada ET (ed) Cognitive studies of human-artifacts interaction: what is 'usability?'. Tokyo, Kyoritsu-syuppan, pp 119–138. (In Japanese)
Hara N, Naka T, Harada ET (2009) How can we make IT appliances easy for older adults? Usability studies of Electronic Program Guide system. In: Maurtua I (ed) Human-computer interaction. In-Tech, Rijecka, pp 369–388
Harada ET (2008) Problems in memory in modern society. In: Ohta N (ed) Psychology of human memory, pp 208–221. (in Japanese) (現代社会における記憶の問題.太田信夫(編著) 記憶の心理学(放送大学教材) 放送大学教育振興会)
Harada ET (2009) What do cognitive aging studies tell us? Some implications from aging investigations on human memory and cognitive engineering studies. Jpn Psychol Rev 52(3):383–395. (In Japanese with English abstract)
Harada ET, Hashimoto E (2016) Understandablity of signboard direction and ageing: the universal design principle is there. Paper presented at the 31st international congress of psychology. Yokohama, Japan
Harada ET, Oishi Y (2014) Robustness of crystallized intelligence: ageing effects on the process to use knowledge. Paper presented at the 31st annual meeting of the Japanese cognitive science society
Harada ET, Suto S, Asano A (2012) Error repetition phenomena and its relation to cognitive control, memory and ageing: why does it happen outside the psychology lab? In: Naveh-Benjamin M, Ohta N (eds) Perspectives on human memory and aging. Psychology Press, New York, pp 49–67
Hasher L, Zacks RT (1988) Working memory, comprehension, and aging: a review and a new view. In: Bower GH (ed) The psychology of learning and motivation, vol 22. Academic Press, New York, pp 193–225
Ikai S (2010) The theory of the hospital century. Yuhikaku, Tokyo. (in Japanese, 猪飼周平,病院の世紀の理論,有斐閣)
Kaiho H, Harada ET (1993) Introduction to verbal protocol analysis: reading from uttered data. Tokyo, Shin-yo-sha. (in Japanese: 海保博之・原田悦子編(1993). 『プロトコル分析-発話データから何を読むか』新曜社)
Mather M, Carstensen LL (2005) Aging and motivated cognition: the positivity effect in attention and memory. Trends Cogn Sci 9(10):496–502
McCoy SL, Tun PA, Cox LC, Colangelo M, Stewart RA, Wingfield A (2005) Hearing loss and perceptual effort: downstream effects on older adults' memory for speech. Q J Exp Psychol A 58(1):22–33
Norman DA (1991) Cognitive artifacts. In: Carroll JM (ed) Designing interaction: psychology at the human-computer interface. Cambridge University Press, New York, pp 17–38

Ophir E, Nass C, Wagner AD (2009) Cognitive control in media multi-taskers. Proc Natl Acad Sci USA 106(37):15583–15587
Salthouse TA (1991) Theoretical perspectives on cognitive aging. Lawrence Erlbaum Associates. Inc., Hillsdale
Sayeki Y (1988) Information processing between human and machine: an introduction to cognitive engineering. In: Takeuchi K (ed) Semantics and informatics. Tokyo University Press, Tokyo. (in Japanese, title translated by the authors)
Schaie KW (1994) The course of adult intellectual development. Am Psychol 49(4):304–313
Suto S (in press) Cognitive psychological view on usability of home-use medical equipment and ageing: a case report. In: Harada ET (ed) Cognitive psychology for better medical care. Seishin Shobo. Co., Tokyo
Suto S, Harada ET, Tanaka S, Adachi Y, Hine K (2014) When older adults met tablet computers: a study of learning process to use newly innovated equipment by young - and older adults. Cogn Stud 21(1):62–82. (in Japanese, with English abstract)
Tanaka S, Harada ET (2016) Older adults' timidness to use new artifacts: analyzing human-artifact interaction in experimental lab and in community activities. Trans Hum Interface Soc 19:1–12
Tanaka S, Harada ET, Fujiwara K, Suto S (2017) Constructing a scale of timidity to use artifacts for healthy older adults. Tsukuba Psychol Res 53:41–50
Verhaeghen P, Steitz D, Sliwinski M, Cerella J (2003) Aging and dual-task performance: a meta-analysis. Psychol Aging 18(3):443–460
Wingfield A, Tun PA, McCoy SL (2005) Hearing loss in older adulthood: what it is and how it interacts with cognitive performance. Curr Dir Psychol Sci 14(3):144–148

FSC
www.fsc.org
MIX
Papier aus verantwortungsvollen Quellen
Paper from responsible sources
FSC® C105338

If you have any concerns about our products,
you can contact us on
ProductSafety@springernature.com

In case Publisher is established outside the EU,
the EU authorized representative is:
Springer Nature Customer Service Center GmbH
Europaplatz 3, 69115 Heidelberg, Germany

Printed by Libri Plureos GmbH
in Hamburg, Germany